13-50

D0716227

INSECTS ON PLA

COMMUNITY PATTERNS A

MECHANISMS

D.R. Strong PhD *Associate Professor of Biology*
Florida State University, Tallahassee

J.H. Lawton PhD *Reader in Biology*
Department of Biology, University of York

Sir Richard Southwood PhD DSc FRS
Linacre Professor of Zoology
University of Oxford

Blackwell Scientific Publications
OXFORD LONDON
EDINBURGH BOSTON MELBOURNE

56600 17312

© 1984 by
Blackwell Scientific Publications

Editorial offices:
Osney Mead, Oxford OX2 OEL
8 John Street, London WC1N 2ES
9 Forrest Road, Edinburgh EH1 2QH
52 Beacon Street, Boston
Massachusetts 02108, USA
99 Barry Street, Carlton
Victoria 3053, Australia

All rights reserved. No part of this publication may be reproduced, stored in a retrieval system, or transmitted, in any form or by any means, electronic, mechanical, photocopying, recording or otherwise without the prior permission of the copyright owner

First published 1984

Printed and bound in Great Britain
by the Camelot Press, Southampton

KING'S
COLL. LIBR.
CAMB.

British Library
Cataloguing in Publication Data

Strong, D.R.
Insects on plants.
1. Insect societies
I. Title II. Lawton, J.H.
III. Southwood, T.R.E.
595.7'0524 QL496

ISBN 0-632-00907-1
ISBN 0-632-00909-8 Pbk

Contents

WITHDRAWN
King's College Library
Cambridge

Preface

Three questions dominate community ecology at the present time. How predictable are natural communities? How important is competition between component species in determining community structure? And what proportion of coexisting species in contemporary communities are coevolved? To a greater or lesser extent the answer to each question is dependent on answers to the other two, and each leads logically to a series of more detailed, related questions.

This book addresses itself to each of these three major questions, and many of their logical subsidiaries. The themes will be familiar to most ecologists. Our choice of organisms will probably be much less familiar. Why pick insects on plants?

At one level the answer is that we have studied insect communities on plants — for an aggregate total of 65 years. More fundamentally we believe insects have certain advantages for the community ecologist. In contrast to birds their communities are rich in species and individuals, whilst unlike many plants and corals the individuals are distinct and yet not so small as invariably to demand microscopic examination. These features make them ideal experimental material and we believe the experimental approach is important in testing ideas in community ecology.

The relative neglect of phytophagous insects by community ecologists baffles us. Paradoxically, one reason may be their extraordinary diversity. Non-entomologists may feel overwhelmed by the very profusion of insects. It is our hope this book will encourage more community ecologists to study insects on plants.

However, a sense of awe at the diversity of phytophagous insects is perfectly reasonable. At least half, and probably more, of all the known macroscopic species in the world are plants and the insects that feed on them. Not surprisingly, the literature on insects and their host plants is vast and growing rapidly. Nobody can hope to read, let alone summarize, all the relevant papers in biochemistry, behaviour, physiology, population dynamics, community ecology, evolution, entomology and taxonomy. In this book we have tried to summarize our collective impressions of this enormous field as they affect the structure and dynamics of communities of insects feeding on plants. That is, we have focused on interspecific interactions, predation, parasitism, competition and, most important, the interaction of insects with their host plants. Undoubtedly the greatest challenge in writing has been to restrain ourselves from tackling more facets of even this relatively restricted focus. Certainly, ecological communities of phytophagous insects could validly be approached with a point of view more autecological, more coevolutionary, or more ecosystem oriented than ours. So, the experience of writing a short book on a broad topic is one of hopeful frustrations; we have continually wanted to write more, about different organisms and different relationships, and have discovered how much more needs to be known about insects on plants.

The book draws on literature published up to the end of 1982. A small number of papers and reviews due to appear in 1983 are referred to, but inevitably we failed to include some work that appeared late in 1982. Most glaring of these omissions is our failure to incorporate material from the Proceedings of the 5th International Symposium on Insect–Plant Relationships, held at Wageningen in Holland in 1982 and published towards the end of that year.

It is symptomatic of the rapidity with which this subject is developing that, as authors, we have sometimes disagreed among ourselves about the significance of particular studies, about the strength of evidence for or against an hypothesis, and even about what to include and what to discard. The result, we hope, is a balanced compromise.

Our own work on insects on plants, on which we have drawn extensively in parts of the book, has been generously supported by research grants from the National Science Foundation (D.R.S.) and by research grants and studentships from the Natural Environment Research Council (J.H.L. and T.R.E.S.). To both grant-giving bodies we are extremely grateful.

We have also benefited enormously when planning and preparing the book from discussions with a large number of friends and colleagues. In particular we would like to thank Lawrence Abele, Valerie Brown, Mick Crawley, Simon Fowler, Larry Gilbert, Mike Hassell, Malcolm MacGarvin, Stuart McNeill, Cliff Moran, Patrice Morrow, Peter Price,

Daniel Simberloff, Peter Stiling, Joseph Travis and Mark Williamson. Ann Thistle typed and edited the prose on numerous drafts, and throughout, Bob Campbell and Simon Rallison at Blackwell's were generous with their patience, tact and help.

Last but not least we owe a debt of gratitude to our families, for support and encouragement.

January 1983

D.R.S. — Tallahassee
J.H.L. — York
T.R.E.S. — Oxford

Chapter 1
Introduction

1.1 Phytophagous Insects

A clergyman once asked J.B.S. Haldane what the scientist had learned about the Lord from a lifetime of studying His creations. 'His inordinate fondness for beetles', Haldane replied (Hutchinson 1959).

Beetles do indeed come in a bewildering variety of shapes and forms. They are one of nine orders of insects some or most of whose members feed on the living tissues of higher plants. These phytophagous insects include representatives of familiar groups — Haldane's beetles, butterfly caterpillars and grasshoppers for example — as well as less familiar ones

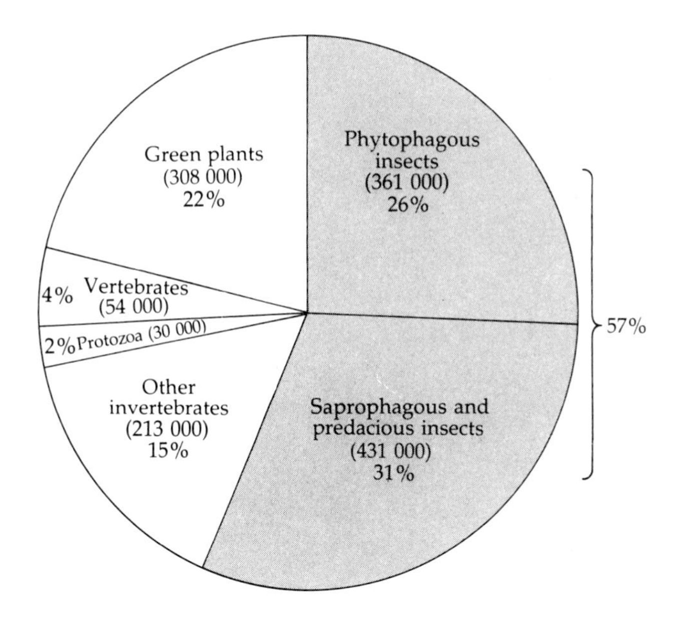

Fig. 1.1. The number and proportions of species in major taxa, excluding fungi, algae and microbes. Numbers are taken from Southwood (1978a). The proportion of phytophages is from Price's (1977) estimates for British insects, assuming that proportions in the world as a whole are broadly similar.

such as sawflies and agromyzid flies. Together they make up the animal communities that form the subject of this book.

The diversity of phytophagous insects is staggering. There are at least one-third of a million species. Contrast this with 8500 species of birds and 4500 species of mammals. There are nearly ten times more species of Lepidoptera (butterflies and moths) than all birds and mammals combined. Herbivorous ungulate mammals (Proboscidea, Perissodactyla and Artiodactyla) muster few more than 200 species.

Current estimates suggest that phytophagous insects make up approximately one-quarter of all living species (Fig. 1.1). Their hosts, the green plants, make up a second quarter. For every species of phytophagous insect there is also approximately one predatory, parasitic or saprophagous insect species, and these comprise a third quarter of macroscopic organisms. The final quarter encompasses all the vertebrates, protozoa, and invertebrates other than insects.

Terrestrial insects that feed directly on the living tissues of higher plants — phytophagous insects — are virtually confined to nine extant orders. They are:

Coleoptera	beetles
Collembola	springtails
Diptera	flies
Hemiptera	sucking bugs
Hymenoptera	sawflies, wasps, etc.
Lepidoptera	butterflies and moths
Orthoptera	grasshoppers
Phasmida	stick and leaf insects
Thysanoptera	thrips.

Appendix I outlines the taxonomy and general biology of the species and higher taxa within these nine orders, dealt with in the text.

The proportion of mainly phytophagous species varies from order to order (Fig. 1.2). Lepidoptera, Hemiptera, Orthoptera and Phasmida, for example, are almost entirely phytophagous, but only about one-third of beetle species and one-tenth of Hymenoptera feed on the living tissues of higher plants. Their non-phytophagous relatives are only of passing concern in this book, except as enemies of plant-feeding insects.

By defining phytophagous insects as species feeding on the living tissues of higher plants we also exclude from consideration those aquatic and terrestrial insects exploiting algae, together with wood-borers,

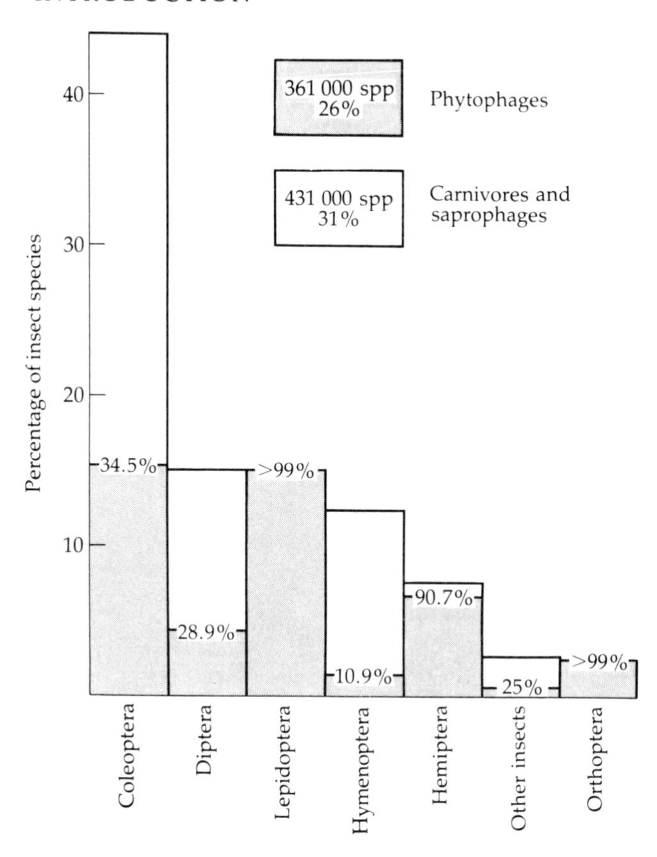

Fig. 1.2. Proportions of insect species that are directly phytophagous as opposed to those that feed upon dead plant material or upon animals. Relative sizes of orders are taken from Southwood (1978a). Feeding modes are based on Price's (1977) estimates for the British fauna, assuming that proportions in the world as a whole are broadly similar.

nectar-feeders and those species that use dead plants and leaf-litter as food. Strict adherence to our definition of phytophagy is occasionally difficult. Seed-feeders in particular pose problems. In general, insects that feed on developing seeds are phytophagous whilst those that exploit shed seeds, granivorous ants for example, are not. This restricts attention to ecologically coherent groups of species that occur together on host plants, chewing leaves, mining stems, boring into developing seeds and so on.

Phytophagous insects exploit their hosts in four ways: some feed externally by biting and chewing, others by sucking from individual cells or from the plant's vascular system, some species mine into their host, and a few form galls. Most parts of a plant can be attacked in any or all of these ways, producing communities that are rich in biological detail, and fascinating to study.

Our objectives are to search for patterns in communities of phytophagous insects, to understand their evolution and to explain their structure.

1.2 The Organization of the Book

It is impossible to understand contemporary communities without knowing something about their evolution. Chapter 2 examines the evolution of phytophagous insects. Eating plants poses a number of unexpected problems — hurdles that must be overcome before insects can exploit plants for food. Hurdles to phytophagy are discussed in the first part of Chapter 2. The second part of Chapter 2 traces the rise of major orders and suborders of phytophagous insects over the broad sweep of the fossil record. The most important conclusion to emerge from this survey is that the number of orders and suborders (and hence probably species) of phytophages has apparently increased steadily from the Devonian onwards.

This conclusion is compatible with a major theme of the book, namely that interspecific competition between phytophagous insects is in general feeble and relatively rare. Arguments to support this view are gathered in Chapter 5. Competition between phytophagous insects does not appear to have been a serious constraint in the past on the numbers of coexisting species. Nor are, it seems, the numbers of species in contemporary communities limited significantly by interspecific competition.

Processes determining how many species of insects are found on particular species of plants are explored in Chapters 3 and 4. Size of host plant range is a major factor. Common and widespread plants have more associated insects than rare plants. Effects of host range and other characteristics such as growth form (whether the plant is a tree, bush or herb) and taxonomic affinity are dealt with in Chapter 3. Accumulation and speciation of phytophagous insects on new hosts are reviewed in Chapter 4.

From a search for broad patterns in Chapters 2, 3 and 4, emphasis shifts in Chapters 5 and 6 to the ecological processes underpinning these patterns. Populations of many species of phytophagous insects fluctuate markedly from generation to generation; hence local communities are rather unpredictable in their composition and in the relative abundance of component species. We have already remarked that competition between species is relatively unimportant in structuring these communities. Natural enemies, in contrast, play a much more important role, summarised in Chapter 5. Chapter 6 is devoted to the ways in which host plants influence the population dynamics of phytophagous insects. As might be expected, host plant effects are profound. Thus the major forces acting on populations and communities of phytophagous insects work vertically through the food web (plants–insects–enemies) rather than horizontally through competition.

A consideration of these vertical influences leads logically to the problem of coevolution, the long-term reciprocal interaction between insect and host plant. Chapter 7 explores coevolution, and in so doing returns full circle to the problems of evolution, species richness and hurdles to phytophagy.

1.3 Community Ecology

Communities are groups of species that interact, or have the potential to interact, with one another.

Delineating a group of phytophagous insects, their host plants and their enemies as a community is to some extent arbitrary. Animals other than insects also feed on plants, and certain of the enemies also attack other taxa. Problems with delimiting the system are common to all studies of ecological communities. The hope is that interactions within the chosen system are more important than the unstudied and often unknown interactions outside the system.

Ecology textbooks often discuss communities as though a uniform set of processes determined the behaviour of all such systems. However, it is becoming increasingly obvious that communities built up from different sorts of species behave in rather different ways. The rules for birds (e.g. Lack 1971; Cody 1974) are not necessarily the same as those for intertidal marine organisms (Paine 1980), freshwater invertebrates (Zaret 1980), tropical coral reef fish (Sale 1977) or parasites (Price 1980), and each of these systems may differ from all the others. Of major concern are: (i) the relative roles of interspecific competition and predation in determining community structure (Connell 1975), and (ii) the degree of determinism and predictability of structure displayed by communities (Cody & Diamond 1975: Price *et al.* 1983; Strong *et al.* 1983). Hence there is an urgent need for studies in depth of a range of community types. Phytophagous insect communities, because they embrace so many of the world's species, deserve particular attention.

1.4 Three Examples of Phytophagous Insect Communities

Each of the authors has studied communities of insects on plants over a number of years, in different parts of the world and with emphasis on different questions. We can best introduce the study of phytophagous insect communities by describing some of the systems most familiar to us, and some of the questions we have tried to answer.

1.4.1 Insects on Spartina alterniflora *(D.R.S.)*

Spartina alterniflora is a common grass of saltmarshes along protected seashores on the Atlantic and Gulf coasts of North America. The local community of insects associated with *S. alterniflora* off the Gulf coast of Florida consists of phytophages of two kinds (leaf-feeding species and stem-miners or borers), saprophages or 'decomposers', which live in and eat dead plant tissues, and the predatory and parasitic insects that attack the phytophages and saprophages (Fig. 1.3).

The assemblage of phytophagous insects, together with its principal parasitoids, constitutes a relatively discrete system. The principal phytophages are illustrated in Fig. 1.3. Larvae of the leaf-miner *Hydrellia valida* live and feed within leaves, in the epidermis towards the tip. Pupae remain within the leaf; adults fly freely among many plants. The plant-hopper *Prokelisia marginata* lives externally on the plant in the nymphal and adult stages, feeding upon the vascular sap by means of stylets inserted into the leaf. The grasshopper *Orchelimum fidicinium* feeds by rasping tissue away from the leaf surface.

The stem-boring phytophages frequent different portions of *Spartina* stem at different phenological stages of plant growth. First, in the spring, the long-legged fly *Thrypticus violaceus* hatches, and larvae begin boring upward in tiny emergent shoots of the plant. In spring and early summer, caterpillars of the moth *Chilo plejadellus* also feed upon young stems.

The remaining three species bore downwards, hatching from eggs laid much later in the season, when the plant is mature. The cecidomyid midge *Calamomyia alterniflorae* restricts larval activity to a single area of shoot, where it lives communally in a plug of fungus which fills the stem section. The tumbling flower beetle *Mordellistena splendens* bores downward from the inflorescence, as do larvae of the lizard beetle *Languria taedata*.

A diagram like Fig. 1.3 gives no clues as to whether horizontal, competitive interactions between members of the same trophic level are significant, or whether plant–phytophage–enemy interactions operating vertically in the food-web provide the key to understanding the structure of the community, or if both, or neither, are of great importance. These can only be discovered by careful observation and experimentation. In this book we have tried to use detailed case histories, like that of the *Spartina* system, to determine what sorts of interactions appear to be common and widespread in phytophagous insect communities and what sorts are generally rare and feeble.

1.4.2 Insects on Pteridium aquilinum *(J.H.L.)*

Bracken fern (*Pteridium aquilinum*) grows naturally on all continents except Antarctica, and may well be one of the five commonest plants in the world. It therefore provides an excellent system to study replicated communities, exploiting the same food plant, in different parts of the world and at different places within one country.

Table 1.1 lists all the phytophagous insects feeding on two patches of bracken roughly comparable in size and other characteristics, one in the mountains of New Mexico at Sierra Blanca, the other at Skipwith Common near York in northern England. Also shown are all the species known to feed on the plant in New Mexico and Arizona on the one hand, and Britain on the other. Two things are immediately obvious. First, not all the species known to occur in the regional 'pool' are found on the food plant at one locality within that region. Secondly, the regional pools differ markedly in species richness. This prompts questions of what determines the size of the regional pool of species exploiting one species of plant, and what determines the species richness of local communities.

Bracken is interesting in other ways. No beetles feed on the plant at Skipwith, and only one species of beetle feeds on the plant in Britain (it is too rare to include in Table 1.1). No beetles exploit the plant in the American south-west. Haldane would be disappointed! Sawflies

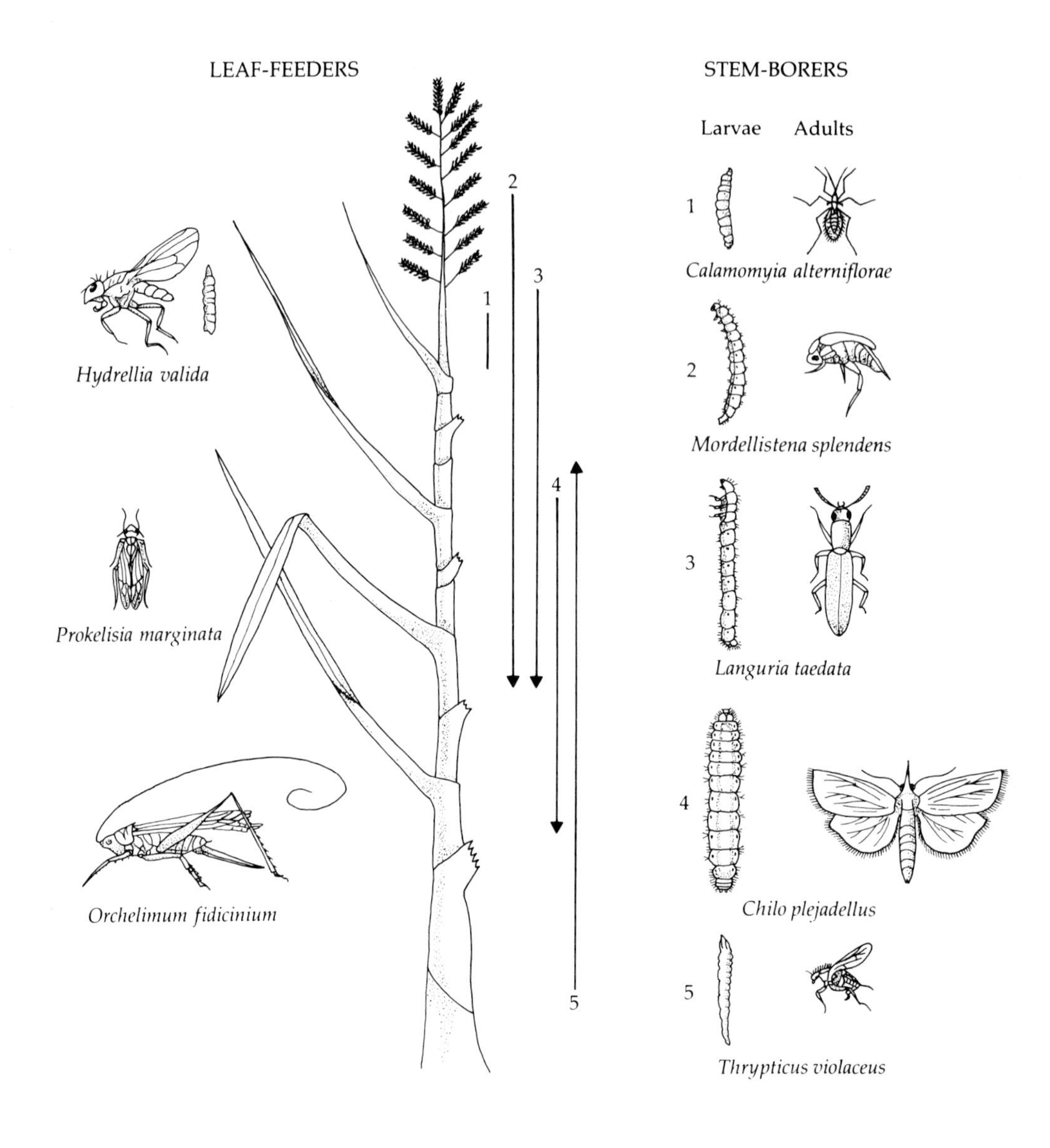

Fig. 1.3. The phytophagous insects, parasitoids and saprovores associated with the saltmarsh grass *Spartina alterniflora* on the Gulf coast of North America. Only the phytophagous species are illustrated. The stem-borers occupy positions indicated by the numbered lines to the right of the grass stem. (After a drawing by Peter Stiling.) Feeding relations listed on facing page.

LEAF-FEEDERS		PARASITOIDS (HYMENOPTERA*) not illustrated
Leaf-miner *Hydrellia valida* (Diptera: Ephydridae)	Larvae	*Opius* sp. (Braconidae)
	Larvae	*Pteromalus* sp. (Pteromalidae)
	Larvae	*Sympiesis* sp. (Eulophidae)
Plant-hopper *Prokelisia marginata* (Homoptera: Delphacidae)	Eggs	*Anagrus delicatus* (Mymaridae)
	Nymphs	*Elenchus koebelei* (Strepsiptera)
	Larvae	*Pseudogonatopus* sp. (Dryinidae)
Grasshopper *Orchelimum fidicinium* (Orthoptera: Tettigoniidae)	Eggs	*Macroteleia surfacei* (Scelionidae)
	Eggs	Eulophid (Eulophidae)

STEM-BORERS	LARVAL PARASITOIDS (HYMENOPTERA)
Calamomyia alterniflorae (Diptera: Cecidomyidae)	*Inostemma* sp. (Platygasteridae)
	Platygaster sp. (Platygasteridae)
	Meromyzobia sp. (Encyrtidae)
Mordellistena splendens (Coleoptera: Mordellidae)	*Cephalonomia hyalinipennis* (Bethylidae)
	Heterospilus languriae (Braconidae)
	Eurytoma sp. 1 (Eurytomidae)
Languria taedata (Coleoptera: Languriidae)	*Cephalonomia hyalinipennis* (Bethylidae)
	Heterospilus languriae (Braconidae)
	Eurytoma sp. 1 (Eurytomidae)
Chilo plejadellus (Lepidoptera: Pyralidae)	*Gambrus bituminosus* (Ichneumonidae)
Thrypticus violaceus (Diptera: Dolichopodidae)	*Eurytoma* sp. 2 (Eurytomidae)
	Eurytoma sp. 1
	H. languriae

SAPROVORES (DIPTERA) not illustrated	LARVAL PARASITOIDS (HYMENOPTERA)
Chaetopsis apicalis (Otitidae)	*Cothonapsis* sp. (Eucoilidae)
	Hexacola sp. (Eucoilidae)
C. aenea (Otitidae)	*Cothonapsis* sp. (Eucoilidae)
	Hexacola sp. (Eucoilidae)
Incertella insularis (Chloropidae)	
Elachiptera sp. (Chloropidae)	

* Except *Elerchus koebelei* (Strepsiptera)

Britain		New Mexico and Arizona	
	Collembola		
+ *Bourletiella viridescens*	CP		
	Diptera		
+ *Chirosia albifrons*	MP		
0 *C. albitarsis*	MR		
+ *C. histricina*	MP		
+ *C. parvicornis*	MP		
+ *Dasineura filicina*	GP		
+ *D. pteridicola*	GP		
+ *Phytoliriomyza hilarella*	MP	0 *Phytoliriomyza clara*	MP
0 *P. pteridii*	MP		
	Hemiptera: Homoptera		
+ *Macrosiphum ptericolens*	SP	+ *Macrosiphum clydesmithi*	SP
+ *Ditropis pteridis*	SR, P		
+ *Philaenus spumarius*	SR, P		
	Hemiptera: Heteroptera		
+ *Monalocoris filicis*	SP		
	Hymenoptera: Symphyta		
0 *Aneugmenus padi*	CP	0 *Aneugmenus scutellatus*	CP
+ *A. fürstenbergensis*	CP		
+ *A. temporalis*	CP	+ *Eriocampidea arizonensis*	CP
+ *Stromboceros delicatulus*	CP		
+ *Strongylogaster lineata*	CP		
+ *Tenthredo ferruginea*	CP		
+ *Tenthredo* sp.	CP		
	Lepidoptera		
0 *Ceramica pisi*	CP		
+ *Euplexia lucipara*	CP		
0 *Lacanobia oleracea*	CP	0 Tortricid sp.	CP
+ *Olethreutes lacunana*	CP		
+ *Paltodora cytisella*	GR	0 *Monochroa placidella*	MR
+ *Petrophora chlorosata*	CP		
+ *Phlogophora meticulosa*	CP		
	Thysanoptera		
		+ *Frankliniella occidentalis*	SP
Total species			
22	27	3	7
Skipwith	Regional pool	Sierra Blanca	Regional pool

(Symphyta), however, are surprisingly rich in species, although they are not usually a dominant group of phytophages. Sawflies are regarded as a rather primitive group of insects and ferns are primitive plants, so the association between a fern, bracken, and several species of sawflies inevitably raises questions about the evolution of members of this community.

1.4.3. *Insect Communities on Trees (T.R.E.S.)*

Our third example comes from a study of the insect fauna of trees: six species in Britain and an equal number in South Africa (Moran & Southwood 1982; Southwood *et al.* 1982a and b). On all plants there are insects other than herbivores; this is particularly true of trees, with their trunks, branches, dead wood and epiphytes providing shelter and food for many insect species. These insects (and other macroinvertebrates, especially spiders) can be divided into six groups, or major guilds, according to their life-styles: phytophagous species (herbivores, sub-divided into chewers and sap-suckers), the fauna associated with epiphytes, scavenging fauna, predators, parasitoids and 'tourists'. Tourists are non-predatory species that have no lasting association with the plant, but may be attracted for shelter, sun-basking or sexual display; they may fall victim to local predators and so become part of the trophic

Table 1.1. Phytophagous insects feeding regularly on the above-ground parts of bracken (*Pteridium aquilinum*) in Britain and in New Mexico and Arizona. Within these large geographic regions, two patches of bracken of broadly similar size and structure were studied. Species from the regional pool present on bracken growing at these sites are indicated by +, species that were absent by 0.
Indicated after each species are its mode of feeding and the part of the plant it usually attacks (see below for key). Very rare species and species of uncertain status have been omitted. (After Lawton 1982.)

- C = chews externally
- M = mines inside tissues
- S = sucks from individual cells or the vascular system
- G = makes galls
- P = attacks pinna ('leaves') and/or associated leaf-veins (the costa and costule)
- R = attacks rachis (the main 'stem' of the plant)

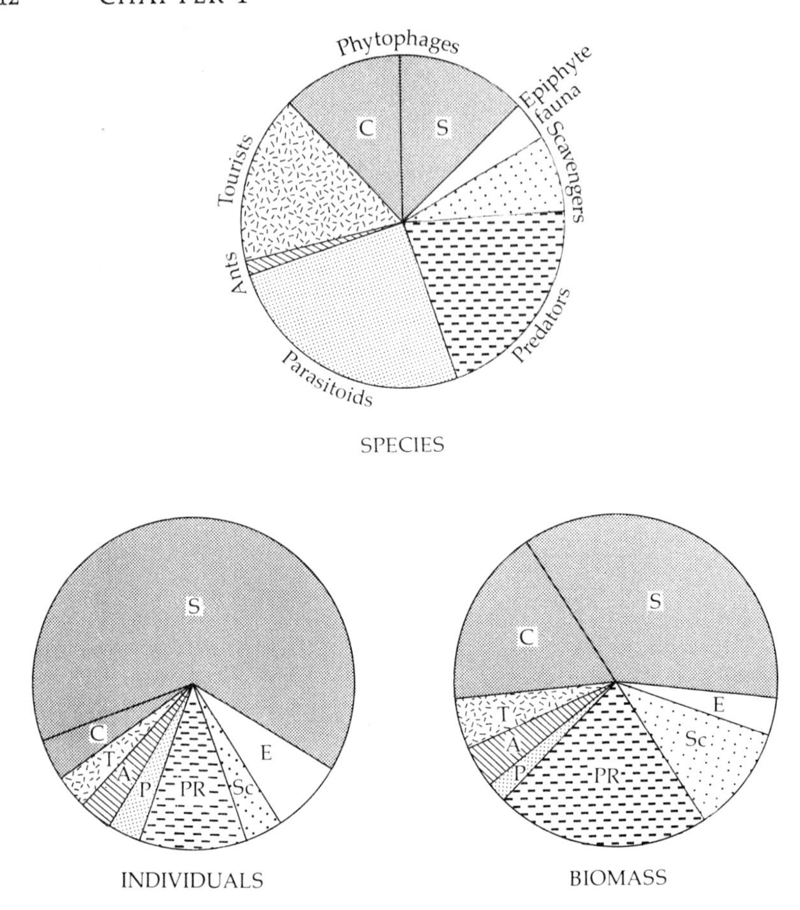

Fig. 1.4. Composition of the arthropod fauna of trees in terms of major guilds, expressed as mean number of species, number of individuals and biomass, based on comparable samples from six species of tree in Britain and six in South Africa. C = chewers, S = sap suckers, E = epiphyte fauna, Sc = scavenging (dead wood, etc.) fauna, PR = predators, P = parasitoids, A = ants, T = tourists. (After Moran & Southwood 1982.)

web of the arboreal community. Ants, which have a peculiar and varied role, are most simply treated as a seventh group.

The average proportional make-up of the insect fauna in terms of these various groups did not vary greatly from tree to tree or from continent to continent and is shown in Fig. 1.4 as the numbers of species,

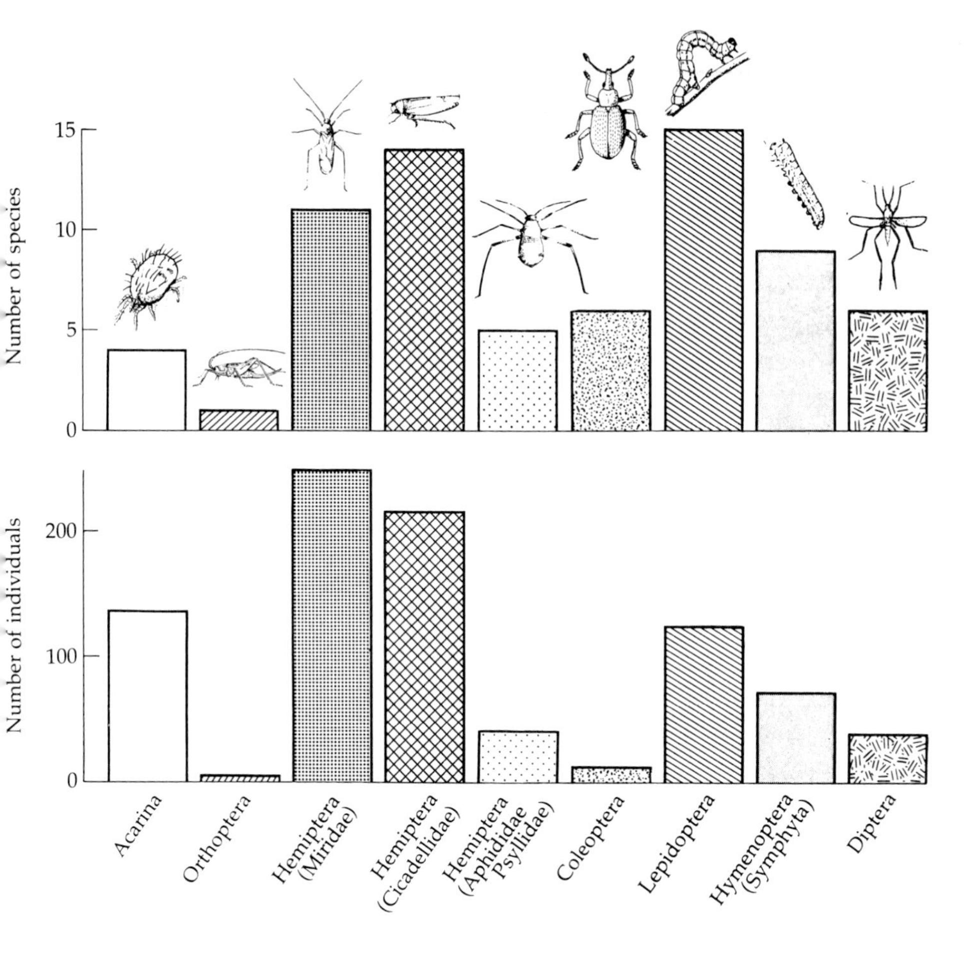

Fig. 1.5. The community of phytophages found on sallow trees (*Salix cinerea*) in Southern England in late summer. Note that some of the groups, Acarina, Miridae and Orthoptera, feed on both plant material and other insects.

the numbers of individuals and the biomass (weight) for each guild. It will be seen that the phytophages constituted about one-quarter of the species, one-half of the weight and two-thirds of the individuals. However, these samples, which were taken by 'knocking down' the insects from the trees with an insecticide spray, were all taken in late summer; they therefore represent a 'snap-shot' of the communities at that season. In the spring, soon after leaf-burst, the pattern would be different, with more chewers and fewer sap-suckers.

Although representatives of 23 different orders of insect or other major invertebrate groups were found on the trees, the herbivore community was drawn from eight orders of insect (and the phylum Mollusca). The taxonomic composition of the herbivore community in this temporal 'snap-shot' for one tree species, sallow (*Salix cinerea*), in Britain is shown in Fig. 1.5, but these major groups of plant-feeding insects were present on all the other trees studied. Why so few insects are phytophagous is a point that we address at the start of the next chapter.

These are the sorts of communities we wish to study, and some representative questions we can ask of them. Our exploration starts, as it should, at the very beginning, with the evolution of phytophagous insects. It is to these problems that we must now turn.

Chapter 2
The Evolution of Phytophagous Insects

2.1 The Hurdle of Phytophagy

2.1.1 The Nature of the Problem

As we saw in Chapter 1, insects of only nine orders exploit the living tissues of higher plants for food. Yet there are 29 living orders of insects (Richards & Davies 1977). Compared with phytophages, there are one-and-a-half times as many orders of insects in which predatory species are important or dominant (14 orders), and a partially overlapping set of 17 orders whose members are mainly or commonly scavengers or microbial or fungus feeders. Thus the exceptional species diversity of plant-feeding insects (see Fig. 1.1) is concentrated in rather few higher taxa (Southwood 1973), a point clearly shown in Fig. 2.1.

The small fraction of insect orders that have adapted to feed on green plants is remarkable because plants are the most obvious and readily available source of food in terrestrial communities. The biomass of invertebrate prey available for predators, for example, is typically at least an order of magnitude less than the biomass of plants available for herbivores.

These observations prompted Southwood (1973) to argue that life on higher plants presents a formidable evolutionary hurdle that most groups of insects have conspicuously failed to overcome. Once the hurdle is cleared, however, radiation may be dramatic. Southwood suggested that insects had at least three major difficulties to overcome before they could exploit plants successfully: the problems of desiccation, attachment, and food.

2.1.2 Desiccation

Although living plant material contains a high proportion of water, air movement often causes a considerable saturation deficit at only a very small distance from the leaf surface (Ramsay *et al.* 1938). Thus even

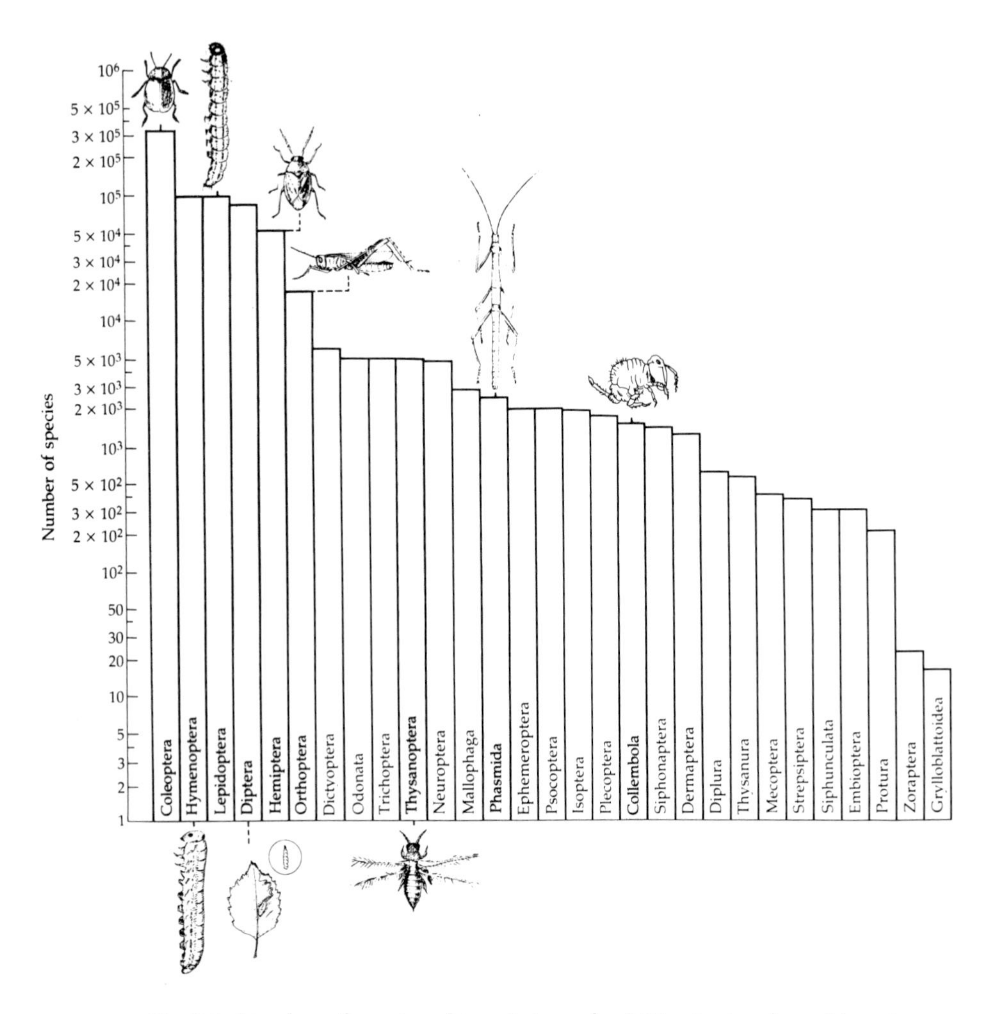

Fig. 2.1. Number of species of insects in each of 29 extant orders of insects (Richards & Davies 1977). Note the large numbers of species in each of the orders (shaded) with members that feed on the living tissues of higher plants (phytophagous *sensu* Chapter 1). (N.B. Ordinate scale is logarithmic.)

though the active stages of insects gain moisture by feeding on foliage, those that live in exposed situations still need to reduce the risk of desiccation. Three methods may be adopted, and these are often combined. First, insects may supplement their water intake by drinking (Mellanby & French 1958). Secondly, they may rest on those parts of the leaf where the boundary layer of air circulation is deepest and the humidity highest, as Fennah (1963) found in cacao thrips (Fig. 2.2). Thirdly, and most commonly, water loss is reduced by modifications and development of the cuticle and tracheal system (Wigglesworth 1972), combined with tolerance and control of the changes in haemolymph osmolality that inevitably accompany water loss (Willmer 1980). Other alternatives are to secrete a cover of 'spittle', as do cercopid homopteran larvae, or to

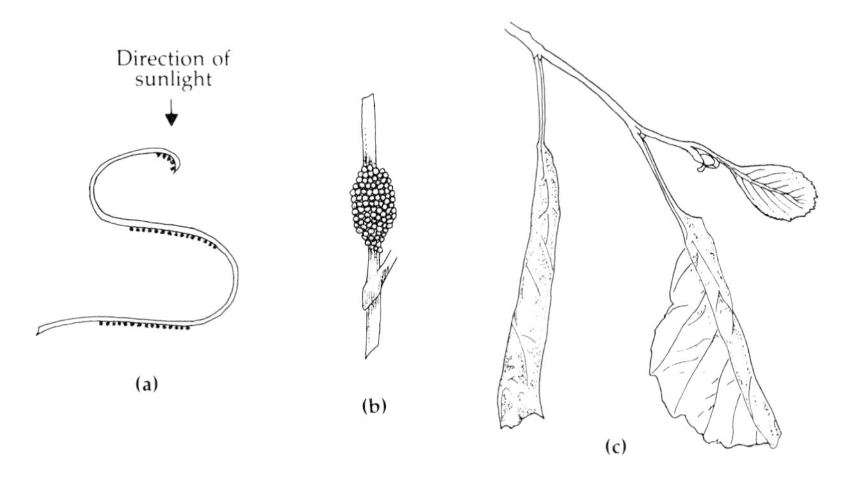

Fig. 2.2. Methods of reducing desiccation. (**a**) The position of cacao thrips (*Selenothrips rubrocinctus*) on a leaf (after Fennah 1963). (**b**) 'Spittle-mass' of a cercopid larva. (**c**) Leaf-roll of tortricid moth (*Epinotia*) on alder (after Bradley *et al.* 1979).

reduce air flow across the body with a web, case or leaf-roll, as do many Lepidoptera (Fig. 2.2). Insects living in galls, mines and flower buds are well protected from desiccation, and often die quickly if they are removed from the safety of their mines (Willmer 1980; Fig. 2.3).

Even when insects have overcome the problems of immediate desiccation, the water content of foliage is often sufficiently low to reduce the growth rates of lepidopteran and sawfly caterpillars (Scriber 1979; Scriber & Feeny 1979; Scriber & Slansky 1981; Fig. 2.4). Hence caterpillars are forced to spend more time on the host plant and are exposed to greater risks of predation, parasitism and other hazards. Hemiptera presumably reduce such problems by feeding directly on the phloem, xylem or liquid cell contents.

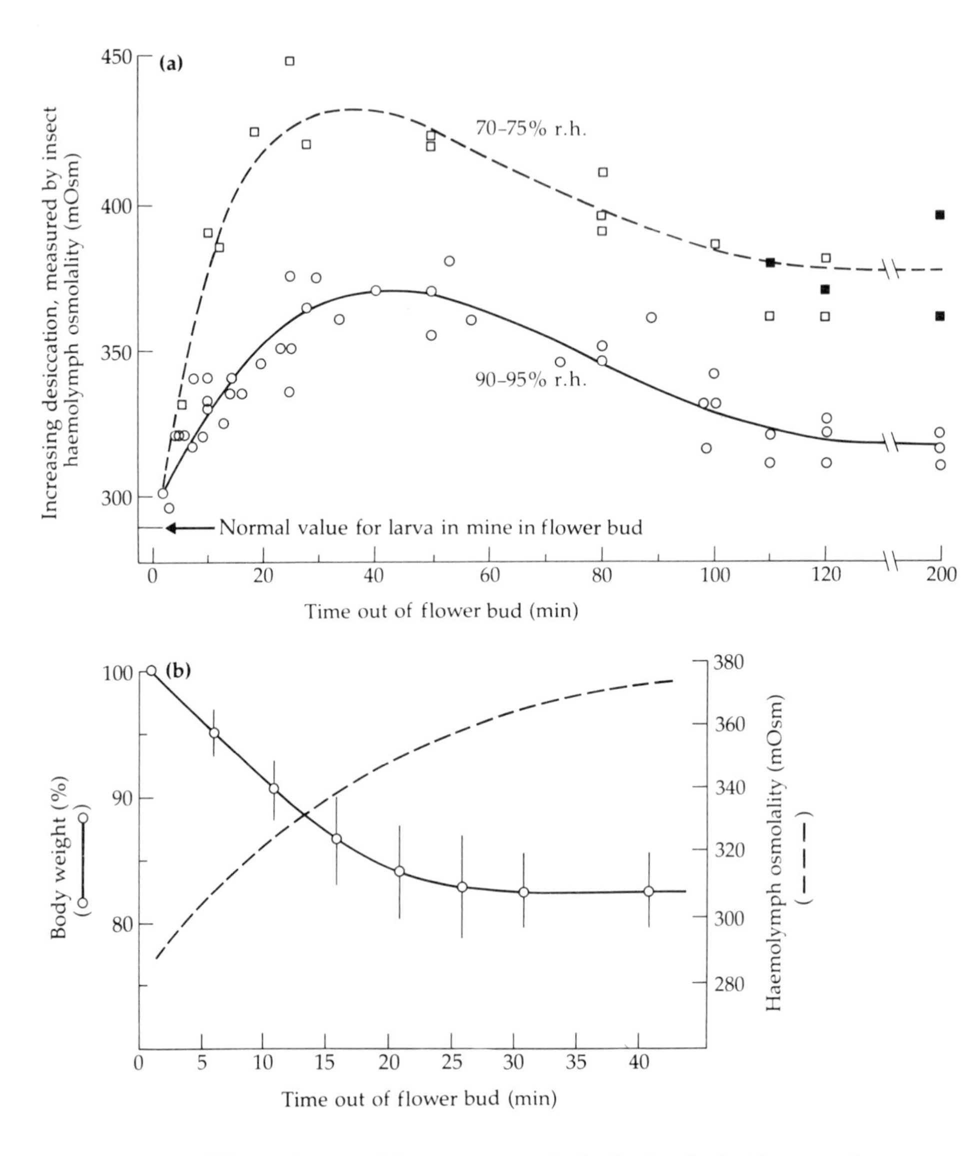

Fig. 2.3. Effects of removal from mines on body fluids of a lepidopteran larva. Larvae of the plume moth *Platyptilia* sp. removed from mines (in tightly enclosed flower buds) and maintained at artificial relative humidities (as labels). (**a**) Effect on haemolymph osmolality. Solid symbols indicate larvae approaching death. (**b**) Changes in body weight over the first 40 minutes, due to water loss, compared with those in haemolymph osmolality for 90–95% r.h. Solid line, percentage bodyweight with ± 2 SE; broken line, osmolality from (a). (After Willmer 1980.)

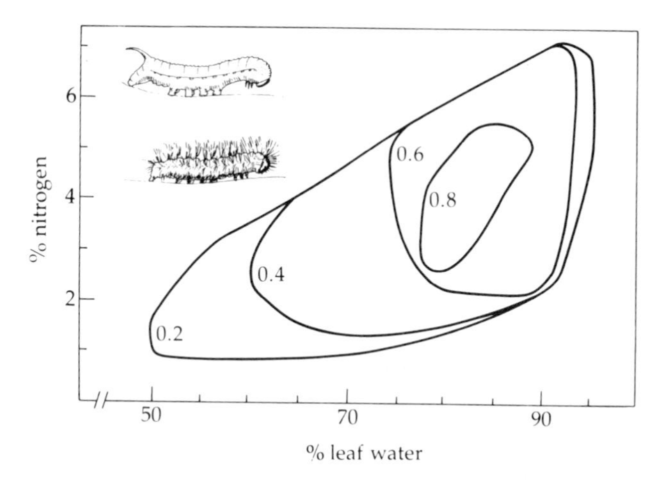

Fig. 2.4. Relative growth rates for 25 species of penultimate instar lepidopteran caterpillars and four species of sawflies as a function of the nitrogen and water content of their host plants. The 'contours' enclose all values equal to, or greater than, the specified values but do not exclude lower values. The figure is compiled from 284 separate experiments, involving nearly 2000 caterpillars, assembled by Scriber and Slansky (1981). Relative growth rates are given in mg growth per mg body weight per day.

The dangers of desiccation are particularly severe for insect eggs because of their high surface-to-volume ratios and the fact that eggs cannot replace lost water. Some groups overcome this problem by inserting the eggs into plant tissue, but eggs with long development times need protection from the crushing, enclosure or suffocation caused by the surrounding plant tissue. Eggs of the dipterous leaf-miner *Phytomyza rufipes* can be literally 'squeezed' out of the leaf by the growth of cells at the point of attachment (van Emden & Way 1973). The Heteroptera and Lepidoptera in particular have relatively slowly developing eggs exposed on the surface of plants, and these insects have many adaptations to reduce water loss from eggs (Southwood 1955; Kayumbo 1963; Bartell 1966).

2.1.3 *Attachment*

Insects can have great difficulty in holding on to smooth, waxy or hairy plants. In general, the problems of staying on the host have been solved in two ways (Southwood 1973):

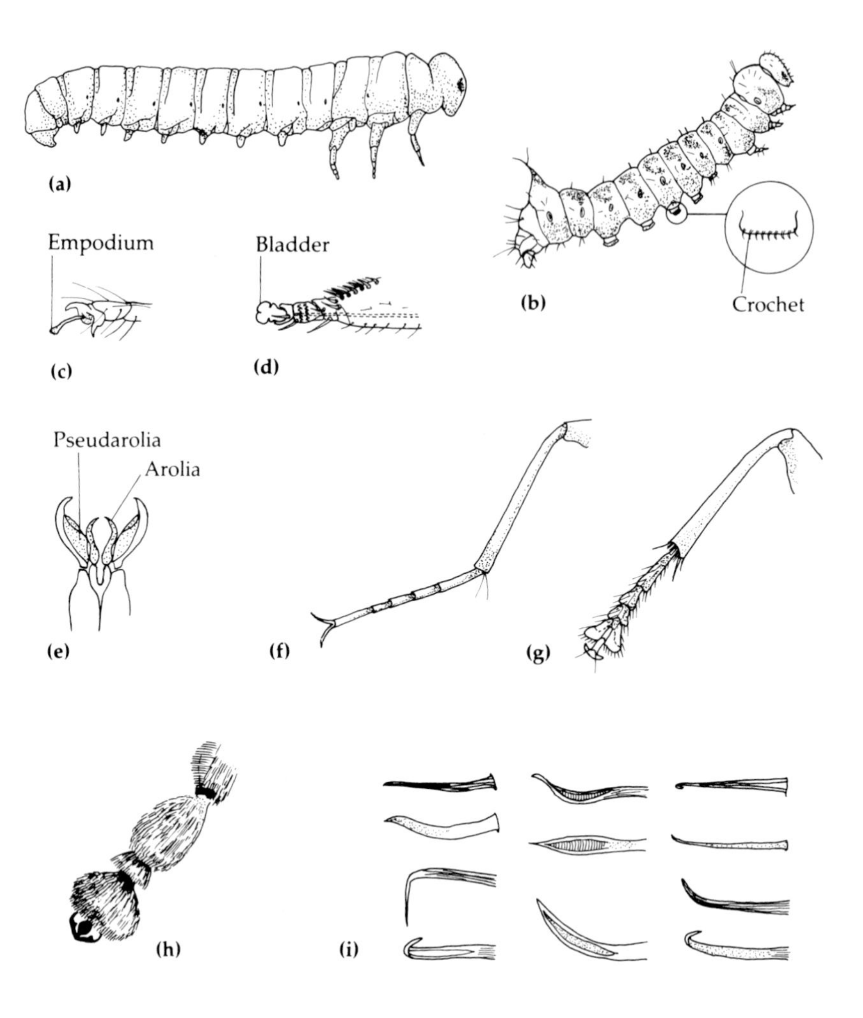

Fig. 2.5. Some adaptations for 'holding on' to plant surfaces. (**a**) Sawfly larva with abdominal legs (after Waterhouse 1970). (**b**) Moth larva with abdominal legs bearing crochets (after Waterhouse 1970). (**c**–**e**) Pretarsal structures. (**c**) Lacewing (Neuroptera) larva, a plant-dwelling predator (after Waterhouse 1970). (**d**) Thrips (after Jones 1954). (**e**) Mirid (after Southwood 1953). (**f**) Hind leg of a ground-dwelling predatory carabid beetle for comparison with (**g**) hind leg of *Demetrias*, a carabid that lives on vegetation (note enlarged fourth tarsal segment) (after Lindroth 1974). (**h**) Leaf beetle (Chrysomelidae), underside of tarsus showing broad segments covered with adhesive setae (after Waterhouse 1970). (**i**) Adhesive tarsal setae of various chrysomelid beetles, ventral and lateral views (after Stork 1980).

(1) by structural modifications or secretions that increase the insect's adhesion; and
(2) by living inside the plant, as do leaf-miners, stem-borers and gall-formers.

The pre-tarsal structures of many insects aid in gripping plant surfaces (Gillet & Wigglesworth 1932; Holway 1935; Edwards & Tarkanian 1970; Bauchhenb & Renner 1977; Stork 1980). However, to be capable of walking about, insects must not only hold on, but also let go. The Coleoptera use small and numerous tarsal adhesive setae for climbing (Fig. 2.5), and these tarsi are peeled off a substrate rather like the material known as 'Velcro' (Stork 1980). (Setae used for amplexus and copulation, found only in the male, are larger and usually fewer than plant-adhesive setae.)

Sawfly and lepidopteran caterpillars have independently evolved very similar sucker-like abdominal prolegs (Fig. 2.5); in Lepidoptera their grip is enhanced by crochets, small hooked hairs. Many caterpillars also spin silk threads to aid attachment as they move about, much as rock climbers use ropes, pitons and belays. In a similar vein, some chrysomelid beetle larvae (e.g. *Galerucella*) use the end of the abdomen for an extra grip, and some mirid larvae extrude the last part of the rectum, which adheres to the smoothest surface, when the plant is jolted. Many Homoptera help anchor their body to the plant with the rostrum, but this is a delicate structure that can break! Another method available to sessile feeders is to attach with a secretion, more or less permanently. This method is used by whitefly larvae and scale insects.

2.1.4 *Food*

All insects living on plants, be they predators or phytophages, have overcome the hurdles of desiccation and attachment. But plants as food present problems only for phytophages. The additional hurdles confronting phytophagous insects are of two sorts — problems of nourishment and problems of physical and chemical defences.

Nourishment

Problems of obtaining adequate nourishment arise because plants and animals differ markedly in chemical composition. Insects have a high

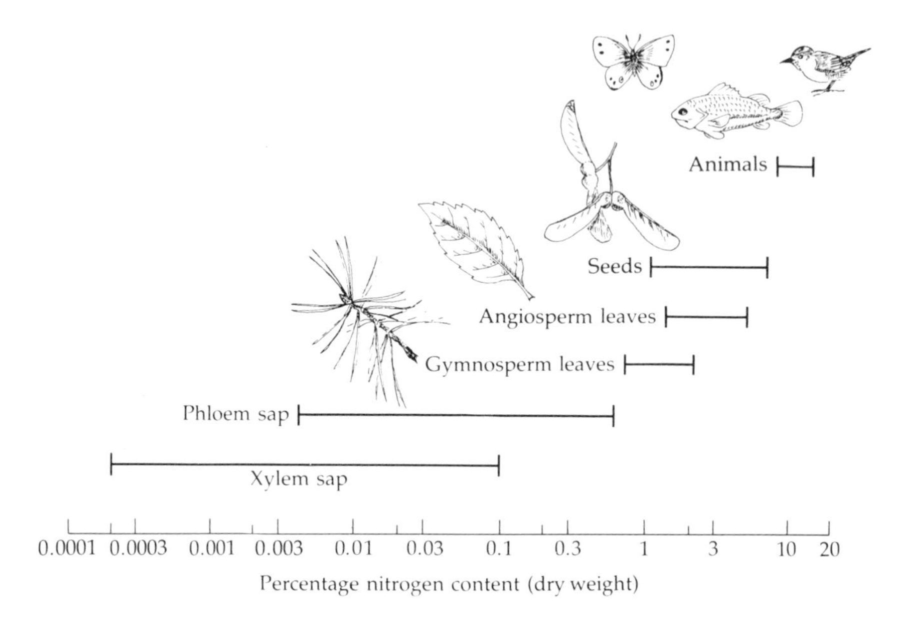

Fig. 2.6. Variation in nitrogen concentration (% dry weight) of different plant parts and of animals. The bars span the range of typical values. (After Mattson 1980.) Percentage nitrogen × 6.25 gives an estimate of the protein content of plant and animal tissues (e.g. Golterman 1969).

Fig. 2.7. Effects of nitrogen content of the host plant on assimilation and growth efficiencies of insects, within (**a**–**c**) and between (**d**) species. (**a**) Assimilation efficiency in relation to total nitrogen levels in food for the plant bug *Leptoterna dolabrata* (from McNeill & Southwood 1978). Assimilation efficiency is: food energy or biomass assimilated (A)/food energy or biomass consumed (C). (**b**) Assimilation and (**c**) growth efficiencies in third-instar larvae of the chrysomelid beetle *Paropsis atomaria,* as a function of the nitrogen content of the leaves of its *Eucalyptus* hosts (from Morrow & Fox 1980). Growth efficiency is: biomass gained (G)/biomass consumed (C). Vertical bars are ± 1 SE. (**d**) Growth efficiencies of various invertebrate phytophages (mainly, but not entirely, insects) as a function of the nitrogen contents of their food (from Mattson 1980).

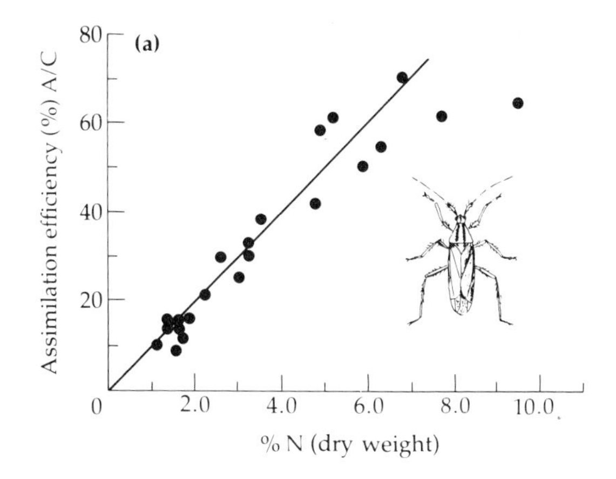
(a)
Assimilation efficiency (%) A/C
% N (dry weight)

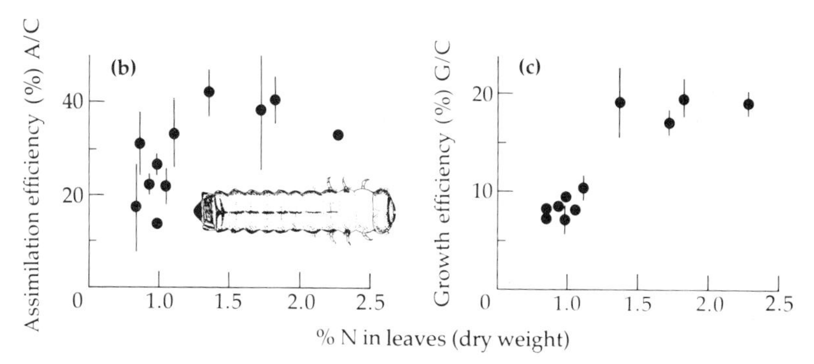
(b)
Assimilation efficiency (%) A/C
(c)
Growth efficiency (%) G/C
% N in leaves (dry weight)

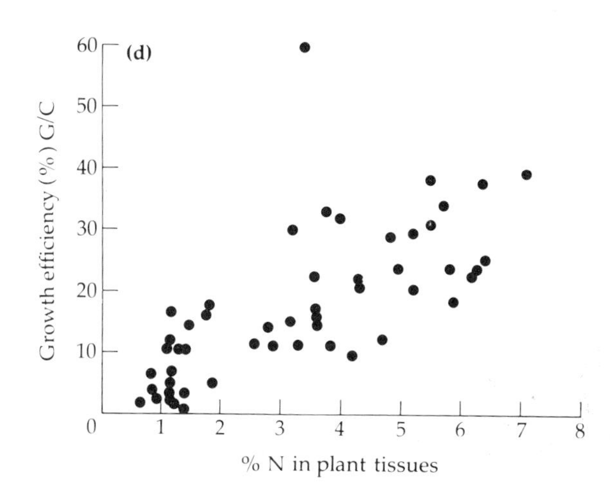
(d)
Growth efficiency (%) G/C
% N in plant tissues

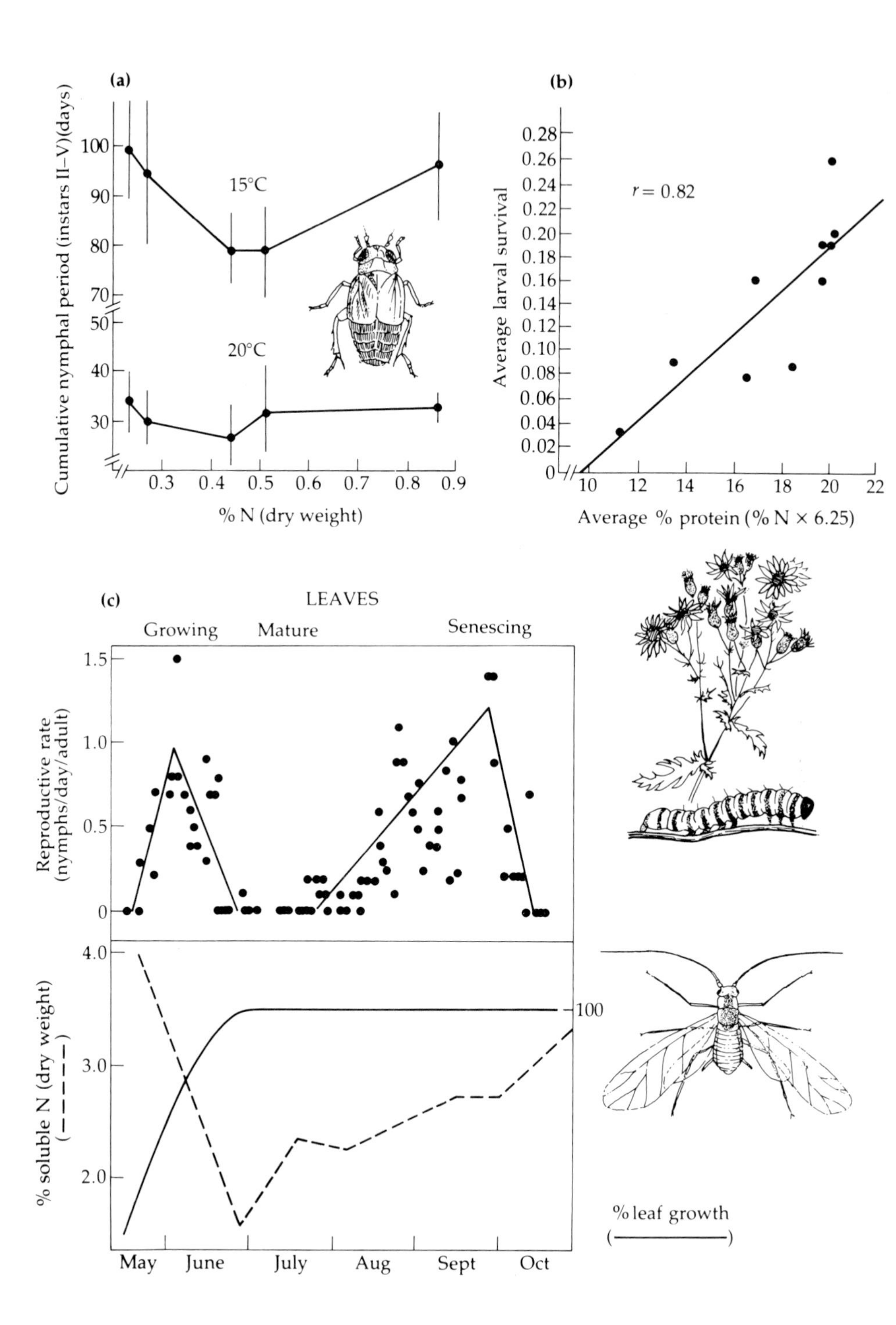

(a)
Cumulative nymphal period (instars II–V)(days)
100
90
80
70
50
40
30
15°C
20°C
0.3
0.4
0.5
0.6
0.7
0.8
0.9
% N (dry weight)
(b)
Average larval survival
0.28
0.26
0.24
0.22
0.20
0.18
0.16
0.14
0.12
0.10
0.08
0.06
0.04
0.02
0
r = 0.82
10
12
14
16
18
20
22
Average % protein (% N × 6.25)
(c)
LEAVES
Growing
Mature
Senescing
Reproductive rate (nymphs/day/adult)
1.5
1.0
0.5
0
% soluble N (dry weight) (– – – – –)
4.0
3.0
2.0
100
% leaf growth (———)
May
June
July
Aug
Sept
Oct

protein content, but plants are composed predominantly of carbohydrates (Southwood 1973; Mattson 1980; Fig. 2.6). Furthermore, the proportions of the various amino acids differ between animal and plant proteins, and insect tissues have a higher energy content than those of land plants. (The mean energy content of insects is 22.8 joules mg^{-1} dry weight, that of land plants is 18.9 joules mg^{-1} dry weight (Cummins & Wuycheck 1971).) This differential is reflected in the low assimilation and growth efficiencies of insect phytophages compared with those of predatory insects; for example, one study found that phytophages convert ingested food into insect biomass with efficiencies between 2 and 38 per cent, while predators attain efficiencies between 38 and 51 per cent (Southwood 1973).

A component of plant tissue critical for phytophages is protein nitrogen. Plant nitrogen is variously measured as 'total nitrogen', 'soluble nitrogen', 'amino acid levels', or 'protein nitrogen', depending upon the author. As the nitrogen content of the food plant increases, assimilation and growth efficiencies often improve (McNeill & Southwood 1978; Mattson 1980; Fig. 2.7); rates of growth, survival and reproduction of individual insects and populations rise; and average population sizes increase (Onuf 1978; Prestidge 1982b). Examples are presented in Figs 2.4, 2.8 and 2.9. These studies leave very little room to doubt that the nitrogen content of host plants is often crucial, even for phytophages closely adapted to their hosts.

However, the problem is not completely straightforward. Enhanced host nitrogen levels do not always improve larval growth rates (e.g. Auerbach & Strong 1981), and Stark (1965) concluded from a survey of the forestry literature that adding fertilizers to forest trees often leads to a

Fig. 2.8. Effects of nitrogen content of the host plant on (a) growth rate, (b) survival, and (c) fecundity of three species of phytophagous insects. (**a**) Leafhopper *Dicranotropis hamata* (Delphacidae). Notice that the nitrogen content of the host (the grass *Holcus lanatus*) appears to be optimal for this species at intermediate concentrations (0.4–0.5%). Vertical bars are 95% CI. (After Prestidge 1982a.) (**b**) Larval survival in the cinnabar moth, *Tyria jacobaeae*, feeding on ragwort, *Senecio jacobaea* (Myers & Post 1981). (**c**) Sycamore aphid, *Drepanosiphum platanoides*, feeding on sycamore, *Acer pseudoplatanus* (Dixon 1970). In (a) host nitrogen levels were manipulated by addition of fertilizer; in (b) they were naturally occurring variations at different sites; in (c) they were naturally occurring seasonal changes at one site.

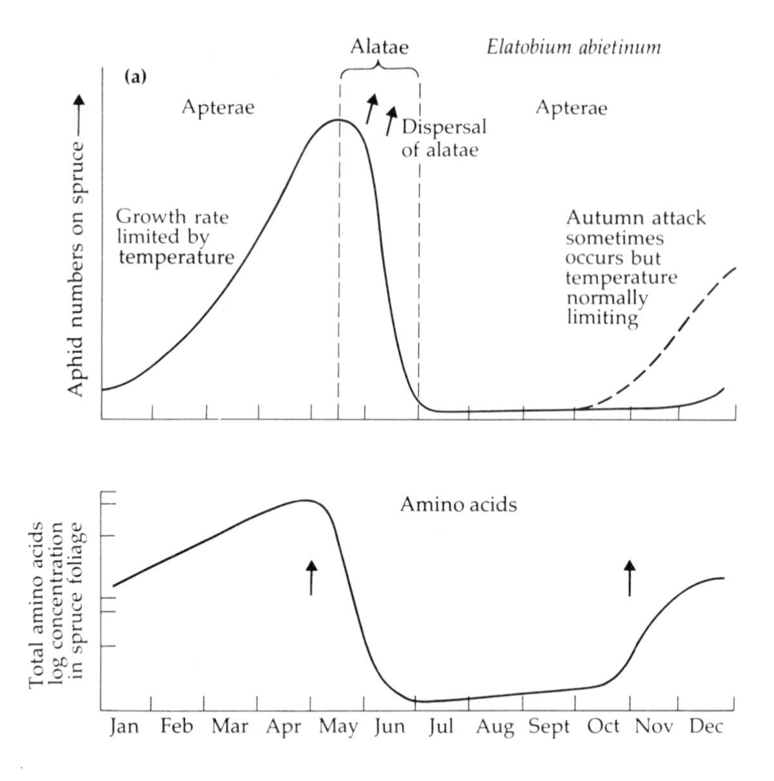
Alatae
Elatobium abietinum
(a)
Apterae
Dispersal
of alatae
Apterae
Aphid numbers on spruce
Growth rate
limited by
temperature
Autumn attack
sometimes
occurs but
temperature
normally
limiting
Amino acids
Total amino acids
log concentration
in spruce foliage
Jan
Feb
Mar
Apr
May
Jun
Jul
Aug
Sept
Oct
Nov
Dec

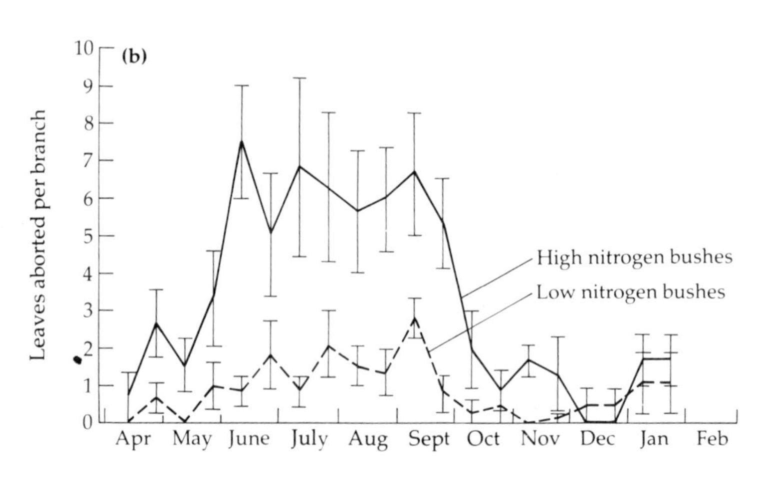
(b)
Leaves aborted per branch
0
1
2
3
4
5
6
7
8
9
10
High nitrogen bushes
Low nitrogen bushes
Apr
May
June
July
Aug
Sept
Oct
Nov
Dec
Jan
Feb

reduction in the abundance of many kinds of insects. Most probably, insects respond not only to total host nitrogen levels but also to the form and availability of nitrogen, for example to the balance of amino acids (Prestidge & McNeill 1983), as well as to the mixture of protein nitrogen and other essential components of the diet, not least the water content of the host (see Fig. 2.4). These are difficult problems that deserve further study.

However, given the importance of protein-nitrogen levels in the diets of many species, it is interesting that insect palaeontologists have suggested two main routes for the evolution of phytophagous insects; both begin with high-nitrogen food sources (Section 2.2.1). One route was via pollen feeding (Malyshev 1968; Zherikhin, in Rohdendorf & Raznitsin 1980). The proportion of protein in fresh pollen is high and is similar to that in insects. The other route enlists the aid of external microorganisms in digesting dead and dying plant tissue (e.g. Mamaev 1968). One contemporary version of this mode of phytophagy is practised by the larvae of *Drosophila* that feed on yeasts in the rot-pockets of cacti (Fellows & Head 1972; Starmer *et al.* 1976). Another phytophage that relies upon microorganisms in plant wounds is the onion fly *Delia antiqua*, which eats cellulytic bacteria in feeding lesions in the host plant. An evolutionary shift to more intimate mutualism between microbes and phytophages is seen in plant-feeding insects that culture yeasts or bacteria within their own guts (Richards & Davies 1977; Mattson 1980; Jones *et al.* 1981; Jones 1983). Gut microorganisms are transmitted from one generation to the next through the egg, and not only supply essential amino acids and proteins but also synthesize

Fig. 2.9. Effects of host nitrogen levels on population abundance. (**a**) Influence of seasonal changes in leaf amino acid levels on population growth of the spruce aphid (*Elatobium abietinum*) (from McNeill & Southwood 1978, data of C.I. Carter). Apterae are wingless aphids. As amino acid levels fall in May and June, alatae (winged aphids) are produced which disperse from the host. (**b**) Effects of feeding by caterpillars of the olethreutid moth *Ecdytolopha* sp. on leaf loss in red mangrove (*Rhizophora mangle*). The feeding and tunnelling of the larva inside a bud causes abortion of the leaves. 'High-nitrogen' mangroves have egret and pelican colonies, and the mud beneath them is enriched with droppings. Several species of phytophagous insect, including *Ecdytolopha*, are appreciably more abundant on the high-nitrogen, bird islands (after Onuf *et al.* 1977). (Data are means ± 1 SE.)

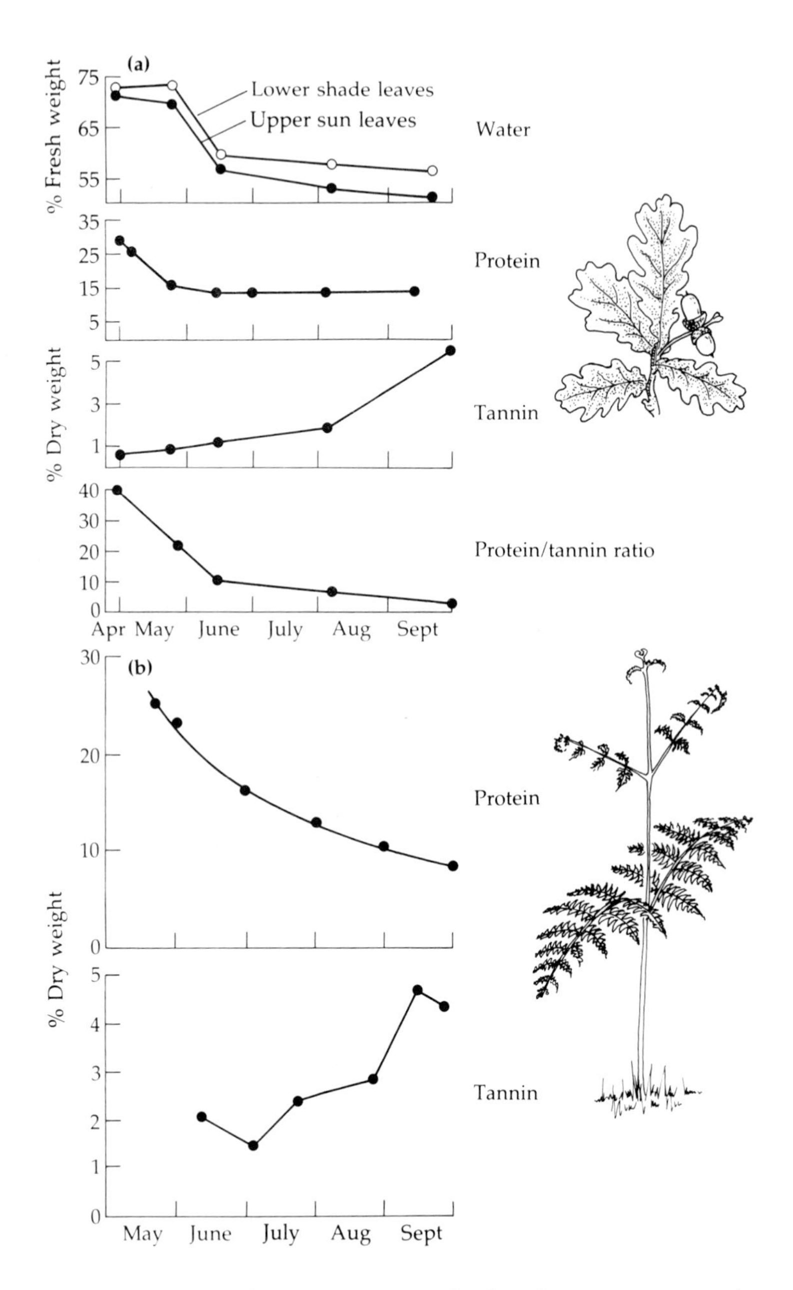

Fig. 2.10. Seasonal changes in nutritional value of two temperate-region plants. (**a**) Seasonal variation in water, protein and tannin content of oak, *Quercus robur*, leaves (after Feeny 1970). (**b**) Seasonal changes in protein and tannin content of bracken fern fronds (after Lawton 1976).

vitamins and sterols. They are housed in special cells, mycetocytes, which may be grouped into organs called mycetomes. Symbiotic micro-organisms are found in a wide range of Coleoptera and in many groups of Hemiptera (particularly Homoptera, but also some Heteroptera).

The evolution of such specialized means of obtaining essential nutrients is good evidence for the existence of nutritional hurdles that have to be overcome before insects can successfully exploit green plants for food.

Host Plant Defences

Once evolved, phytophagous insects inevitably trigger selection pressures on their hosts to evolve defences. We will have cause to consider this evolutionary 'arms race' between plants and insects in several later parts of the book. Here we simply outline the main characteristics of plant defences.

Plant defences against herbivores of all varieties, not only insects, broadly fall into three categories.

(i) Physical defences — hooks, spines, trichomes, tough leaves, etc.

(ii) The low levels of available nitrogen, or low water/nitrogen ratios, that we have already touched on.

(iii) Noxious phytochemicals that are repulsive, unpalatable or poisonous, or that interfere with assimilation by insects. These are the so-called defensive 'allelo-chemicals' of plants, which function against other organisms (Whittaker & Feeny 1971).

Most wild plants maintain high levels of soluble nitrogen for only limited periods of time (McNeill & Southwood 1978; Mattson 1980; Figs 2.8c, 2.9a and 2.10). Other plants store nitrogen as non-protein amino acids, many of which are poisonous to animals (Rosenthal & Bell 1979). Canavanine, for example, is stored in the seeds of many legumes (Janzen 1978; Rosenthal & Bell 1979) and is poisonous to animals, thus combining a reduction in the level of available nitrogen with chemical defence.

Non-protein amino acids are but one class in the broad spectrum of noxious phytochemicals, which range from simple molecules, like oxalic acid and cyanide, to complex cyanogenic glycosides, alkaloids, toxic lipids, terpenoids, saponins, flavonoids, tannins and lignins (Rosenthal & Janzen 1979; Futuyma 1983). Many of these compounds may originally have served an autecological function in the plant, for example as defence against desiccation, cold, salinity or ultraviolet light, and some still play a role in metabolism (Beck & Reese 1976; Seigler & Price 1976; Jones 1979). However, many now have no obvious autecological or metabolic functions within the plant (Jermy 1976), and their principal role appears to be in defence, though not necessarily against insects. Of course, a defensive role may complement an autecological role (Chew & Rodman 1979; Harborne 1982). Many noxious phytochemicals are unequivocally toxic or repellent to one or more species of phytophages (e.g. Beck & Reese 1976; Levin 1976; McKey 1979; Rhoades 1979). Some cost the plant much to produce (Chew & Rodman 1979; Rhoades 1979) or necessitate additional mechanisms for the plant to avoid autotoxicity (e.g. Fowden & Lea 1979). A number of these substances are particularly effective agricultural insecticides (e.g. pyrethrum).

Noxious phytochemicals have been classified broadly into two groups on the basis of their inferred biochemical actions: (i) 'qualitative' or 'toxic' and (ii) 'quantitative' or 'digestion reducing' (Feeny 1975, 1976; Rhoades & Cates 1976; Cates & Rhoades 1977; Rhoades 1979). The former are poisons, such as alkaloids and cyanogenic glucosides, toxic in small quantities. However, some insects have evolved mechanisms to detoxify such poisons, just as they have in response to insecticides. Detoxification mechanisms can render harmless even large concentrations of a poison. Thus many plants have specialist phytophages that are resistant to qualitative defences. Indeed, some phytophages may depend on the presence of a 'defensive substance' to recognize their host plant, they may use the substance as a phagostimulant, or even sequester it to render themselves toxic (Schoonhoven 1973; Rothschild 1973; Chapman & Blaney 1979).

Quantitative or digestion-reducing substances act in proportion to their concentrations; more is better. These are substances like tannins, resins and silica. Until recently, it was generally held that insects could not evolve resistance to quantitative defences. However, it is now clear that this generalization is false, at least for one group of such compounds, the hydrolysable tannins. Some insect species have not only evolved resistance to hydrolysable tannins (e.g. Fox & McCauley 1977), but are actually stimulated to feed by their presence (Bernays 1981). Moreover, the pH of the midgut of lepidopteran caterpillars exploiting plants that contain high concentrations of the second major group of tannins, condensed tannins, may be sufficiently alkaline (an average pH of 8.8) to eliminate the digestion-inhibiting properties of these compounds (Berenbaum 1980). In short, although the distinction between qualitative and quantitative defences was a reasonable theoretical beginning for this field, current discoveries are making it increasingly difficult to maintain this distinction for specialized insects.

Difficulties of classification aside, it is clear that allelochemicals are a major hurdle for phytophages. None provides absolute defence; probably every species of plant is attacked by at least one species of insect, but most species of plants are inedible and unavailable to most phytophagous insects most of the time. It is a guess, but a reasonable one, that the first primitive land plants were less well defended against herbivory than are modern plants; but the chemical and physical defences of plants have undoubtedly been one of the major hurdles to phytophagy throughout most of the evolution of terrestrial ecosystems.

It is to the evolution of land plants and phytophagous insects, a process that started several hundred millions of years ago, that we must now turn.

2.2 Fossil Insects

Land plants and insects appear in the fossil record 380–400 million years ago, in the Devonian Period (see Fig. 2.13). Unfortunately, the fossil record for insects is discontinuous and fragmented, the majority of specimens coming from a very few productive localities (Hughes & Smart 1967; Hennig 1981; Wootton 1981). In spite of much new evidence, particularly from Russian studies (Rohdendorf & Raznitsin 1980), our historical knowledge of insect evolution in general and of plant-feeding forms in particular is scanty and tentative.

2.2.1 Insects of the Devonian, Carboniferous and Permian

The Devonian

The earliest fossil insect is *Rhyniella praecursor*, a collembolan from the Devonian of Scotland 380 million years ago (Whalley & Jarzembowski

1981). Although most living Collembola are not herbivorous, Kevan *et al.* (1975) suggested that *Rhyniella* may have fed on 'plant juices obtained through puncture wounds', perhaps on *Rhynia*, a primitive fossil pteridophyte from the same geological formation. However, it was so long ago, and fossils are so scarce, that good evidence for *Rhyniella* as a herbivore is impossible to obtain.

The Carboniferous

The Carboniferous saw the rise of all modern plant divisions except the angiosperms, and without doubt their main herbivores were insects (Swain 1978). It is tempting to speculate about two major influences on insects that the evolution of giant arborescent plants may have exerted immediately prior to the Carboniferous, in the Devonian. First, arborescences may have favoured the evolution of winged insects. Secondly, arborescence greatly elaborated plant architecture, and the diversity of modern insects is strongly correlated with this elaboration, as we will show in Chapter 3. It therefore seems likely that the appearance of large, complex plants at the end of the Devonian may have been an important factor fostering the evolution of a diverse insect fauna during the Carboniferous.

Most of the insect orders that flourished in the Carboniferous are now extinct. Three are particularly important. These are the Dictyoneurida* (= Palaeodictyoptera) (Fig. 2.11), the Mischopterida (= Megasecoptera) and the Diaphanopterida (Fig. 2.11). The Dictyoneurida are the most abundant terrestrial insects preserved as fossils from the Carboniferous. All three of these orders had beak-like, apparently piercing mouthparts reminiscent of the mouthparts of modern Hemiptera. An expanded clypeal region implies well-developed sucking abilities (Carpenter & Richardson 1978; Wootton 1981), and Russian palaeontologists in particular believe that these insects were phytophagous, feeding on the reproductive organs and fruits of Carboniferous and Permian plants (V.V. Zherikhin, in Rohdendorf & Raznitsin 1980; see also Smart & Hughes 1973; Carpenter 1977).

* Here we follow the Russian usage for names of the extinct insect orders. Brief accounts of these extinct orders are given in this section and are not repeated in Appendix 1.

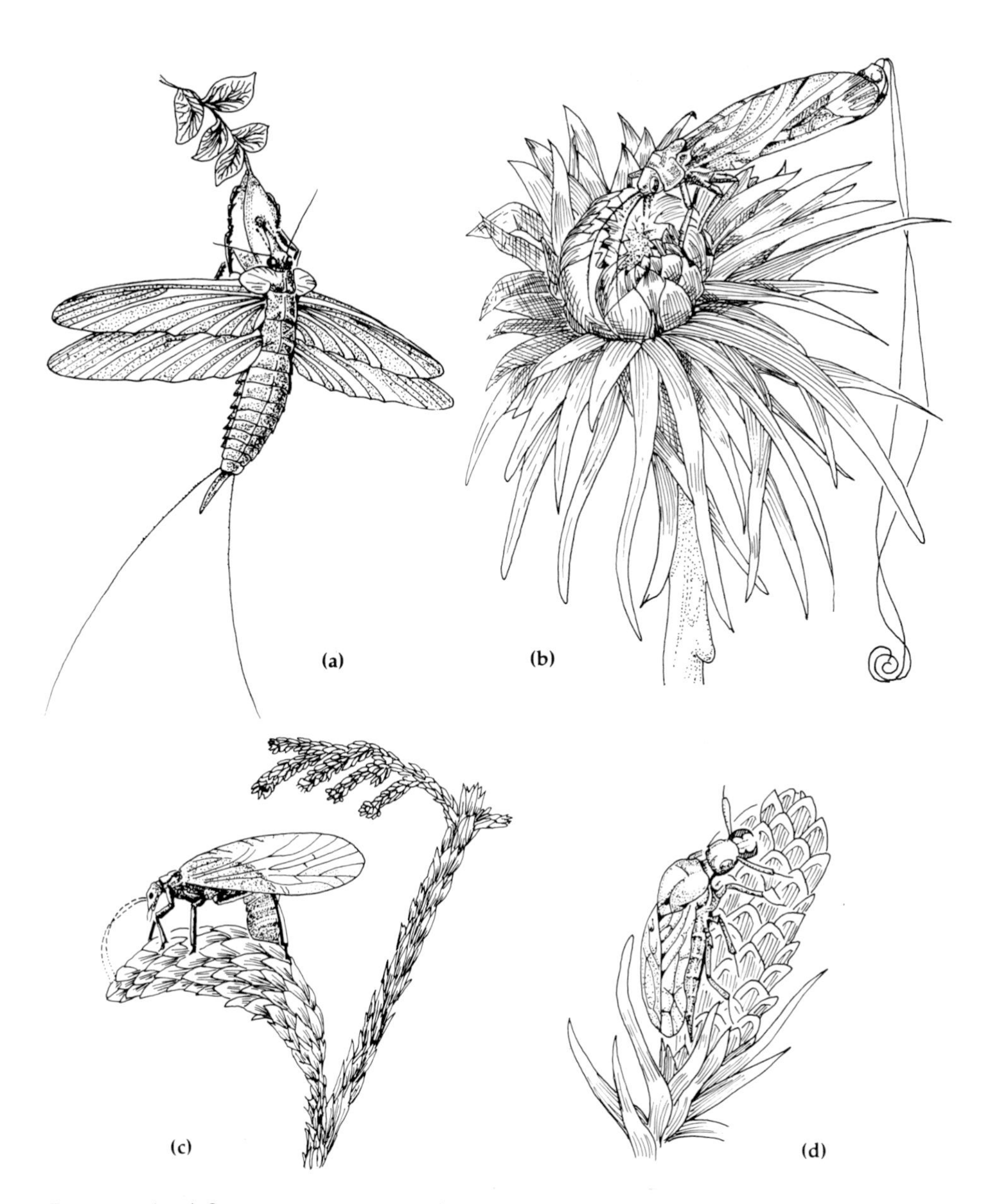

Fig. 2.11. (**a–c**) Some representatives of extinct orders of supposed phytophages from the Carboniferous, Permian and Triassic. (**a**) Dictyoneurida — *Goldenbergia*. (**b**) Diaphanopterida — *Uralia*. (**c**) Palaeomanteida — *Palaeomantina*. (**d**) Early fossil sawfly, *Archexyela smilodon* (Xyeloidea). (From Ponomarenko, in Rohdendorf & Raznitsin 1980.)

It is interesting to speculate about the evolution of these first phytophages. The earliest insects and other primitive arthropods were probably scavengers, rather like the larvae of present-day scorpion flies (Mecoptera). Malyshev (1968) found fossil evidence that these larvae would feed voraciously on pollen on the soil surface. In the Upper Carboniferous and Permian the spores and pollen falling from various primitive plants may have sometimes formed a layer on the soil surface, and Malyshev concluded that it was but a small evolutionary step from feeding on shed spores and pollen to feeding on the concentrated sources, the 'fruiting bodies' on the plants. Recent work has strengthened this view. V.V. Zherikhin (in Rohdendorf & Raznitsin 1980) has pointed out that whereas very few fossil leaves from the Carboniferous show signs of damage (and this damage seems more like that caused by worms, snails or millipedes on the fallen leaves), reproductive bodies of *Samaropsis*, a fossil gymnosperm from Mongolia, show evidence of damage apparently caused by piercing mouthparts. On present evidence, Zherikhin concludes that leaf-feeding insects did not evolve until after the Carboniferous.

Other insects of the Upper Carboniferous include the Palaeomanteida (= Miomoptera) (Fig 2.11), which were some of the earliest insects undergoing a complete metamorphosis; their larvae are believed to have developed in the fruiting bodies of gymnosperms (Rohdendorf & Raznitsin 1980). The 'Protorthoptera' (Carpenter 1977), a group of mixed pedigree, include fossils of true Orthoptera. Although modern Orthoptera are predominantly herbivorous, these Upper Carboniferous forms are assigned to an extinct family whose present-day relatives are largely predaceous, but also eat plant material. We may conjecture that some of these Carboniferous orthopteroids may have chewed tissue of green plants, but possibly foliage that was more dead than alive. The earliest fossil leaves showing damage attributed to chewing insects are found in the early Permian; the incisions on the lamina are reminiscent of those chewed out by a modern orthopteran, and are probably not the bite of a lizard or other reptile (Smart & Hughes 1973).

The Permian

The insect fauna of the Permian was diverse, with 19 known orders. Ten of these orders appeared then for the first time and include

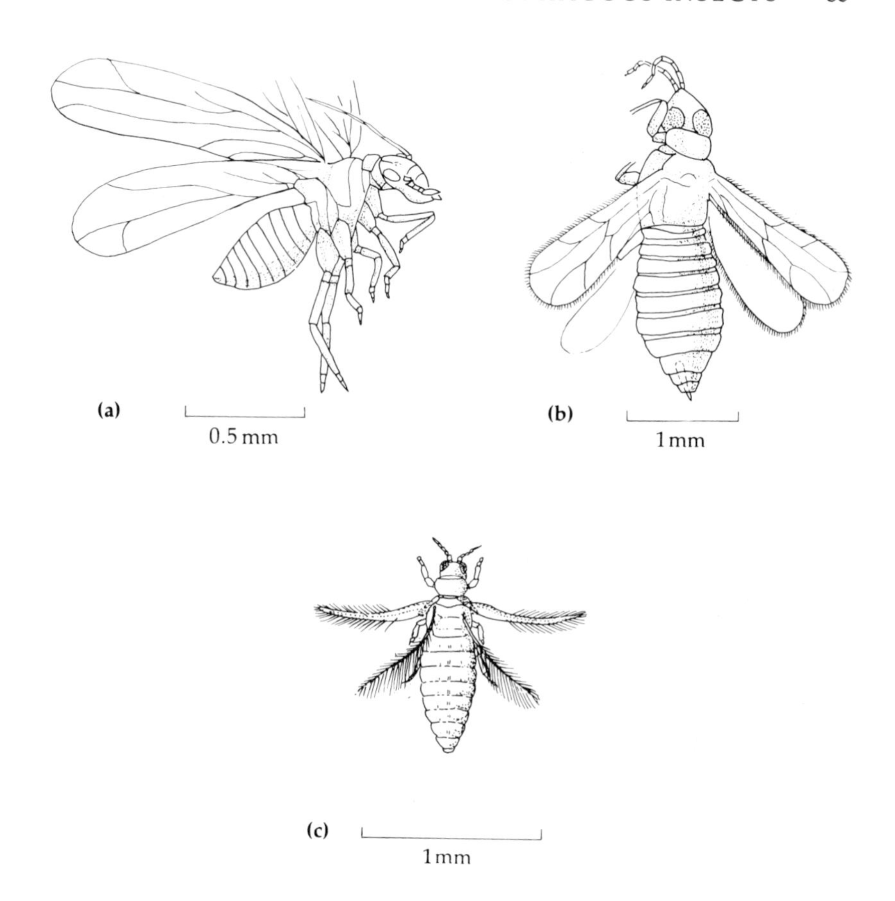

Fig. 2.12. (**a**, **b**) 'Fossil Thysanoptera'. (**a**) Lophioneurid — *Jantardachus reductus* (after Vishnyakova, in Rohdendorf & Raznitsin 1980).
(**b**) *Karatothrips jurassicus* (after Sharov 1972). (**c**) A typical recent Thripidae. Note narrowing of wing membrane.

the Hemiptera-Homoptera in the Lower Permian and Hemiptera-Heteroptera in the upper part of the same period (Harland *et al*. 1967). It is not clear upon what the Permian Heteroptera fed. Phytophagous Heteroptera are not known until the early Jurassic (Popov & Wootton 1977). However, all the evidence suggests that Permian Homoptera were then, as now, predominantly phytophagous, with fossils attributed to two modern groups, the Auchenorrhyncha (present 'in profusion and confusion' (Wootton 1981)) and Sternorrhyncha (Harland *et al.* 1967). Plants with a thin cortex and phloem close to the outer surface of the stem, suitable for exploitation by sucking insects, did not become abundant until the late Carboniferous. The evolution of Homoptera may have been made possible by the appearance of this major new food resource (Smart & Hughes 1973).

Haldane's beloved beetles also appeared in primitive form (Protocoleoptera) in the Permian. These first Coleoptera have some affinities with the extant suborder Archostemata (Carpenter 1977; Crowson 1981; Hennig 1981) and probably fed on fungi under bark on rotting wood. Such micro-saprophagy also seems likely for certain other early groups of insects and may have been an important alternative route to herbivory (e.g. Cecidomyidae, p. 37; see also Section 2.1.4). Those groups of beetles that are today predominantly phytophagous, the Curculionoidea and the Chrysomeloidea, do not appear in the fossil record until the Upper Jurassic (Crowson 1981).

The extinct family Lophioneuridae (Fig. 2.12), first known from the Lower Permian, is considered by Vishnyakova (quoted in Rohdendorf & Raznitsin 1980) to be a primitive group of the Thysanoptera. Lophioneuridae may have been pollen feeders, living within the pollen sacs of their host plants. This habitat would have created evolutionary pressures toward small body size and the narrow, less delicate membranous wings that are the main hallmarks of modern thrips. The first undoubted thrips, *Karatothrips* (Fig. 2.12), is found in the late Jurassic.

2.2.2 *The Mesozoic*

Few fossil insects of any kind are known from the Triassic, though the chewing stick insects, Phasmida, have been recorded (Harland *et al.* 1967; Carpenter 1977).

The most important groups to emerge during the Mesozoic were the Hymenoptera, Cecidomyidae (Diptera) and Lepidoptera.

Hymenoptera

Primitive Hymenoptera were present in the early Triassic. These fossils are assigned to an extant superfamily of sawflies, the Xyeloidea (Fig.

2.11). In the next geological period, the Jurassic, fossils of other sawfly superfamilies have been found (Rohdendorf & Raznitsin 1980); their larvae may well have fed on plant tissues, as do their present-day relatives (Hennig 1981).

Malyshev (1968) argues that phytophagous sawflies evolved from pollen-feeders. If the genesis of sawflies was in the early Permian, the dominant plants would have been gymnosperms, the Pteridospermae and the Cordaitae, which often 'cradled' reproductive organs on the surfaces of their leaves. Malyshev suggests that the early sawflies used natural plant cavities for shelter, rest and feeding, as larvae of the archaic *Blasticotoma filiceti* do today in the stems of the fern *Athyrium filix-femina*. From these habits it is not difficult to imagine how stem-boring developed, as seen in members of the modern sawfly family Cephidae; gall-forming sawflies presumably evolved from borers and miners.

The other main suborder of the Hymenoptera also contains phytophages, particularly the Cynipidae, many of which are specialist gall-makers on oaks (*Quercus*) and Rosaceae. In addition, some members of the superfamily Chalcidoidea are seed parasites. Malyshev (1968) argued that these habits arose from parasitism of the embryos of plants ('phytoophagy'), which is so well developed in the fig wasps (Agaonidae) today. Subsequent evolutionary steps for these Hymenoptera involved the stimulation of galls, first in reproductive organs, then in growing points, and last in normal tissue. Cynipids appear in the fossil record in the Cretaceous.

Cecidomyidae

The probable evolutionary path of the gall midges (Cecidomyidae) has been reconstructed in detail by Mamaev (1968). Fossils of Diptera are known from the Lower Triassic, but specimens do not include any phytophages. By the Upper Eocene, fossils in Baltic amber include a large number of adult gall midges in at least 16 genera. Mamaev considers that the ancestors of gall midges diverged in the Upper Jurassic or Lower Cretaceous and had larvae that fed on detritus in the litter of wet forests. The path to phytophagy may have begun with species exploiting fungal hyphae among dead leaves (cf. the comments on Coleoptera on p. 36, and Section 2.1.4), proceeding via fungal necroses on living plants to true herbivory. The diverse range of species present in the Eocene argues strongly for evolution of gall-forming species before the end of the Cretaceous.

Lepidoptera

The Lepidoptera also diversified in the Cretaceous (Carpenter 1977), the earliest fossils being primitive Micropterygidae from Lower Cretaceous amber (Whalley 1977; Skalski 1979; see also Shields 1976). Caterpillars of extant micropterygids feed on and among mosses and liverworts and the adults feed on pollen. A more typically modern form has been found in mid-Cretaceous Canadian amber. This fossil is thought to represent an external feeder, though whether it chewed foliage, detritus or fungus is unclear (Powell 1980). By the early Caenozoic, beautifully fossilized caterpillar leaf-miners are known (Opler 1973; Hickey & Hodges 1975; Crane & Jarzembowski 1980). Powell (1980) suggests that the two major extant suborders of Lepidoptera, Monotrysia and Ditrysia, evolved after the radiation of angiosperms in the Cretaceous. In contrast to the situation with sawflies, there is no evidence for an initial evolution by early Lepidoptera on primitive plants, pteridophytes for example.

2.2.3 The Rise of the Angiosperms

The origins of the angiosperms (flowering plants) have now been pinpointed near the Jurassic–Cretaceous boundary (Doyle 1978; Knoll & Rothwell 1981). Angiosperms dominated terrestrial ecosystems by the mid-Cretaceous. The radiation of the Lepidoptera, Cecidomyidae and Cynipidae followed.

As more primitive groups of plants disappeared, many insects must have become extinct along with their hosts. Among important extant phytophages only the Hemiptera, sawflies and Orthoptera certainly predate the rise of the angiosperms. Moreover, some of the most successful phytophagous superfamilies within these three early groups did not appear until after the evolution of angiosperms, in the Cretaceous. Among groups appearing during the Cretaceous were the Aphidoidea in the Hemiptera and the Acridoidea in the Orthoptera. However, several other superfamilies of the Hemiptera that are today successful had definitely appeared in the Jurassic or even earlier, well before the angiosperm explosion. These insects now have many hosts among higher flowering plants and include the leaf-hoppers (Cicadelloidea, Fulgoroidea), plant-hoppers (Psylloidea), shield bugs (Pentatomoidea), and the group that includes the modern plant-bugs (Miridae), the Cimicoidea.

We can only speculate as to why some insect groups successfully survived the host plant revolution caused by the evolution of angiosperms. Modern Psylloidea, for example, are particularly abundant in Australia, and it is possible that isolation may have played some role in their survival. For other groups, alternative explanations must be sought. Many mirids are partially or entirely predatory, giving this group a valuable 'food refuge'; shield bugs feed predominantly on fruiting bodies, and so may have been less affected than leaf-feeders; sawflies may have survived because they fed on primitive plants that persisted, and so on. In each of these cases, however, we can do little more than guess. The greatly varying success of insect groups in surviving and adapting to the angiosperm explosion is striking, and worthy of further study.

2.3 Patterns of Diversity through Time

2.3.1 Data and Interpretations

The detailed fossil histories of the last section are brought together in Fig. 2.13, where it is clear that from the Carboniferous onwards, the number of orders and major suborders of phytophagous insects, and hence presumably the number of species, appears to have increased continuously.

Interpretations

The impression of continuous increase in diversity goes straight to the heart of a fundamental issue in evolutionary ecology, the extent to which evolutionary adaptation has led to a progressive rise in the number of species and higher taxa. We can encompass this issue, in broad strokes, by two extremes in perspective. On the one hand, proponents of the *ecological-saturation hypothesis* (Raup 1972; May 1974a, 1981; Gould 1981) argue that while 'actors' — species, orders or even phyla —

enter and leave the ecological stage, the total number of 'roles' in the play — the number of niches — stays roughly constant. Under this hypothesis, interspecific competition prevents more than a roughly constant number of species from coexisting on a finite supply of resources. As new species evolve, others are excluded by competition and become extinct.

An opposite perspective might be termed the *expanding-resources hypothesis*, and would consider that both the quantity of resources and the ratio of niches (and hence the number of species) per resource can change through evolution (Whittaker 1969, 1977; Whittaker & Woodwell 1972).

This issue has two main components, the number of species exploiting each major resource, and the constancy of resources. Resources are difficult to define, but if we ignore this difficulty for the moment, we can offer no fewer than five alternative, though not mutually exclusive, explanations for the patterns revealed by Fig. 2.13.

(i) The impression given by Fig. 2.13 is an illusion, because sporadic preservations and discoveries make the fossil record on insects seriously misleading (see Gould 1981).

(ii) The rise in insect diversity is not a palaeontological artefact, but mirrors a rise in plant diversity, that is, a rise in the number of major types of resources in the form of species of host plants.

(iii) The variety of resources made available by an 'average plant' has tended to increase during evolution.

(iv) The number of ways in which insects exploit plants has similarly increased; new ways of exploiting resources enhance the variety of exploiters (e.g. Schoener 1974; Gatz 1980; Ausich & Bottjer 1982).

(v) Phytophagous insects are not limited to any significant extent by interspecific competition. Hence species have evolved largely unconstrained by competition with other species.

Fig. 2.13. The evolution of major orders and suborders of fossil insects. Data on individual groups are summarized in the text (Sections 2.2.1, 2.2.2). The top (right-hand) panel is obtained by summing across the definite and possible records of orders and suborders numbered 1–15 in the body of the figure. Both 'minimum' and 'maximum' records imply a steady increase in the number of major taxa of insects exploiting plants during the Phanerozoic (the total fossil record). Note also the progressive addition of major feeding modes (sucking, chewing, mining and gall-forming).

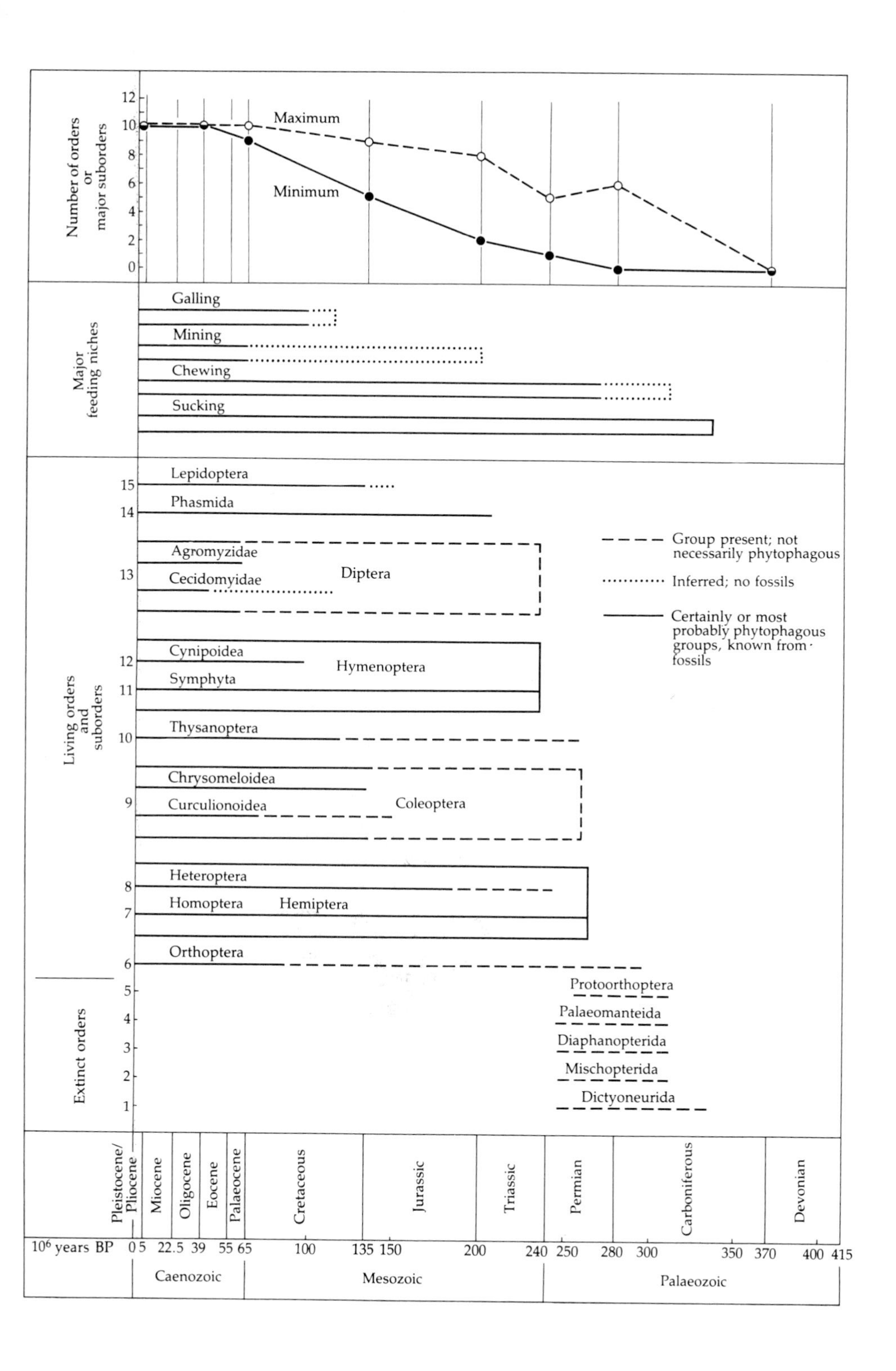

Number of orders or major suborders
12
10
8
6
4
2
0
Maximum
Minimum
Major feeding niches
Galling
Mining
Chewing
Sucking
Living orders and suborders
15
Lepidoptera
14
Phasmida
13
Agromyzidae
Cecidomyidae
Diptera
12
Cynipoidea
11
Symphyta
Hymenoptera
10
Thysanoptera
9
Chrysomeloidea
Curculionoidea
Coleoptera
8
Heteroptera
7
Homoptera
Hemiptera
6
Orthoptera
Extinct orders
5
Protoorthoptera
4
Palaeomanteida
3
Diaphanopterida
2
Mischopterida
1
Dictyoneurida
Group present; not necessarily phytophagous
Inferred; no fossils
Certainly or most probably phytophagous groups, known from fossils
Pleistocene/ Pliocene
Miocene
Oligocene
Eocene
Palaeocene
Cretaceous
Jurassic
Triassic
Permian
Carboniferous
Devonian
10⁶ years BP
0 5 22.5 39 55 65 100 135 150 200 240 250 280 300 350 370 400 415
Caenozoic
Mesozoic
Palaeozoic

Tests of the Hypotheses

Leaving aside (i), which is an important possibility that cannot easily be tested, evidence can be mustered in favour of all these hypotheses.

The fossil history of North American land-plant communities (Knoll *et al.* 1979), and presumably those of other continents, suggests a series of plateaux in plant diversity, with major periods of increase through the Devonian, in the mid-Carboniferous, and in the Upper Cretaceous and Tertiary. It is reasonable to suppose that much of the increase in phytophagous insect diversity reflects this increase in plant diversity.

Not only plant species diversity has increased during evolution, so also have plant size and structural complexity. Sufficient is known of the morphology of the plants of the Devonian and Carboniferous to show that the overall architecture of these plant communities would have been less diverse than that of today's forests (Sporne 1975). As we will show in Chapter 3, in contemporary communities there are more species on structurally diverse plants. An increase in the architectural complexity of plants during evolution should therefore have contributed to an increase in the diversity of phytophagous insects.

Figure 2.13 also suggests that the insects themselves evolved several new ways of exploiting plants. The Dictyoneurida and their contemporaries probably used their piercing and sucking mouthparts to feed on the reproductive organs of plants. In chronological order, foliage feeding by external chewing, sucking from vascular systems and/or leaf tissue, and mining and gall-forming were added to the repertoire of phytophagous insects. The major increases in phytophage diversity were correlated with these new ways of exploiting plants.

Finally, we can ask to what extent insect diversity was, and is, constrained by interspecific competition, at any resource level determined by plant species diversity and architecture, and at any point in the development of insect adaptations to exploit plants. We will argue in Chapter 5 that contemporary communities of phytophagous insects do not appear to be strongly constrained by interspecific competition, and that past communities probably behaved in a similar way. Although relevant fossil data are few, we do know that later North American insect faunas of the Oligocene contained many more families than the earlier Eocene faunas, despite apparently similar environments (Wilson 1978). The number of lepidopteran families, in particular, appears to have expanded continuously since the Eocene, although part or all of the increase could be due to poorer preservation of older fossils. With this important caveat in mind, Wilson's data imply that more species came to exist on roughly the same number of resources, and are consistent with the hypothesis that species richness of phytophagous insects is not seriously constrained by interspecific competition.

If this is true, and with an increasing variety of resources and ways of exploiting resources developing over the long run of the fossil record, it is little wonder that the trend in numbers of major insect taxa (and presumably also species) in Fig. 2.13 is steadily upwards. Most important of all, the available data firmly reject the ecological saturation hypothesis for the evolution of phytophagous insect communities.

2.3.2 *Consequences for later chapters*

An obvious corollary of the conclusions in Section 2.3.1 is that most contemporary communities of phytophagous insects are unlikely to be saturated with species; in the long run we see no reason why many more species would not evolve to share existing resources. Hence questions about what determines the absolute species richness of present-day communities are probably meaningless: 'Why are there about 60 species of butterflies in Britain and not 6 or 600?', has no simple answer.

What we will try to answer in later chapters are questions about *relative* differences in the diversity of phytophagous insect communities and about the fraction of some existing pool of species found on one species of plant, or within one habitat. We may never be able to say why the pool itself is not much larger or smaller than it actually is. As Fig. 2.13 makes plain, for the world as a whole the diversity of major taxa of phytophagous insects has apparently never been constant.

2.4 Reciprocal Evolutionary Influences between Insects and Plants

We draw this chapter to a close by reiterating a simple but crucial message. Obviously, plants have fundamentally influenced the evolution of phytophagous insects. However, phytophagous insects have probably played a very important part in the evolution of plants. It is a guess, but a reasonable one, that there were fewer hurdles to the colonization of plants by insects in the Carboniferous and Permian than there are today (Section 2.1.4), and that phytophagous insects had a strong hand in generating the bewildering biochemical diversity and the rich variety of growth forms, leaf shapes and seasonal phenologies of modern land plants.

These reciprocal influences between plants and insects — 'coevolution' — form the subject of Chapter 7. Much of the fascination of coevolution centres on understanding the detailed responses of individual plant taxa to their herbivores and vice versa, although, as we shall see in Chapter 7, there are problems in deciding just how important is coevolution between individual species of plants and single species of phytophagous insects. In very general terms, the process that has been called 'diffuse coevolution' by Janzen (1980) and Fox (1981) (see Section 7.2.3) may be an important selective force generating plant defences, and ultimately responsible for the steady rise in insect diversity displayed in Fig. 2.13.

There are two reasons for saying this. First, differences in defences among plant taxa will promote insect diversity. Secondly, insects can markedly reduce the abundance of the plants on which they feed (Chapter 7; and see Caughley & Lawton 1981). In so doing, they may reduce interspecific competition between plants, and nurture an increase in the diversity of plant species (e.g. Janzen 1970; Harper 1977; May 1981).

Hence, it is not surprising that a recent palaeobotanical review concluded: 'The dramatic and continual increase in fossil angiosperm diversity observed throughout the Tertiary .... has defied any attempt to model Cenozoic plant diversity as an equilibrium phenomenon. Insect angiosperm coevolution may contribute to an explosive diversity increase (in plants), whose ceiling is not obvious.' (Knoll & Rothwell 1981). In other words, a significant part of the rise in insect diversity depicted in Fig. 2.13, particularly in more recent times, can be attributed to an increase in plant diversity; and plant diversity increases in response to the pressures imposed by herbivores, not least to attacks by phytophagous insects.

These considerations bring us to the present day, and to the problem of what determines the number of species of phytophagous insects exploiting modern plants. It is these questions that we consider in the next chapter.

2.5 Summary

The first land plants evolved in the Devonian, and the first undoubtedly phytophagous insects are known from the Carboniferous. Thereafter, modern orders of phytophagous insects appear steadily throughout the fossil record. The last to appear were the Lepidoptera, coincident with the rise of the angiosperms in the Cretaceous. Several other important groups evolved after the appearance of the angiosperms, and a number of more primitive insects became extinct.

The number of major orders and suborders of phytophagous insects has apparently increased continuously in the fossil record, implying concomitant increase in the number of species. There is no reason for believing that the numbers of extant species and higher taxa of phytophagous insects are constant or in unchanging equilibrium through evolutionary time. The complementary and parallel evolution of host plants, themselves a major fraction of all species, makes the idea of a simple set point of insect diversity illogical.

Most insect orders are not phytophagous (20 of 29), but over half of insect species are phytophagous. This pattern implies major evolutionary barriers to eating plants. Since the barriers were breached, however, the great diversity and biomass of plants have been a spectacular resource. We suggest that the major hurdles to phytophagy for insects are desiccation, attachment, nourishment, and plant chemical defences.

Reciprocal adaptation and counteradaptation between plants and insect phytophages has been an important mechanism driving a steady increase in plant and insect diversity over the broad sweep of the fossil record.

Chapter 3
The Major Determinants of Diversity

3.1 Introduction

How many species of phytophagous insects exploit particular kinds of plants within one geographic region? How, for example, does the number of insects on birch in Britain compare with the number of species known to attack broom, or bracken? More generally, how do insect diversities on herbs compare with those on trees from the same region, or diversities on rare plants compare with those on common plants, and so on? What are the major correlates of diversity for contemporary insect–plant associations, and what in the way of mechanism and cause can we infer from these major patterns?

Our emphasis in this chapter is upon total numbers of phytophage species attacking single species or genera of plants within large geographic regions. We will deal in later chapters with how such insects came to colonize the plants, their taxonomic composition, their ecological roles and what happens on a local scale.

3.2 Species–Area Relationships on a Geographic Scale

3.2.1 *Examples*

The total number of insect species associated with plants is strongly influenced by how common and widespread or local and rare the plants are. Very simply, geographically widespread species of plants have more sorts of insects feeding on them than similar but less widespread species, a phenomenon first documented for Hawaiian trees by Southwood (1960a; Fig. 3.1). Southwood (1961a) also observed how several species of conifer are more extensively distributed, and host more insect species, in Russia than in Britain. We now also know that within the British Isles themselves widespread species of trees are hosts for more species of insects than are rare trees (Fig. 3.2a). A growing number of other, similar examples are available, summarized in Table 3.1 from which a sample is illustrated in Fig. 3.2b–f.

Often a double-logarithmic plot of the number of species of insects associated with each species of plant against the size of the geographic range of that plant (a 'species–area' relationship) provides a good working fit to the data. The pattern holds good for both genera (Fig. 3.2a) and species (Fig. 3.2b–h) of plants; for entire faunas (Fig. 3.2a–c); or for some taxonomically defined subportions, such as leaf- and stem-mining agromyzid flies on British umbellifers (Fig. 3.2d), leaf-mining Lepidoptera on Californian oaks (Fig. 3.2h), or American gall-forming cynipid wasps (Fig. 3.2e–g). The double-log plot also gives a good fit for the same species of plant in different countries (Fig. 3.2c).

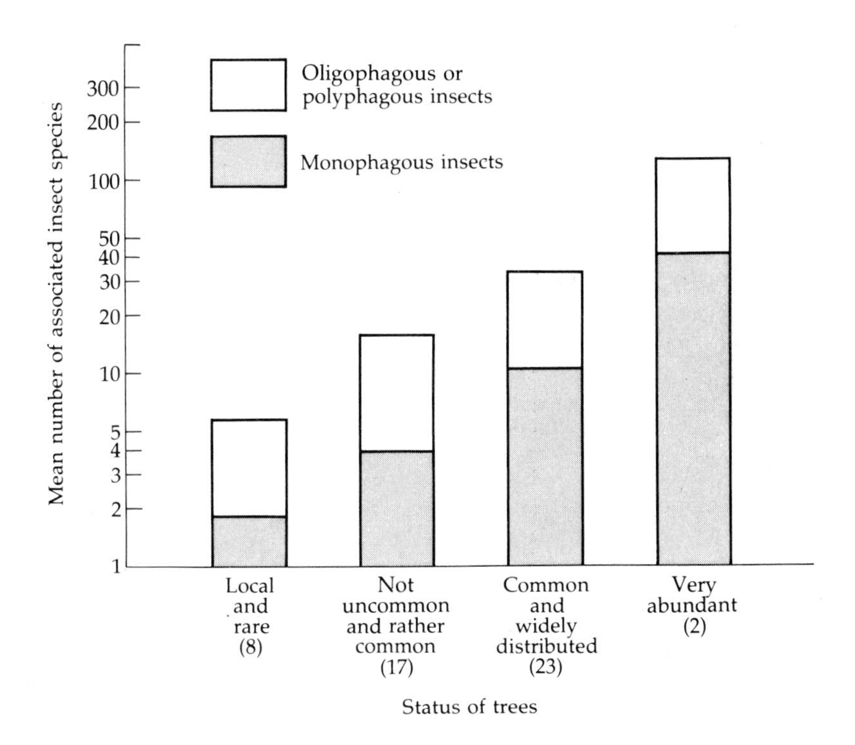

Fig. 3.1. Hawaiian trees grouped according to abundance (pre twentieth century) with the mean number of insect species associated with each group. The number of species of tree is given below histograms. Note that the insect-richness scale is logarithmic. Monophagous insects feed on only one species of tree, oligophogous and polyphagous insects on more than one species. (From data in Southwood 1960a.)

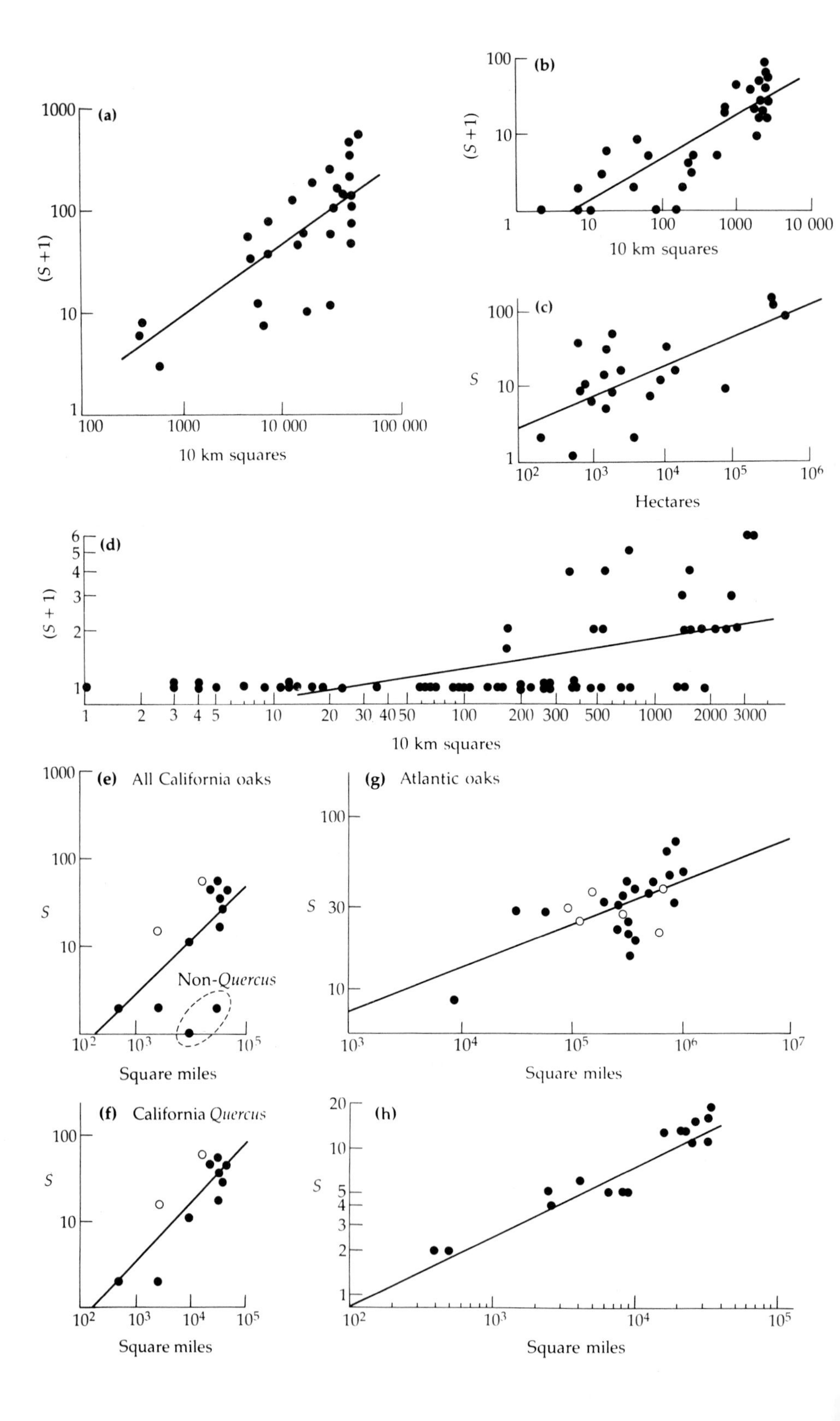

(a)
1000
100
10
1
(S + 1)
100
1000
10 000
100 000
10 km squares
(b)
100
10
1
(S + 1)
1
10
100
1000
10 000
10 km squares
(c)
100
10
1
S
10^2
10^3
10^4
10^5
10^6
Hectares
(d)
6
5
4
3
2
1
(S + 1)
1
2
3
4
5
10
20
30
40
50
100
200
300
500
1000
2000
3000
10 km squares
(e) All California oaks
1000
100
10
S
Non-Quercus
10^2
10^3
10^5
Square miles
(g) Atlantic oaks
100
30
10
S
10^3
10^4
10^5
10^6
10^7
Square miles
(f) California Quercus
100
10
S
10^2
10^3
10^4
10^5
Square miles
(h)
20
10
5
4
3
2
1
S
10^2
10^3
10^4
10^5
Square miles

The amount of variation explained by these species–area correlations varies markedly (Table 3.1). Possible reasons for the residual scatter (unexplained variation) around species–area regression lines are considered in subsequent sections and in Chapter 4.

3.2.2 *Explanations*

Species–area relationships are common in ecology, be they for insects on plants, birds on islands, or snails in ponds (Strong 1979; Connor & McCoy 1979; May 1981). A number of hypotheses have been advanced to describe the mechanisms involved. One of these was developed specifically for associations between insect species and plants. In a form appropriate to insect–plant interactions, the more interesting of these hypotheses are summarized below. They are not mutually exclusive explanations.

(i) The *habitat-heterogeneity* hypothesis was first developed by Williams (and summarized in Williams 1964), who proposed that large areas embrace more kinds of habitat and hence support more species; more habitats support more species because habitat requirements vary among species.

Fig. 3.2. Plots of number of insect species (S or $S + 1$) against host plant range for various associations.
(**a**) All species of phytophages on genera of British trees (C.E.J. Kennedy & T.R.E. Southwood, unpublished data).
(**b**) All species of phytophages on British perennial herbs (Lawton & Schröder 1977).
(**c**) Insect pests on cacao — each point is for a different country (Strong 1974a).
(**d**) Agromyzid flies on British Umbelliferae (Lawton & Price 1979).
(**e–g**) Cynipid gall wasps on North American oaks; various regions and oak species combinations. The open circles are shrub species, the remainder trees (Cornell & Washburn 1979).
(**h**) Leaf-mining Lepidoptera on North American oaks (Opler 1974).

Ranges measured in '10 km squares' refer to records of host presence over the British National Grid, from Perring & Walters (1962). Table 3.1 summarizes the slopes and correlation coefficients of the fitted regression lines, with the exception of the data on British trees (a), discussed in Section 4.1.3.

Table 3.1. Summary of large-scale species–area relationships for insects and host plants. S = number of insect species, A = size of geographic range of the host plant, c and z are constants. The data are divided into two groups according to the nature of the statistical transformations used. Authors who included host plants with no records of phytophagous insects in the group(s) in question have used the transformation $(S + 1)$ to avoid taking the logarithm of zero. All the regressions are statistically significant.

Authors	Insect–plant system	Slope of $\log S$ (or $\log (S + 1)$ where used) against $\log A$ z	Proportion of variation in insect diversity explained by size of geographic range r^2	
$(S = cA^z)$				
Auerbach & Hendrix (1980)	All taxa on ferns, world review	0.29	0.18	
Banerjee (1981)	Pests on tea, world review	0.34	—*	
Cornell & Washburn (1979)	Cynipid gall wasps on oak trees			
	Atlantic region	0.25	0.41	Fig. 3.2g
	California region	0.63	0.33	Fig. 3.2e
Opler (1974)	Leaf-mining Lepidoptera on American oaks	0.47	0.90	Fig. 3.2h
Strong (1974a)	Pests on cacao, world review	0.90	0.43	Fig. 3.2c
Strong & Levin (1979)	All taxa, British plants			
	Herbs	0.49	0.28	Fig. 3.7b
	Shrubs	0.66	0.49	Fig. 3.7b
	Trees	0.89	0.58	Fig. 3.7b
Strong *et al.* (1977)	Pests on sugar cane, world review	0.45	0.50	
$(S + 1 = cA^z)$				
Claridge & Wilson (1981)	Leaf-hoppers (Typhlocybinae) on British trees	0.34	0.16	
Claridge & Wilson (1982)	Leaf-miners on British trees	0.34	0.19	
Lawton (1983b)	All taxa on bracken (*Pteridium*), four world regions	0.70	0.96	

Lawton & Price (1979)	Agromyzids on British Umbelliferae	0.14	0.32	Fig. 3.2d
Lawton & Price (1979)	Microlepidoptera on British Umbelliferae	0.16	0.24	
Lawton & Schröder (1977)	All taxa, British plants			
	Perennial herbs	0.54	0.71	Fig. 3.2b and Fig. 3.7a
	Shrubs	0.45	0.85	
	Weeds and other annuals	0.47	0.59	
	Monocots	0.39	0.51	
Neuvonen & Niemela (1981)	Macrolepidoptera on Finnish deciduous trees and shrubs	0.30	—*	

* Multiple regression with other variables; individual correlation coefficients not given.

Applied to host plants, this hypothesis has two important facets that we will concentrate upon: (a) the variety of habitats occupied by a widely distributed plant, and (b) looking ahead to Section 3.3, the greater variety of microhabitats present in large, structurally diverse plants such as trees, compared with structurally simple plants, for example rushes.

(ii) The *encounter-frequency* (or *frequency of exposure*) hypothesis was developed specifically to account for the larger numbers of insect species on widespread plants (Southwood 1961b). The probability of an insect species encountering a potentially suitable new host plant depends on the frequency of exposure or encounter. Other things being equal, encounter frequency depends on the distribution and abundance of the potential host plant — more widespread plants receive greater exposure. Of course the rate of encounter between plants and insects is also influenced by the abundance of the phytophagous insects present in the region and their dispersal behaviour. Once encountered, the chances that an insect colonist will be physiologically adapted to a new host plant depend on the biological (physical, chemical, phenological, etc.) similarity between the new plant and the insect's normal host(s) — its predilection. Southwood illustrated this mechanism by arguing that the recruitment of insects by a new host is similar to the accumulation of insects resistant to an insecticide; breadth of application increases the chance of finding resistant insects, just as the distribution and abundance

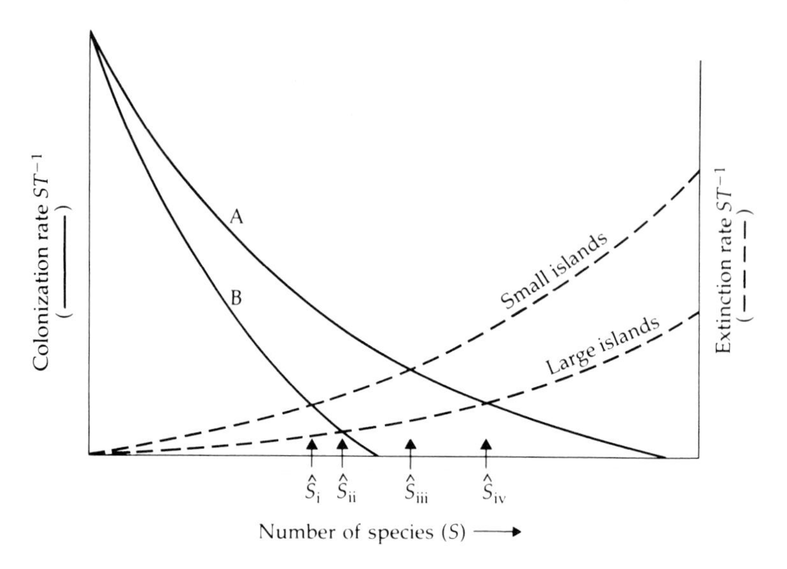

Fig. 3.3. The Equilibrium Theory of Island Biogeography. The number of species on a host plant 'island' is set by a dynamic balance between the rate at which species colonize that host, and the rate at which established species become extinct. Colonization rates maybe higher (curve A) on large 'islands' (hosts with large ranges) and/or on plants with many close relatives; colonization rates are lower (curve B) on hosts with small ranges and/or few or no close relatives (taxonomically isolated hosts), and/or on biochemically unusual or particularly toxic hosts. Extinction rates are influenced by island size (see text). $\hat{S}_i$–$\hat{S}_{iv}$ are various possible equilibrium species numbers under this hypothesis. For example, theory predicts that there will be fewer species at equilibrium $\hat{S}_i$ (a plant with a small total range and/or one that is taxonomically and biochemically isolated) than at equilibrium $\hat{S}_{iv}$ (a widespread non-noxious host with many relatives).

of new hosts determine the chances of recruiting phytophages. Widespread plants with high exposure therefore recruit more species of herbivores than rare plants.

(iii) *The Equilibrium Theory of Island Biogeography* was first proposed by Preston (1960, 1962) and elaborated by MacArthur and Wilson (1967). Janzen (1968a, 1973) first suggested its application to insect plant interactions by treating species of host plants as 'islands' in a 'sea' of other vegetation. The theory of island biogeography proposes that the number of species inhabiting an island is set by a dynamic balance between rates of immigration (colonization) and extinction (Williamson 1981; Fig. 3.3). MacArthur and Wilson suggested that colonization rates depend upon the physical distance of the island from a pool of potential colonists. For example, host plants of British Lepidoptera are isolated from the same species on mainland Europe, and this geographic isolation greatly affected recolonization of the island after the Pleistocene glacial periods (Dennis 1977). By analogy, a species of plant may be 'isolated' from insects because it is anatomically, morphologically, biochemically or otherwise biologically unusual. (Southwood 1961a and b, 1977a; Janzen 1968a, 1973).

In MacArthur and Wilson's original form, the theory of island biogeography envisaged colonization rates to be independent of island size. However, larger 'targets' will almost certainly have higher colonization rates (Connor & McCoy 1979; Strong 1979; Fig. 3.3). So modified, the immigration (colonization) component of the Equilibrium Theory is similar to hypothesis (ii), the encounter-frequency hypothesis (Southwood 1977a).

Extinction rates are assumed in the Equilibrium Theory to be inversely proportional to total population size, and hence inversely proportional to island area; extinctions will therefore occur more often on small islands or on plants with small ranges.

Drawing together these effects (Fig. 3.3), the theory predicts an equilibrium number of species $\hat{S}$ that is higher for plants with larger ranges, lower on geographically isolated or rare plants, and lower on plants that are physically, morphologially, anatomically, biochemically, or otherwise biologically 'isolated'.

3.2.3 *Some Tests of These Hypotheses*

Habitat Heterogeneity

Habitat heterogeneity (hypothesis (i)) undoubtedly contributes strongly to species area relationships for insects on plants. Thus the presence alone of a suitable species of host is but the first of many habitat requirements to be met before a population of phytophagous insects can establish itself. In consequence, most plants have geographic ranges

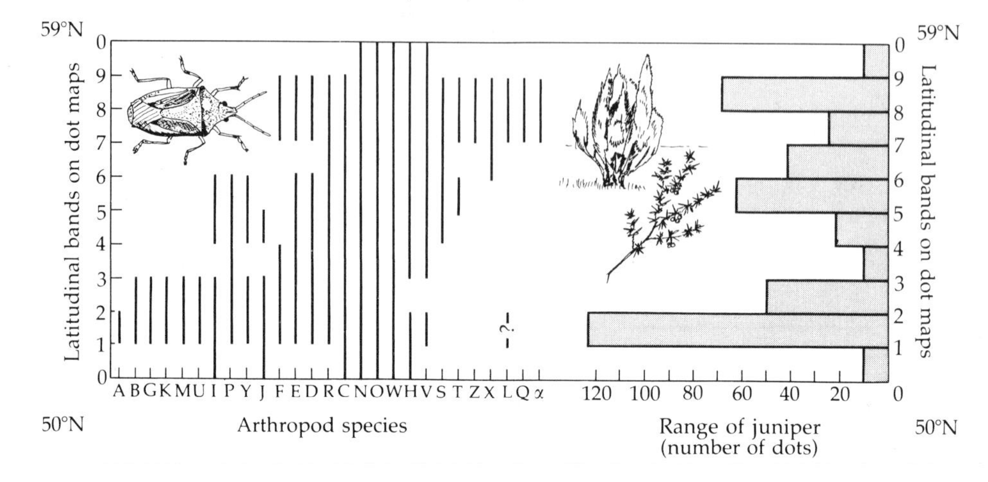

Fig. 3.4. Different arthropods (mainly but not entirely insects) inhabit different sections of their host's distribution. Each vertical line indicates the meridional distribution of a specific arthropod species associated with juniper in Britain. (From Ward 1977, after Strong 1979.) (The distributions of both the host plant and its arthropods were mapped using 10 km square 'dot maps'.)

larger than the ranges of their insect consumers, and different species of insects exploiting the same species of host occupy different, idiosyncratic portions of the host's total range (Strong 1979). Examples are provided by Ward (1977) and elaborated by Strong (1979) for the insect fauna of juniper (*Juniperus communis*) in Britain (Fig. 3.4), by McClure and Price (1976) for North American leaf-hoppers of the genus *Erythroneura* attacking sycamore* (*Platanus occidentalis*), and by Denno *et al.* (1981) for the Auchenorrhyncha (Delphacidae and Cicadellidae) feeding on *Spartina patens* in salt marshes along the eastern seaboard of America. Some examples for British Lepidoptera are shown in Fig. 3.5.

Climate undoubtedly plays an important role in limiting the distribution of phytophagous insects within the ranges of their hosts, although critical studies are few.

* In North America *Platanus occidentalis* is known as the sycamore, but in Britain this tree is called the plane and *Acer pseudoplatanus* is termed the sycamore.

The white admiral butterfly (*Ladoga camilla*) is confined to southeast England, although the food plant of its monophagous caterpillar, the honeysuckle *Lonicera periclymenum*, is virtually ubiquitous (Fig. 3.6). Pollard (1979) shows that bird predation on the late larvae and pupae is the 'key factor' (*sensu* Varley & Gradwell 1960) in the population dynamics of *Ladoga*. When summer temperatures are low, development is slow, and the birds have more time to find the caterpillars and pupae and so impose heavy mortality. Pollard argues that *Ladoga* is excluded from large parts of the British Isles by summer temperatures too low to allow sufficient numbers to escape predation and maintain the population. The dramatic expansion of range shown by the white admiral in the 1930s and early 1940s (Fig. 3.6) correlates with June temperatures during this period higher than usual in Britain. (Pollard 1979). Several other British butterflies expanded their ranges at this time (Heath 1974), with no appreciable change in the distribution of their host plants.

A second example is provided by the thimbleberry aphid (*Masonaphis maxima*) in North America. This aphid is confined to only a small part of the total range of its sole host, thimbleberry, *Rubus parviflorus* (Gilbert 1980). The host's range extends from Alaska to Mexico and from the Pacific to the Great Lakes. The aphid is found only

Fig. 3.5. (Overleaf) Host ranges and insect ranges in British Lepidoptera. Plant ranges from Perring & Walters (1962); Lepidoptera from Heath (1973, 1983).

(a) *Prunus spinosa* is abundant and widespread throughout England and Wales. The monophagous *Strymonidia pruni* is confined to the south-east Midlands. *Thecla betulae*, which also feeds on birch (a ubiquitous tree), is confined to southern England and Wales.

(b) *Erebia epiphron* occurs in restricted upland areas within the range of its abundant and widespread host grass, *Nardus stricta*.

(c) Two moths that both feed on birch (*Betula*) and alder (*Alnus*), trees that are virtually ubiquitous. *Endromis versicolora* is extremely rare, and is now confined to Scotland; both south of England populations are apparently extinct. *Drepana falcataria* is apparently absent from parts of Scotland where both hosts grow, and is thinly distributed and absent from many areas where both hosts grow in central and northern England.

(d) Three butterflies that all feed on nettle (*Urtica dioica*), a ubiquitous host, reach the northern limits of their British ranges in different parts of the country; *Polygonia c-album* is the most restricted, *Aglais urticae* the least, and *Inachis io* is intermediate in range.

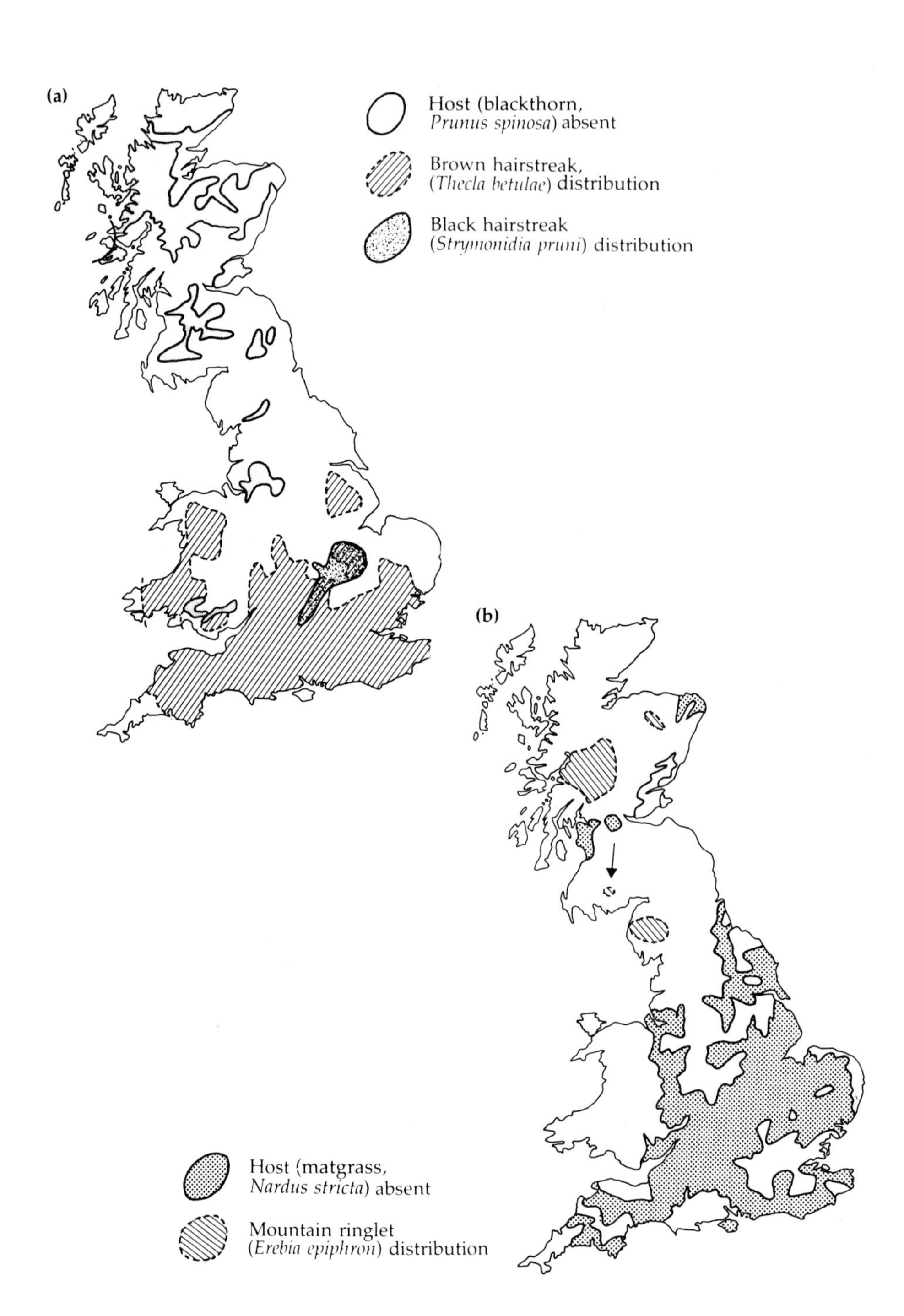
(a)
Host (blackthorn, *Prunus spinosa*) absent
Brown hairstreak, (*Thecla betulae*) distribution
Black hairstreak (*Strymonidia pruni*) distribution
(b)
Host (matgrass, *Nardus stricta*) absent
Mountain ringlet (*Erebia epiphron*) distribution

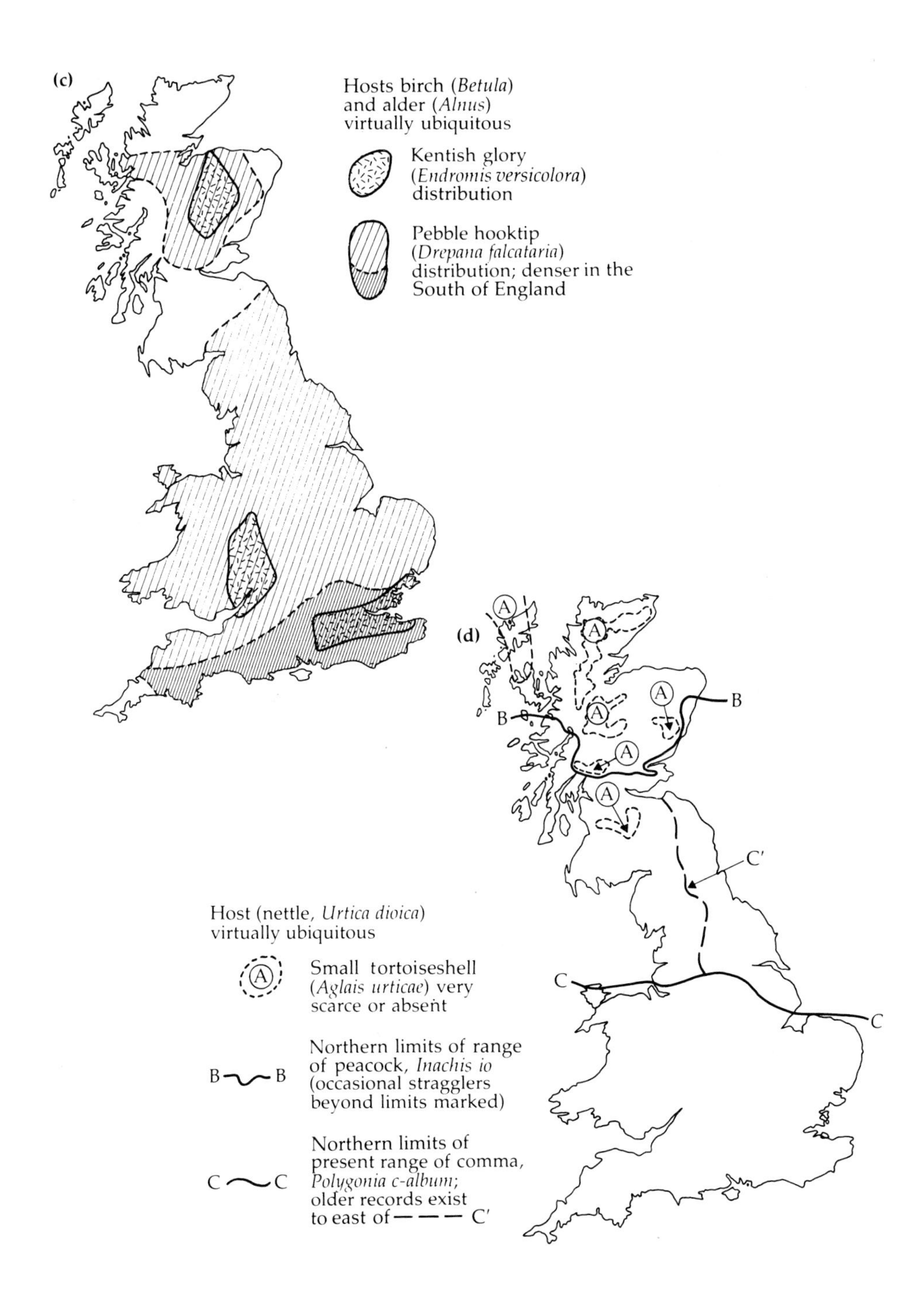
(c)
Hosts birch (*Betula*)
and alder (*Alnus*)
virtually ubiquitous
Kentish glory
(*Endromis versicolora*)
distribution
Pebble hooktip
(*Drepana falcataria*)
distribution; denser in the
South of England
(d)
A
B
B
C'
C
C
Host (nettle, *Urtica dioica*)
virtually ubiquitous
Small tortoiseshell
(*Aglais urticae*) very
scarce or absent
Northern limits of range
of peacock, *Inachis io*
(occasional stragglers
beyond limits marked)
B—B
Northern limits of
present range of comma,
Polygonia c-album;
older records exist
to east of — — — C'
C—C

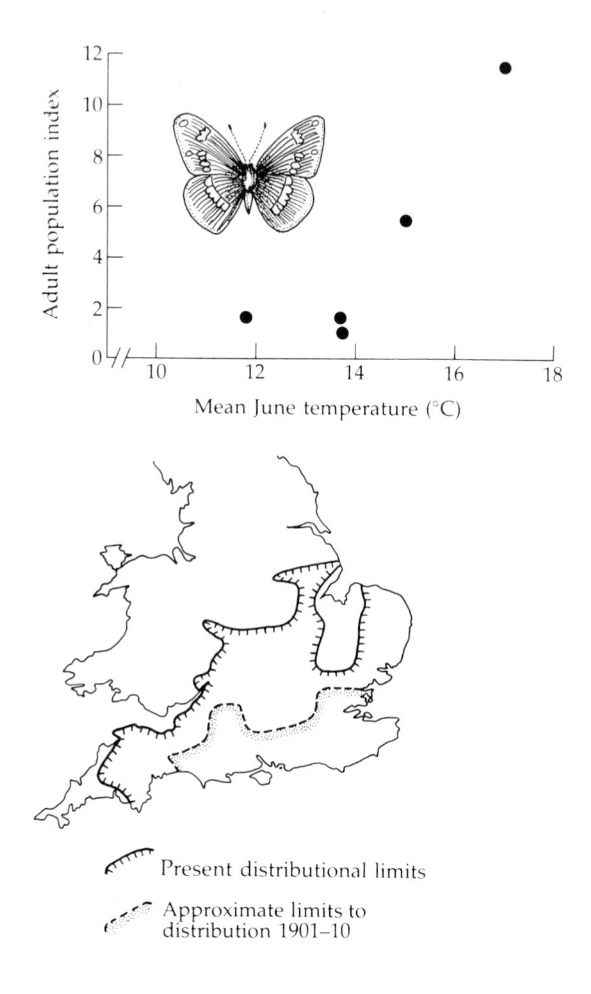

Fig. 3.6. Present and past northern limits to the distribution of the white admiral butterfly (*Ladoga camilla*) in England. Its sole larval host, honeysuckle (*Lonicera periclymenum*), is abundant and widespread throughout England, Wales and southern Scotland, becoming sparse, but not absent in the highlands and the far north. The distribution and abundance of *Ladoga* is much influenced by temperature. At one locality within the present range of the butterfly (Monks Wood), adult numbers are closely correlated with mean June temperature over five years ($r = 0.89$; $p < 0.05$), apparently because lower temperatures lead to poorer late-larval and pupal survival (Pollard 1979).

along the Pacific coast, from northern Vancouver Island to Santa Cruz in California, and inland for some 100 km along the Frazer Valley, but not more than 30 km elsewhere. These limits are, again, apparently climatic, the edge of the aphid's range being determined by the time (three generations) needed to maintain existing populations each season. Colonization of new areas is even more demanding and requires the aphids to complete five generations in a season. If they cannot, colonization fails (Gilbert 1980).

There are no reasons for thinking that these two examples are atypical. We expect climatic heterogeneity to contribute substantially to total habitat heterogeneity, and hence to the generation of geographic species-area relationships for insects on host plants.

The cumulative effects of different sorts of habitat heterogeneity are shown by communities of cynipid gall-wasps on oaks, in the Californian and Atlantic regions of North America (Cornell & Washburn 1979). These two areas are separated by vast regions where oaks are scarce, and have no species of oaks in common. Both show significant species–area relationships for cynipids on oaks (Table 3.1, Fig. 3.2e–g), but the regression for the Californian region has a higher slope and also a higher average diversity of cynipids; and as might be predicted from the habitat heterogeneity hypothesis, topographical diversity, oak varietal diversity, and altitudinal variation are much greater in the Californian region than in the relatively flat and homogeneous Atlantic region.

Finally, focusing down to local habitat types, such as woodlands, pastures and ponds, reveals that British Umbelliferae growing in several habitats support more species of leaf-mining agromyzid flies than umbellifers confined to only one or two habitats (Fowler & Lawton 1982). Indeed, this appears to be the principal mechanism generating the species area relationship for these particular organisms, shown in Fig. 3.2d.

Encounter Frequency

Man's forests, farms and gardens have increased the exposure of many species of plants to potential insect colonists. In the last fifty years, for example, afforestation has greatly increased the abundance of various conifers growing in Britain. A few of these trees have long been present in many of the habitats where they grow now, so the number of habitats occupied has probably increased less than plant abundance in those habitats. As predicted by hypotheses (ii) and (iii), this increase in encounter frequency has led to the colonization of conifers by several new species of insect and to an increase of the range of other insects in Britain (Southwood 1957, 1961a; Hammond 1974; Winter 1974; Entwistle 1978).

Decreasing abundance of native hosts within habitats causing loss of insect species has been observed on juniper in Britain (Ward 1977), and leads us logically to a brief examination of insect extinctions.

Extinctions

Extinctions of phytophages may occur through the complete loss of suitable habitat within the total range of a host plant, in which case they are an illustration of hypothesis (i). Alternatively, extinctions may occur by reduction in the abundance of a host plant within a habitat, as indicated by the case of insects on juniper just referred to; such extinctions are probably generated by chance events operating on small, vulnerable populations, in the manner envisaged by MacArthur and Wilson (hypothesis (iii)).

Extinctions of animals as small as insects may easily go unnoticed, but in Britain some insect extinctions have been recorded. The black-veined white butterfly (*Aporia crataegi*), which feeds on widespread and abundant trees (*Crataegus* and *Prunus*), became extinct at the turn of the century. This butterfly was on the edge of its range in Britain and it is believed that climatic changes interacting with predation or parasitism may have been responsible (just as with the white admiral's expansion described above). In most other cases of historic extinctions of British Lepidoptera (probably 15 species) (Burton 1975; Dennis 1977) or Heteroptera (probably only one) (Southwood 1957), there have been obvious reductions in either the overall abundance of the host plant, as with juniper (Ward 1977), or partial to complete destruction of the insect's particular habitat, complemented by inclement weather (Dempster & Hall 1981; Pyle *et al.* 1981; Robertson 1981). In none of these extinctions have the host plants completely disappeared. For example, remnant British populations of the large blue butterfly (*Maculinea arion*) were pushed to final extinction by drought in 1975 and

1976 followed by disastrously poor summer weather in 1977 and 1978. The weather dealt the *coup de grâce* to a species much reduced in abundance by the loss of suitable habitat, which included, among other things, ants of the genus *Myrmica* to tend the large caterpillars underground over the winter (Brian 1977). The food plant of the large blue's early instars, thyme (*Thymus*), although reduced in abundance because of changes in the large blue's habitat, is still common and widespread in Britain.

Synthesis

Summarizing, species–area relationships on a geographic scale are generated by very simple mechanisms. Widespread plants occur in more habitats and are more exposed to potential colonists. For both reasons, if plant species become more common they recruit phytophagous insects. On the other hand, if the range of a plant contracts it may lose phytophagous insects because it no longer grows in suitable habitats and because small remnant populations of insects are vulnerable to chance extinction.

Other aspects of the hypotheses outlined in Section 3.2.2, particularly the modifying effects of physical and biochemical similarity between host species, are discussed in Section 3.4.

3.2.4 Some Problems of Interpretation and Data

Geographical species–area relationships for insects on plants have recently been the focus of several controversies.

First, Kuris *et al.* (1980) dismissed many of these relationships as sampling artefacts, on the grounds that widespread plants have received more intense investigation, inevitably revealing more sorts of herbivores. This argument overlooks the point that widespread plants must be given more attention if all parts of their ranges are to be properly investigated. It does not follow that numbers of phytophage species sufficient to annihilate reported species–area relationships lurk undiscovered on rare plants. This is particularly true in Western European countries where the insect fauna is very well known. For example, if equal numbers of samples are taken from different species of plants of the same type (trees), the relative species richness of their sampled faunas is similar to that obtained from the published faunal lists on which Fig. 3.2a is based (Southwood *et al.* 1982a). Detailed discussion of the points made by Kuris *et al.* is provided by Lawton *et al.* (1981) and Rey *et al.* (1981).

Secondly, three recent studies (Claridge & Wilson 1976, 1978; Futuyma & Gould 1979; Karban & Ricklefs 1983) found that widespread and common species of trees do not necessarily support the richest faunas of particular groups of herbivores at one restricted locality, or within a small part of a host's total range. Such data could be misinter-

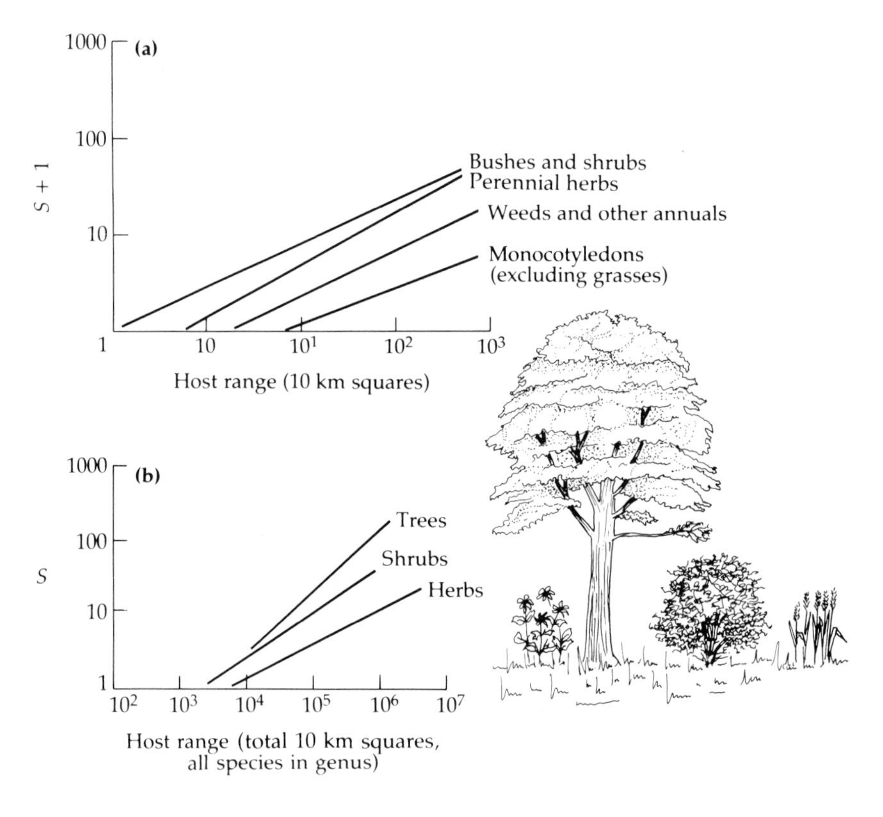

Fig. 3.7. (**a**) Number of insect species on British plant species. The series from bushes and shrubs to monocotyledons in one of decreasing 'architectural complexity' in the above-ground parts of the plants; number of associated phytophagous insect species decreases as architectural complexity declines (Lawton & Schröder 1977). (**b**) As **a** for genera of British plants (Strong & Levin 1979).

preted to mean that size of host plant range has little or no influence on insect species richness. This would be a mistake because, as we have seen, the extensive fauna of a widespread plant invariably contains local species, found only in restricted parts of its range. Furthermore, the insect species found on a particular species of plant at one locality are influenced by several variables: the local abundance of that plant, both its density and its extent, and the presence of other plants, both alternative hosts and non-hosts (Chapter 6). Hence, we cannot expect widespread plants with rich total faunas *necessarily* to host more species of insects at one locality.

3.3 Plant Structural Diversity

3.3.1 Examples

Up to this point we have ignored another major determinant of the diversity of insect communities on plants, namely the size, growth form and variety of the above-ground parts of the host, termed by Lawton and Schröder (1977, 1978) 'plant architecture'. Two major plant attributes are involved: *size*, the spread of plant tissue through different positions in space, and the *variety* of plant structures, both in form and persistence (Southwood *et al.* 1979; Lawton 1983a). The effects of architecture are not discriminated in Fig. 3.2 and Table 3.1, because only plants of a particular type (e.g. trees or perennial herbs) have been grouped together for analysis. The influence of plant architecture is made clearer in Fig. 3.7a, which shows species–area regressions for British woody shrubs, perennial herbs, weeds and other annuals, and monocotyledons (excluding grasses, for which food-plant records tend to be unreliable) (Lawton & Schröder 1977, 1978). The four lines do not differ significantly from one another in slope, only in intercept. British trees (Fig. 3.2a) support even more kinds of insect. Hence comparing plants having similar-sized geographic ranges, these data suggest that phytophagous faunas become progressively more depauperate in the sequence: trees > woody shrubs > perennial dicotyledonous herbs > weeds and other annuals > monocotyledons (excluding grasses). This sequence is one of decreasing size, structural complexity and diversity in the above-ground parts of the plant.

Strong and Levin (1979) in a parallel, but quite independent, study

(Fig. 3.7b) to the one executed by Lawton and Schröder, corroborated most, but not all, of the important details shown in Fig. 3.7a. The differences between the two studies are a salutory reminder of effects of apparently innocent differences in analytical methods. Three methodological differences are apparent. While Lawton and Schröder used only data from the *Biological Flora of the British Isles* (in *Journal of Ecology*), Strong and Levin used other published sources. Did these additional sources sharpen the picture or introduce bias? Lawton and Schröder worked on plant species and included species without insect records; Strong and Levin used plant genera and specifically excluded genera with no records of associated insects. Lastly, and perhaps most significantly, Strong and Levin grouped grasses with monocotyledons, while Lawton and Schröder excluded grasses. The classification of plants as 'weeds' may also pose problems; the weeds of newly disturbed ground are very different in persistence (and many other qualities) from those of a permanent pasture or of an abandoned field.

Both studies confirm that size of the geographic range of plant species (or genera) explains most of the variation in richness of associated phytophagous insects, and that plant architecture accounts for a considerable part of the rest. Strong and Levin were able to quantify this finding, showing that for these data host range has approximately 2.5 times as much influence upon insect species number as division into the three growth forms of trees, shrubs and herbs. The two studies disagree as to whether monocotyledons and weeds have impoverished faunas compared with perennial herbs: this point is clearly where the differences in methods will have the greatest influence.

The influence of plant architecture on insect species richness has been corroborated by several studies, summarized in Table 3.2 (Lawton 1983a). A particularly clear example is provided by Moran (1980), who plotted the number of species of phytophagous insects known to attack each of 28 species of *Opuntia* cacti found in the deserts and grasslands of America against the 'architectural rating' of the plant (Fig. 3.8). Tall species of *Opuntia* (> 5 m when mature) with numerous large, spiney, strongly tuberculate cladodes borne on large, woody stems, host up to 30 species of phytophagous insects. Fewer than five species are found on small, architecturally simple cacti.

To date, only one study has failed to detect significant architectural effects on insect diversity; shrub oaks are not noticeably impoverished in

Table 3.2. Ratios of species richness for phytophages between plants of different architectures. Ratios are: species number on architecturally more complex plants/species on architecturally less complex plants, measured under comparable conditions. Where ranges or minimum values are given, the differences depend in part on size of geographic range. (From Lawton 1983a).

Plants (architecturally more complex plants given first)	Insects	Geographic region	Ratio of average herbivore species richness	Reference
Bushes/Perennial herbs	All phytophages	Britain	1.3–2.5	Lawton & Schröder (1977)
Perennial herbs/Weeds	All phytophages	Britain	1.5–2.7	Lawton & Schröder (1977)
Weeds/Monocots	All phytophages	Britain	1.3–2.9	Lawton & Schröder (1977)
Shrubs/Cynareae (herbs)	All phytophages	Europe	1.3	Lawton & Schröder (1978)
Trees/Shrubs	All phytophages	Britain	2.7	Strong & Levin (1979)
Shrubs/Herbs	All phytophages	Britain	2.1	Strong & Levin (1979)
Trees/Shrubs	Microlepidoptera	Britain	2.0	Price (1977)
Shrubs/Herbs	Microlepidoptera	Britain	3.5	Price (1977)
Largest trees/Shrubs	Macrolepidoptera	Finland	> 1.5	Neuvonen & Niemela (1981)
Largest/Smallest Umbelliferae	Agromyzidae	Britain	> 4.7	Lawton & Price (1979)
Largest/Smallest *Opuntia* cacti	All phytophages	North and South America	5	Moran (1980)
Trees and shrubs/Herbs and grasses	Macrolepidoptera	Finland	> 10	Niemala *et al.* (1982)

their cynipid gall wasp faunas, compared with larger species of oaks (Fig 3.2e–g; Cornell & Washburn 1979).

3.3.2 *Explanations*

Exactly the same explanations suffice to account for plant-architecture effects as account for species–area effects (Section 3.2.2). Larger, more complex plants, with a greater variety and persistence of above-ground

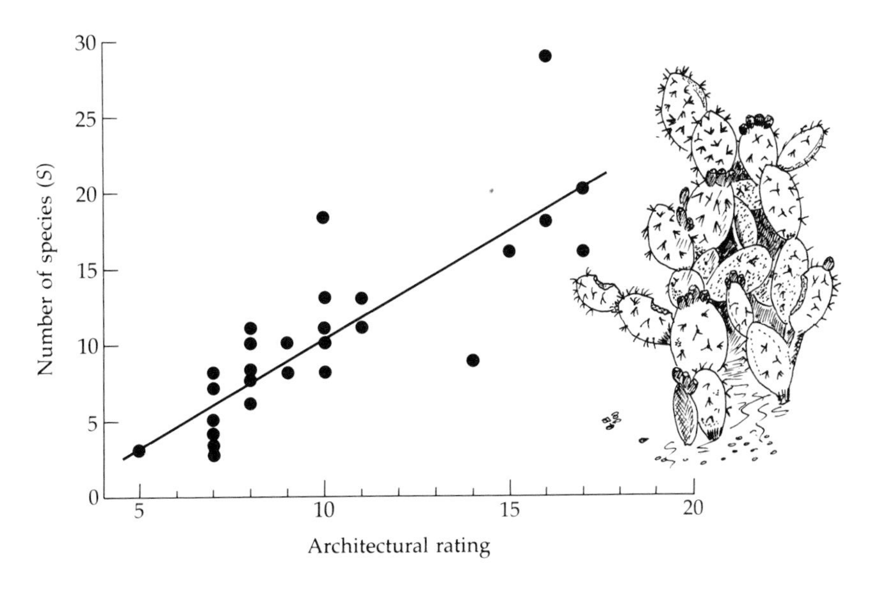

Fig. 3.8. The number of phytophagous insect species associated with 28 *Opuntia* cactus species in North and South America, as a function of the 'architectural rating' of each species of cactus. Architectural rating is the sum of five separate variables, each measured on a score from 1 to 4: height of mature plant, mean number of cladodes on a large plant, cladode size, degree of development of woody stems and cladode complexity from smooth to strongly tuberculate. (After Moran 1980.)

parts, provide a greater range of microhabitats for colonization by insects (hypothesis (i) above) (Lawton 1978; Southwood 1978a), but may also increase plant exposure and support larger total populations, reducing extinction probabilities (hypotheses (ii) and (iii)). We cannot apportion relative roles with any confidence, although we expect that microhabitat heterogeneity plays a major part (Lawton 1983a).

Trees present a greater variety of niches for insects than do, say, herbs in at least three ways. First is microclimate; layers of the canopy often differ markedly in temperature, humidity and insolation, with the result that insects differ in which parts of the canopy they exploit (Askew 1962; Billany *et al.* 1978; Claridge & Reynolds 1972; Darlington 1974; Gross & Fritz 1982; Nielsen & Ejlersen 1977). Second is phenology and age; young trees and mature trees may support different species (Askew 1962; Lemen 1981; Martin 1966; Morris 1974; Niemela *et al.* 1980;

Washburn & Cornell 1981; Watanabe 1981; Zelazny & Pacumbaba 1982). Third and last, insects require more from a host plant than just something to eat; they also require oviposition sites, hiding places, somewhere to overwinter, and so on. In all these respects trees provide more opportunities than small perennial herbs. The importance of overwintering and oviposition sites on the twigs, bark and buds of trees has been shown by Claridge and Wilson (1976), Malicky *et al.* (1970) and Morris (1974).

The importance of some of the more permanent parts of a woody bush for the phytophagous insects that live on it is illustrated in Fig. 3.9. About half of the 22 species found on *Larrea tridendata* in Arizona feed, oviposite or hide on old leaves and woody stems (Schultz *et al.* 1977).

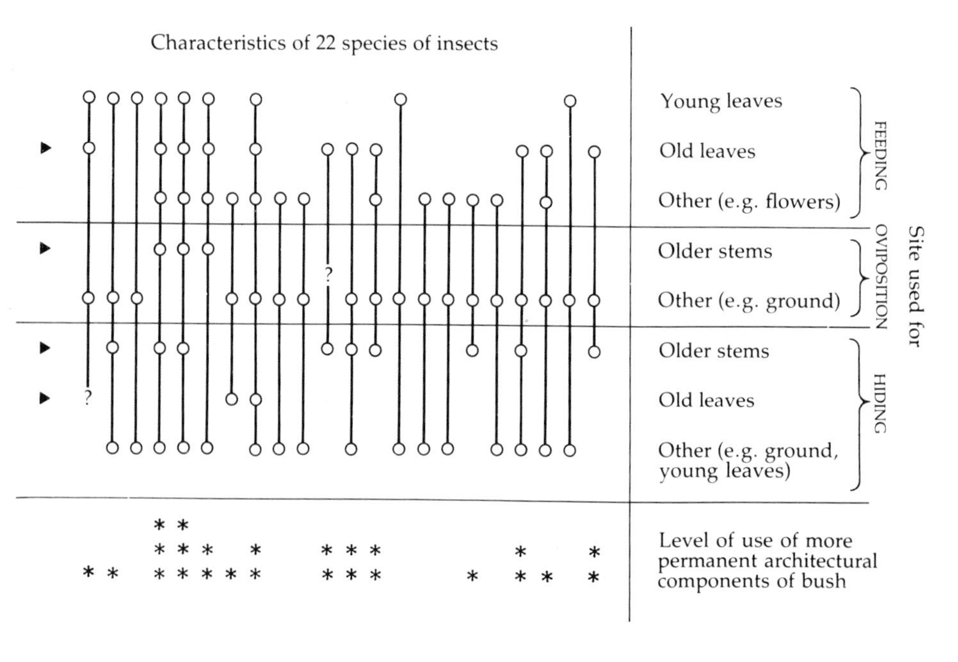

Fig. 3.9. Feeding, oviposition and concealment sites used by 22 species of phytophagous insects on creosote bushes (*Larrea tridentata*) at Silver Bell, Arizona (from Schultz *et al.* 1977). Use of a particular part of the plant is indicated by O. All parts used by each one of the 22 species are connected by a vertical line. Plant parts present on an evergreen woody shrub, but not on a perennial herb, are indicated by ▶ . Use of one, two or three such parts by the insects is indicated by *, * *, or * * * respectively.

Similarly, old leaves and the woody parts of Argentinian *L. cuneifolia* probably accommodate approximately eight of 21 species; these parts of the plant are not available on a herb.

It is difficult without further work to decide how much size *per se* influences the greater species richness of some plants via hypothesis (ii) (bigger plants provide larger 'targets' for colonization) and hypothesis (iii) (larger populations reduce the probability of stochastic extinction). In Moran's (1980) study of *Opuntia*, size alone accounted for 35 per cent of the variance in insect diversity, compared with a total of 69 per cent when all components of plant architecture were considered (Fig. 3.8), but these cacti do not represent a size range as great as the herb–tree contrast. Similar analyses on other groups of plants would be useful.

3.4 Some Residual Effects

3.4.1 Unexplained Variation

Having dealt with the influences of geographical range and plant architecture on the total number of insect species exploiting plants, we are left with varying amounts of unexplained variation, some of it considerable (Table 3.1). The problem is to learn the meaning of this residual variation. Many phenomena *could* contribute to this variation. In no particular order, here is a list of some possibilities.

Local abundance — plants that are scarce within their total geographic range might host fewer species than universally common plants (e.g. Fowler & Lawton 1982), because of their reduced exposure (hypothesis (ii), Section 3.2.2; see also Section 6.2.2 on the Resource Concentration hypothesis).

Leaf form — plants with certain sorts of leaves, finely divided or spiny for example, may consistently have fewer leaf-feeding insects than plants with broad, soft leaves (e.g. Lawton & Price 1979).

Seasonal development — the seasonal development of a plant and the length of its growing season might each influence the number of herbivore species exploiting it (e.g. Strong 1977; Lawton 1978; Niemela & Haukioja 1982; Niemela *et al.* 1982), again because these factors alter exposure (see also Section 6.3.2 on plant phenology).

Taxonomic and biochemical isolation — plants that are taxonomically very different from others in the same geographic region or plants that

are biochemically highly distinct may have impoverished insect faunas, for the reasons outlined in hypothesis (iii) (Southwood 1961a and b, 1977a; Janzen 1968a, 1973; Section 3.2.2).

It is not difficult to think of still other possibilities; the problem is to come up with sensible tests of such ideas. This task is made difficult with insects on plants because variables are often difficult to quantify and are often intercorrelated. For example, biochemical and taxonomic isolation are variables that are correlated, but not perfectly. Taxonomically isolated plants are species with few or no close relatives. These are, by definition, relatively different from one another and in consequence may have unusual biochemistries. However, a plant can be biochemically distinct without being taxonomically isolated and the reverse is also true; often quite unrelated plants contain very similar toxins, cyanogenic glucosides for example (Conn 1979). Moreover, while taxonomic isolation can be reasonably well quantified, biochemical isolation is almost impossible to measure on a linear scale; it is even difficult to rank many biochemical differences.

3.4.2 *Taxonomic Isolation and Plant Toxicity*

Not unreasonably, many entomologists intuitively believe that taxonomically isolated or toxic plants have impoverished insect faunas. That is, such plants should show strong negative residuals in standard species–area plots when compared with less toxic species of similar growth form, or plants with numerous relatives. Despite the reasonable *prima facie* nature of these beliefs, the actual magnitude of any such impoverishment often turns out in practice to be small. Taxonomically isolated, biochemically unusual species of introduced trees, such as *Eucalyptus* spp. in Europe and North America, and *Quercus* spp. in South Africa, Australia and New Zealand, present formidable barriers to colonization by unadapted phytophages and in consequence have impoverished faunas (Connor *et al.* 1980; Moran & Southwood 1982; Southwood *et al.* 1982b). (The colonization of introduced plants is explored in Chapter 4.) Native trees also show some effects. Thus Cornell and Washburn (1979) demonstrate how 'non-*Quercus*' oaks in California support fewer cynipid gall-formers than would be expected on the basis of their range (Fig. 3.2e). Relatively impoverished insect faunas on other taxonomically isolated species of plants have also been reported

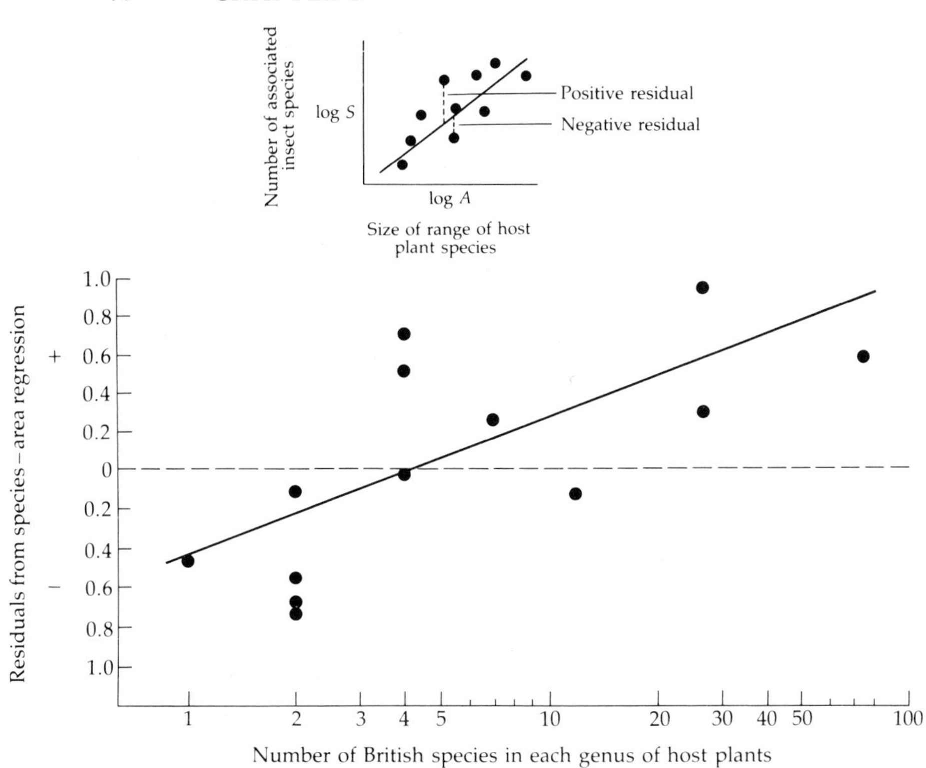

Fig. 3.10. A test of the hypothesis that host plant taxonomic isolation reduces species richness of associated phytophagous insects. Plotted are residuals from the species–area relationship (number of insect species associated with a host plant, and size of host plant range) against the number of congeneric host plants in Britain. Data are for insects on monocotyledons (Fig. 3.7a and Table 3.1) from Lawton and Schröder (1977). If taxonomic isolation acts as predicted under hypothesis (iii) (p. 53), plants with few close relatives should have negative residuals in the species–area plot (i.e. fewer phytophages than expected from size of geographic range), and those with many relatives should have positive residuals (i.e. more phytophages than expected from size of geographic range). For insects on British monocotyledons this hypothesis is supported by the data, with negative residuals on taxonomically isolated plants; the equation is
Residual = 0.3 ln (species in genus) — 0.43 ($.01 > P > 0.005$; $r = 0.68$).

by Neuvonen and Niemela (1981) for macrolepidoptera on Finnish trees, and by Lawton and Schröder (1977) for insects feeding on British

monocotyledons (Fig. 3.10). Other studies have generally found no such effects.

As we have noted, biochemical effects are more difficult to test. One approach is to feed plants to non-adapted, polyphagous insects, under the hypothesis that plants with negative residuals in species–area relationships should be more repellent or more toxic to phytophagous insects than plants with richer faunas. This approach was adopted by Reader and Southwood (1981), who used five taxonomically very different, generalist phytophages to bioassay the palatability of plants from different successional stages. They showed decreasing palatability of plants with increasing successional stage, perhaps related to the nature of the plants' chemical defences. One theory of host-plant chemistry predicts that plants from early stages of succession will be chemically less well defended because they are more able to 'escape' from herbivores in time and space than are long-lived, highly 'apparent' plants from later stages of succession (Feeny 1976; Rhoades & Cates 1976). However, even within the small sample of plants used by Reader and Southwood there were considerable differences in palatability both between plant species from one stage of succession and between the responses of the different species of herbivores used to make the test. Making generalizations in this field is unlikely to be easy.

Different responses by different species of polyphagous herbivores are one possible reason for the surprising results displayed in Fig. 3.11, in which British trees supporting the largest numbers of species of insects are seen to be the most *unpalatable* to generalist herbivores. Wratten *et al.* (1981) explain these results by arguing that selection has favoured investment in defences in those species of plants subject to the greatest levels of herbivore attack. However, more species of herbivores do not necessarily mean greater levels of damage, and since the generalist herbivores used to test the palatability of the trees were snails and not insects, interpretation is made doubly uncertain.

However difficult it might be to show directly the effects of biochemical and/or taxonomic isolation on insect species richness, some plant species that are relatively impoverished in one group of phytophages tend also to be relatively impoverished in other groups, implying a common cause for a reduced diversity of attackers. Thus the residuals from separate species–area plots for two groups of stem- and leaf-mining insects attacking British umbellifers (agromyzid flies and

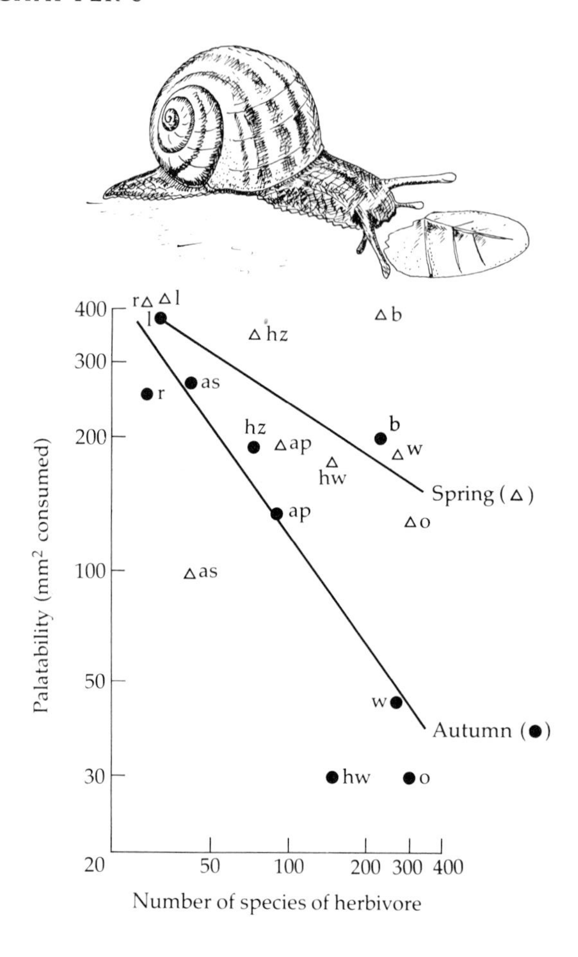

Fig. 3.11. Relationship between palatability of leaves of nine species of British trees and number of phytophagous insect species associated with each species of tree. Palatability measured as mm^2 of leaf consumed by snails under standard conditions, Trees with more species of insects are less palatable than trees with fewer species of insects. (From Wratten *et al.* 1981.) (r — rowan; l — lime; as — ash; hz — hazel; ap — apple; hw — hawthorn; b — birch; w — willow; o — oak).

microlepidoptera) are positively correlated (Lawton & Price 1979), as are the two sets of residuals from species–area relations for leaf-hoppers and leaf-miners (Claridge & Wilson 1981, 1982), and insect species and

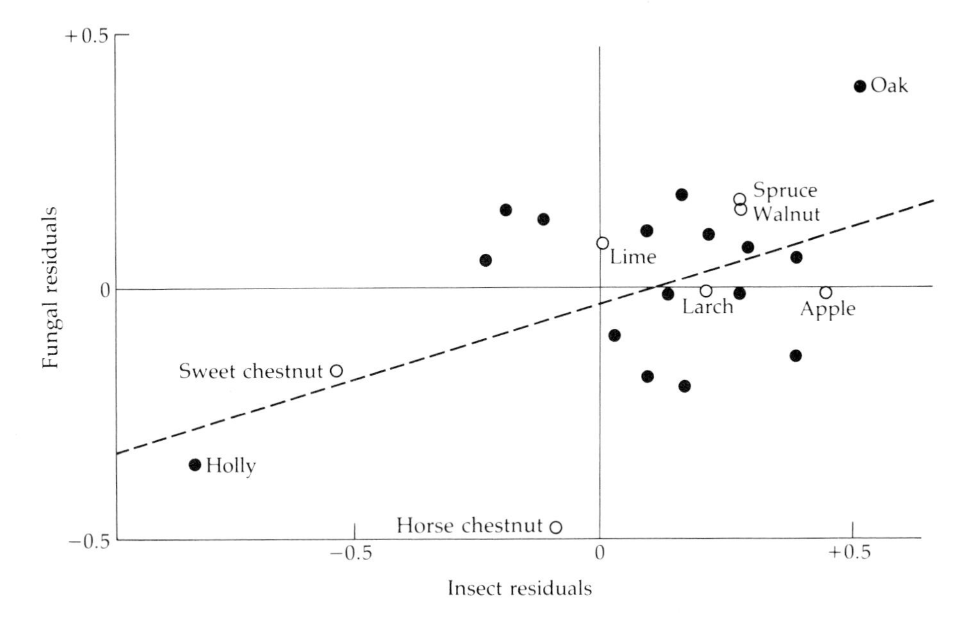

Fig. 3.12. Relationship between residuals from two species–area relationships; one for insects and the other for fungi attacking British trees. Trees relatively impoverished in insect species (negative residuals in the species–area plot) tend also to be impoverished in fungi. The equation is:
Fungal residual = 0.36 (insect residual) + 0.08 ($P<0.001$; $r = 0.65$) (Strong & Levin 1975). Introduced trees are shown as open circles, native taxa as solid circles.

parasitic fungi (Strong & Levin 1975) on British trees (Fig. 3.12). The simplest, but by no means the only, explanation for such results is that each group has independently found the same plants 'easy' or 'difficult' to exploit. If so, species impoverished in both microlepidoptera and agromyzids, leaf- hoppers and leaf-miners, or fungi and insects may well have unusual chemistries, although we know of no detailed studies that show this.

3.4.3 *Other possibilities*

Too little is known of the many other possible influences on insect species diversity for us to say very much. Leaf form has an effect on the agromyzid species richness of British umbellifers, plants with finely divided leaves supporting fewer species and those with broad flat leaves more species than expected from the sizes of their geographic ranges (Lawton & Price 1979; Fowler & Lawton 1982), but similar studies on other plant groups are lacking. Local abundance within each plant's total geographic range influences the species richness of microlepidopterans on trees in Finland (Neuvonen & Niemela 1981), but not that of agromyzids on umbellifers in Britain (Fowler & Lawton 1982). The length of the growing season affects the number of species of hispine beetles attacking *Heliconia latispatha* in Central America (Strong 1977; Fig. 3.13), but again comparable studies are lacking.

Pessimists will note that a considerable amount of the variation in total insect species richness often remains unaccounted for; optimists see the unexplained variation as a challenge for future research and take heart from the fact that a substantial part of the broad patterns in species richness of insects on plants are now understood, at least on a geographic scale. The authors are optimists and look forward to the fruits of future

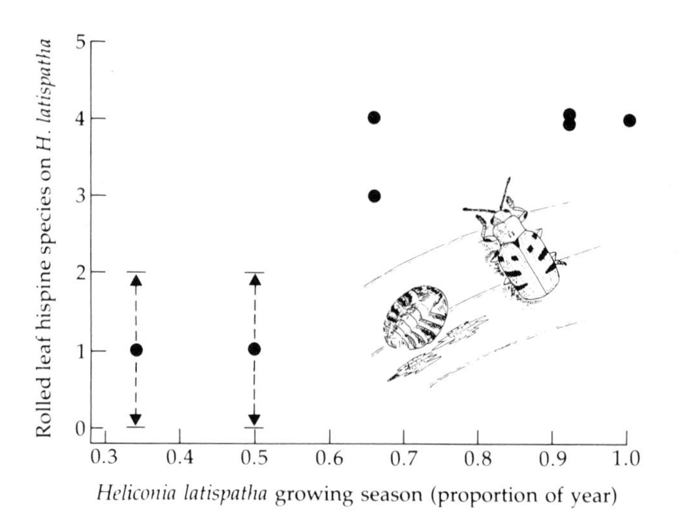

Fig. 3.13. Number of species of rolled-leaf hispine beetles attacking *Heliconia latispatha* at various lowland sites in Costa Rica, as a function of the host's growing season (Strong 1977). The two stations with the shortest growing season support variable numbers of species.

research on factors other than size of geographic range and architecture as determinants of herbivore diversity in contemporary insect–plant associations.

3.5 Summary

Plant species differ markedly in the number of insect species that feed on them. A major determinant of insect species richness is the size of the geographic range of the host plant. Widespread species of plants are hosts for more species of insects than rare plants — the familiar species–area relationship.

We review three general mechanisms that might contribute to species–area relationships for insects and host plants. Widespread species of plants grow in more habitats and over a wider range of climatic zones than rare plants. Hence different species of insects are found in different parts of the ranges of widespread plants. Secondly, widespread plants present more conspicuous 'targets' for colonizing organisms. Thirdly, small populations on plants with restricted ranges may be more prone to extinction.

Phytophagous insect diversity is also influenced by plant 'architecture' — its size and growth form and the variety of resources on the plant. Hence, area for area, trees have more herbivore species than bushes, which in turn have more than herbs.

Finally, some plant characteristics have an enigmatic effect on insect diversity: plant biochemistry, taxonomic affinity and local abundance, to name three. Each of these plant characteristics has been shown to have some effect on insect species richness in at least one study, but on present evidence none appears to be of as consistent importance as size of geographic range and architecture.

A major challenge for future research is the residual variation in insect diversity that remains after geographic range and architecture have been accounted for.

Chapter 4
Community Patterns Through Time: The Dynamics of Colonization and Speciation

4.1 Introduction — Colonization

Every community has a history. In this chapter, we look at patterns of change in phytophagous insect communities over periods between 10 and 10 000 years, a focus intermediate between the contemporary associations discussed in Chapter 3 and the broad sweep of evolution outlined in Chapter 2. Our main questions are: how do newly introduced plant species acquire phytophagous insects, from where, and at what rate?

Man has moved plants round the world for food, for pleasure and by accident, for over 5000 years. In so doing he has unwittingly performed a set of ecological experiments on a grand scale. Insects find and colonize most species of introduced plants remarkably quickly, and the more widely a plant becomes established, the more insect species it recruits (Chapter 3). Some of these colonists are veterans from the plant's original range, dispersed naturally or themselves carried accidentally or deliberately by man, but most are new recruits that add the new plant to their list of hosts. The acquisition of insects by introduced plants is illustrated in Section 4.1.1 with crops and other plants that have been established by man well outside their original ranges, and in Section 4.1.2 with plants that have naturally expanded their geographic ranges. We then move to questions of rates of species accumulation, the kinds of phytophages acquired (Sections 4.1.3–4.1.5), and their subsequent evolution and speciation (Section 4.2).

4.1.1 Insects on Introduced Plants: Some Examples

For obvious reasons, insects colonizing crops have been well studied. Reviews on widely planted crops are provided by Entwistle (1972) for cacao, Turnipseed and Kogan (1976) and Kogan (1981) for soybean, Bournier (1977) for grapes, and Chiang (1978) for maize. Some of the insects exploiting such crops are cosmopolitan, but most are confined to one

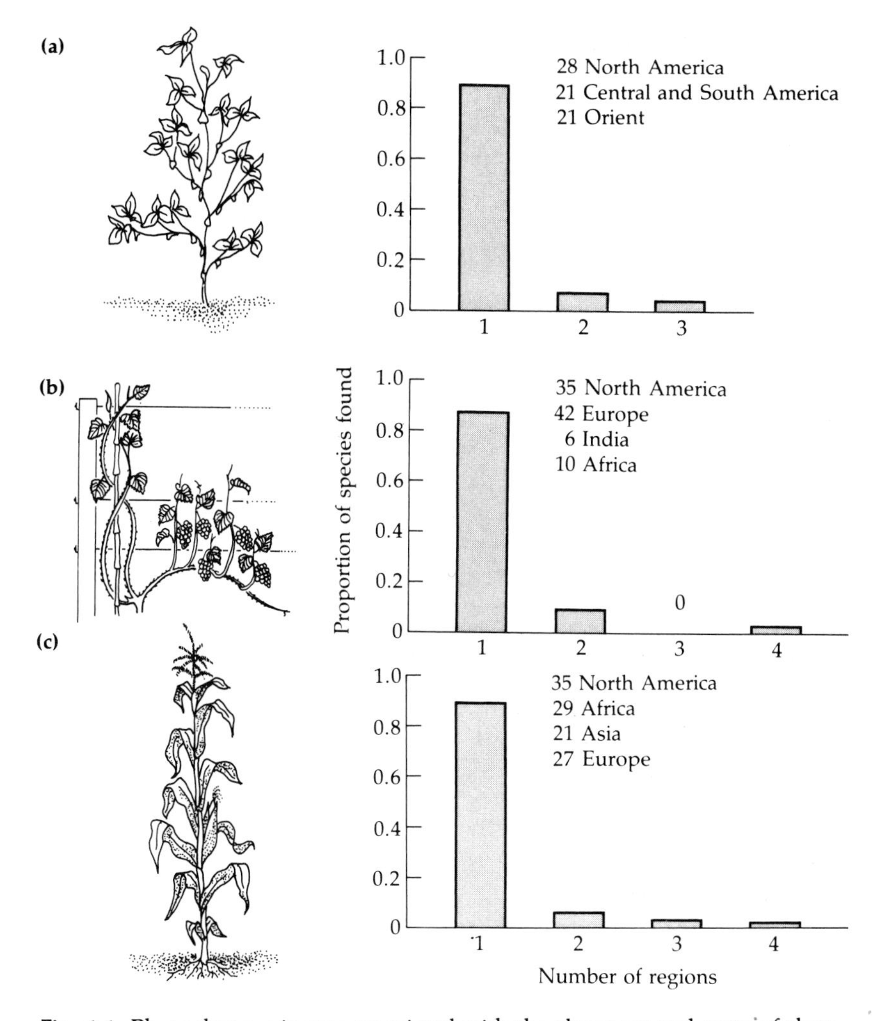

Fig. 4.1. Phytophagous insects associated with the above-ground parts of three cosmopolitan crops (root-feeders, mites, etc., excluded). The histograms show the proportion of species known to feed on the plant in one or more major geographic regions. The regions, together with an estimate of the total number of species (including a few genera or larger 'species groups') found on the plants in each region are given above the histograms. Regions and taxa are as given in the original accounts. (**a**) Soybean (Turnipseed & Kogan 1976); (**b**) grape (Bournier 1977); and (**c**) maize (Chiang 1978). The great majority of phytophagous insects occur in only one region. Recruitment of herbivores has therefore taken place locally; herbivores have not, in the main, been transported round the world with their host plants.

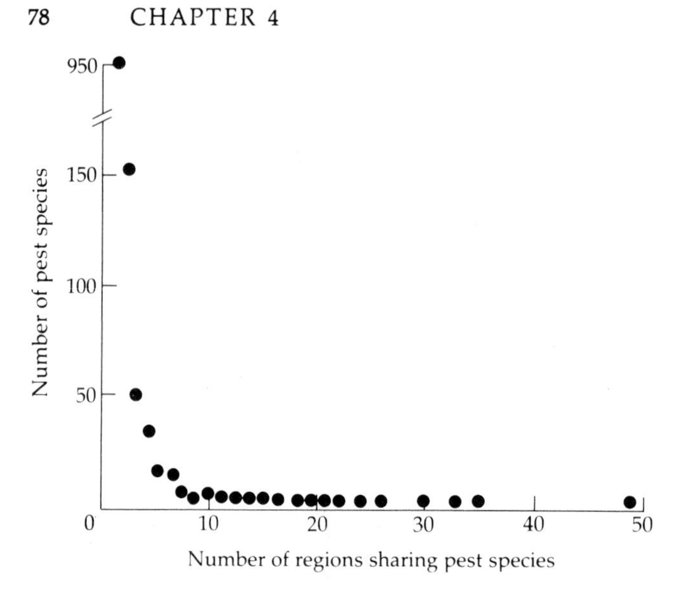

Fig. 4.2. The number of arthropod pest species of sugarcane shared between 81 regions where sugarcane is grown. Of 1645 pests worldwide, 959 occur in only a single sugarcane-growing region. Regions are individual countries, islands, or island groups, throughout the warmer parts of the world. The arthropod pests are mainly but not entirely phytophagous insects. (After Strong *et al.* 1977.)

continent (Fig. 4.1); in general the plant is the traveller and it is native insects that adapt to use new potential hosts. Sugarcane, a widely distributed crop, shows this phenomenon particularly clearly (Strong *et al.* 1977; Fig. 4.2). Nor are crops atypical in this regard. Introduced ornamental bushes (Figs 4.3, 4.10), herbs (Table 4.1) and trees (Fig. 4.10; see also Crooke 1958) all recruit insect phytophages from the local pool of species to which they are exposed.

Many of the insect species that colonize a newly introduced plant are polyphagous, that is they already feed on a wide range of hosts, often members of tribes and families unrelated to the introduced plant (Table 4.1 and Fig. 4.4). This is not particularly surprising, although as we shall see (Section 4.1.5; Fig. 4.10) it is not true of all the insects that colonize new hosts.

Perhaps less obviously, the available data also suggest that chewing and sucking insects feeding externally on the plant are more likely to colonize new hosts than leaf-miners and gall-formers (Fig. 4.4). Most of

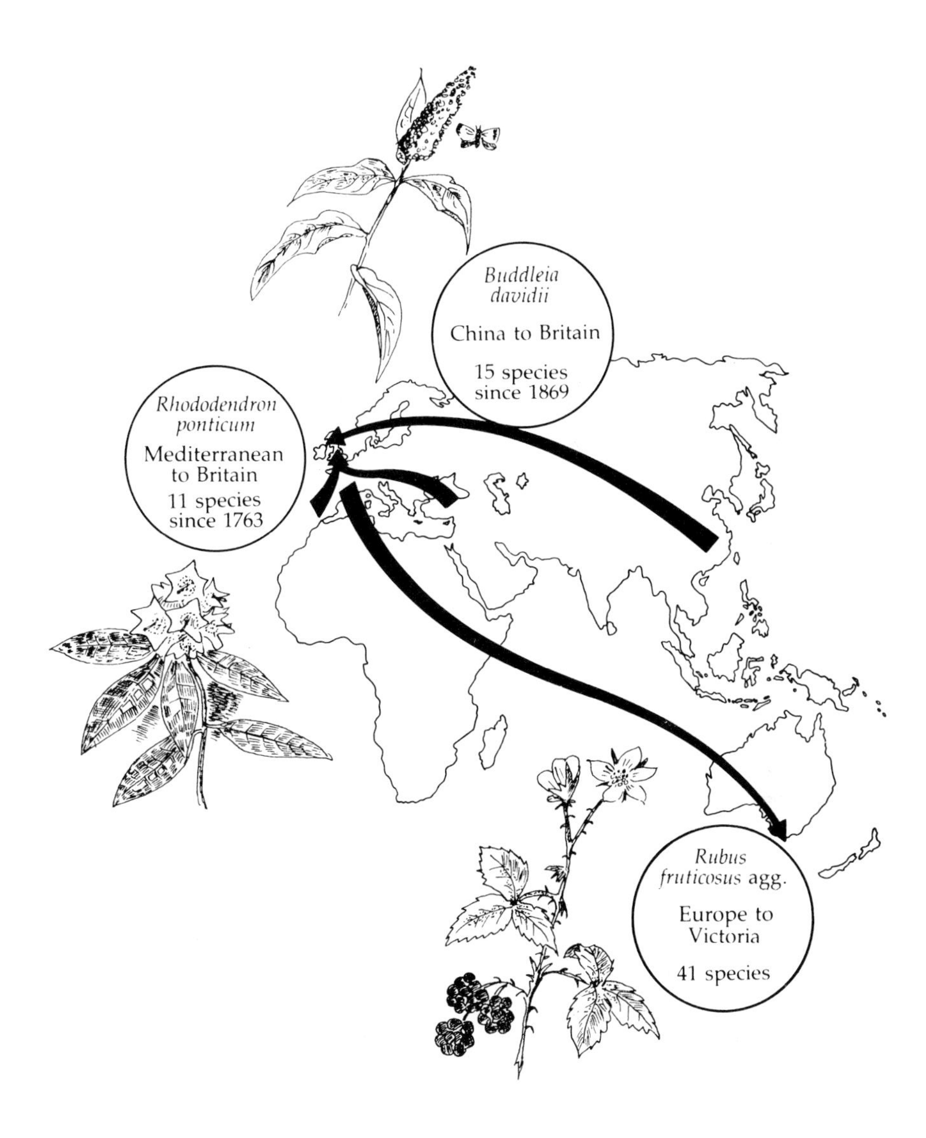

Fig. 4.3. Regions of origin and introduction for three bush species, together with the number of phytophagous insect species recruited in their adopted countries since introduction. (From data in Bruzzese 1980; Cross 1975; Elton 1958; Owen & Whiteway 1980.)

Table 4.1. Some phytophagous insects that attack introduced herbs in the United States.

Author(s)	Plant species, English name and family	Region of origin	Location of survey of introduced plant
Berenbaum 1981c	*Pastinaca sativa* Wild parsnip Umbelliferae	Europe	Tompkins County, New York, USA
Goeden 1971	*Silybum marianum* Milk thistle Compositae	Mediterranean Europe	Southern California, USA
Goeden 1974	*Carduus pycnocephalus* Italian thistle Compositae	Southern Europe	Southern California, USA
Goeden & Ricker 1968	*Salsola kali* Russian thistle Chenopodiaceae	Southeast Russia and western Siberia	Southern California, USA
Root & Tahvanainen 1969	*Barbarea vulgaris* Winter cress Cruciferae	Europe	Ithaca, New York, USA
Wheeler 1974	*Coronilla varia* Crown vetch Leguminaceae	Europe	Pennsylvania, USA

Maximum number of years between introduction and survey	Number of phytophagous insects feeding and/or breeding on plant			Number and proportion of these species that are:	
	Commonly	Occasionally or rarely	Total	Endophagous stem- and leaf-miners and gall formers	Polyphagous (attacking species belonging to other tribes and families)
369	8	—	—	1 (12.5%)	50%
114	19	27	46	2 (4.4%)	'Most'
40	3	>37	>40	5 (<12.5%)	39 (98%)
93	22	49	71	1 (0.4%)	69 (98%)
150	4	31	35	1 (2.9%)	—
83	31	21	52	3 (5.8%)	'Most'

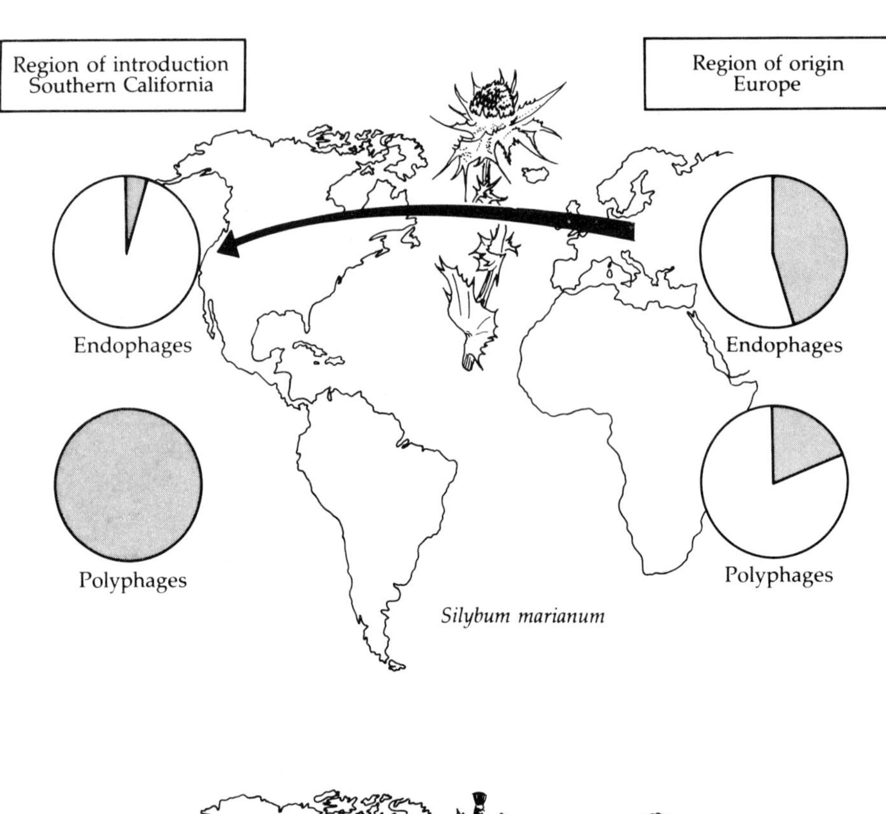
Region of introduction
Southern California
Region of origin
Europe
Endophages
Polyphages
Endophages
Polyphages
Silybum marianum

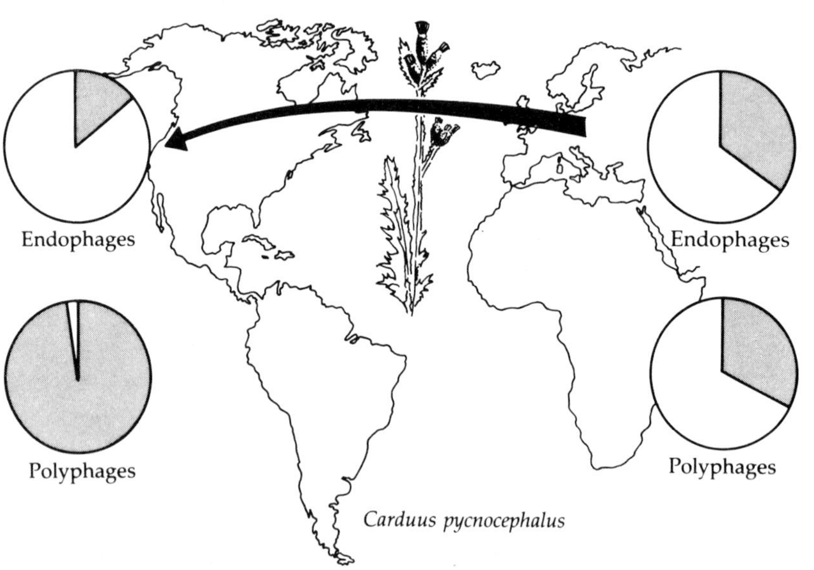
Endophages
Polyphages
Endophages
Polyphages
Carduus pycnocephalus

the species colonizing *Silybum marianum* and *Carduus pycnocephalus* in southern California, where both plants are introduced, feed externally. Endophages (leaf- and stem-miners and gall-formers) make up a much higher proportion of the faunas of these plants in Europe, where they are native. In general, between 10 and 30 per cent of the phytophagous insect species associated with one tribe of native European Compositae, the Cynareae, are miners and gall-formers (Fig. 4.7), a higher proportion than on all but one of the introduced herbaceous plants listed in Table 4.1.

The implication of these data is that colonization or evolution of a varied complement of endophagous, and less polyphagous, species normally takes much longer than the acquisition of external-feeding polyphages. We return to this question later in the chapter (Section 4.1.3).

Intriguingly, not all species of introduced plants recruit insects from the local fauna, despite being exposed to attack for many years. *Opuntia* cacti in southern Africa are notable examples. Natives of South America, two species, *O. ficus-indica* and *O. autantiaca*, have not recruited any African insects, although they have been present in South Africa for 250 years and 150 years respectively (Moran 1980). *Opuntia* in Australia are similarly devoid of native phytophages. The genus does, however, host a rich and varied community of phytophagous insects where it originated in the New World (Moran 1980; Fig. 3.8; Chapter 3).

Gathering these observations together, we see that a comprehensive theory of host-plant colonization must account for the relatively rapid utilization of some introduced plants and plant parts by phytophagous insects, and the marked immunity of others. It must also distinguish

Fig 4.4. Some characteristics of the phytophagous insect faunas associated with two species of thistles, *Silybum marianum* (milk thistle) (above) and *Carduus pycnocephalus* (Italian thistle) (below) in their native Europe, and in southern California where they are introduced. Shaded portions of circles indicate proportions of insect species in particular trophic groups. The introduced plants have a higher proportion of phytophagous species that feed externally on their host, and a higher proportion of non-specialist (polyphagous) species, than plants of the same species in Europe. European data are from Zwölfer (1965), Californian data from Goeden (1971, 1974); the latter paper also gives European data for *C. pycnocephalus*.

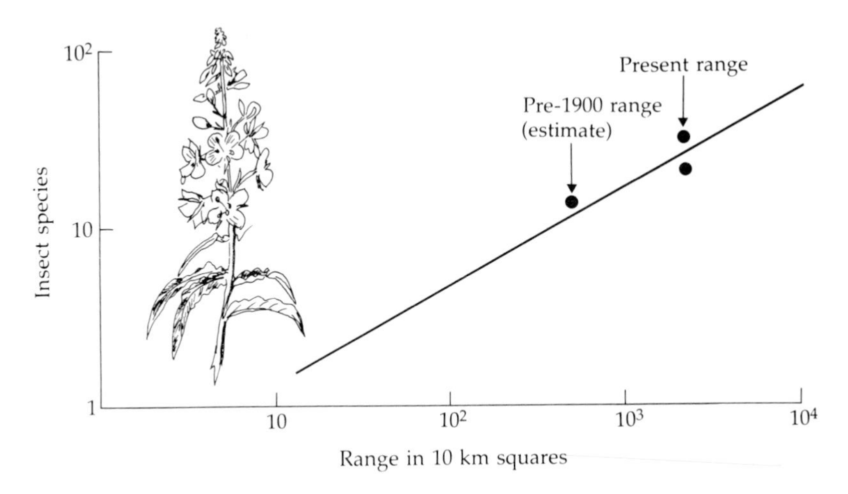

Fig. 4.5. Present and estimated pre-1900 geographic ranges of rosebay willowherb (fireweed), *Chamerion angustifolium*, in Britain; the plant has naturally expanded its range during the present century (pre-1900 range from estimates provided by A.H. Fitter, pers. comm. to J.H.L. and M. MacGarvin; modern range from Perring & Walters 1962). Associated phytophagous insect species are from Horwood (1919), Myerscough (1980) (the lower modern estimate) and MacGarvin (1982, 1983) (the upper modern estimate).
The regression line is not fitted to the points; it is the line
$\ln (S + 1) = 0.54 \ln A - 0.95$
(S = number of species, A = range in 10 km squares) for British perennial herbs from Lawton & Schröder (1977) (see Fig. 3.7 and Table 3.1).

between colonization by generalist, external feeders and by more stenophagous (i.e. more specialised) species, including miners and gall-formers.

4.1.2 *Natural Range Expansions*

Man has caused only a small fraction of the changes in host plant geographic range. Plant distributions vary naturally with climatic and geological changes. The great majority of the species of plants and insects now living in Britain failed to survive the glacial maximum and have colonized the country over the last 10 000 years or so (Birks 1980; Dennis 1977). The species–area relationships for insects on plants shown in Figs

3.2 (a, b, and d) and 3.7 for the British Isles did not exist 10 000 years ago.

A good example of a natural expansion of range by a plant is the recent geographical history of rosebay willowherb (fireweed), *Chamerion angustifolium*, in Britain (MacGarvin 1982, 1983; Fig. 4.5). Formerly a scarce and local plant, with known sites worthy of specific mention in Victorian county floras, it has undergone a remarkable, natural, but poorly understood expansion of range since the First World War. It now occupies 2200 '10 km' squares (Myerscough 1980; Fig. 4.5).

In 1919, Horwood recorded only 13 species of insects from the plant. Recent studies show that at least 30 species now feed on rosebay in Britain (MacGarvin 1982, 1983); a less intense review by Myerscough (1980) records 19 species. Hence some recruitment of phytophages seems likely over the past 60–70 years (Fig. 4.5), although part of the apparent increase in number of species feeding on rosebay is undoubtedly owing to Horwood's list being incomplete. Three colonizations do, however, seem reasonably certain (M. MacGarvin and J.H. Lawton, unpublished). The moth *Scythris inspersella* (Sattler 1981) and the chrysomelid *Bromius obscurus* are rosebay specialists that have recently immigrated into Britain from mainland Europe (although Kendall (1982) disagrees with the view that *Bromius* is a new immigrant, that so conspicuous a beetle can have been overlooked in Britain for so long seems very unlikely). The third recruitment apparently involves a host shift (see p. 97) by the elephant hawk moth (*Deilephila elpenor*) from *Galium* and the related great willowherb (*Epilobium hirsutum*), the only hosts recorded by Scorer in 1913, to rosebay as a third important food plant for this conspicuous moth.

All of these natural and man-made changes in range by host plants and colonizations by phytophagous insects illustrate clearly that species–area relationships are far from immutable. However, it is equally clear that the high rate of insect acquisition after a major expansion of range does not continue indefinitely. If it did, long-established, widespread plants with relatively static distributions would, over several hundred years, build up staggeringly diverse insect faunas. They do not. Hence colonization after a major range expansion must be asymptotic for most species of plants — rapid at first and subsequently slowing to a trickle. Asymptotic recruitment curves are shown by a number of studies, and it is to these that we now turn.

4.1.3 Colonization through Time

Examples from Crops and Other Introduced Plants

Sugarcane (*Saccharum officinarum*), probably native to New Guinea, has been widely introduced to many low-latitude regions of the world. The dates of these introductions range from 1000 BC in India and Pakistan, and 500 BC in Indochina and Egypt, to as recent as 1640 AD in Grenada and 1840 in British Honduras (Strong *et al.* 1977). Recruitment of local insects (Fig. 4.2) has occurred whenever sugarcane has been planted, generating a highly significant contemporary species–area relationship (Table 3.1; Chapter 3). What is particularly interesting in the present context is that the residual variation, not explained by the species–area regression, is not significantly correlated with the length of time sugarcane has been grown in each region. In other words, accumulation of pest species on sugarcane appears to be asymptotic; once the size of the area under cultivation has been allowed for, parts of the world where sugarcane has been grown for over 2000 years do not have noticeably more species attacking the plant than areas first planted only 150 years ago.

Similar comments apply to cacao (*Theobroma cacao*), another widely planted crop (Strong 1974a).

Species accumulation by tea (*Camellia sinensis*) over much shorter time periods has been studied by Banerjee (1981). In 14 areas of Asia and Africa, pest accumulation has continued fairly steadily for periods of between 28 and 153 years. What happens over longer time periods with this crop is not known.

Not all plant species are colonized as quickly as sugarcane, cacao and tea. *Opuntia* cacti in South Africa, as we have just seen (Section 4.1.1), have not been exploited by *any* native insects, even after 250 years, despite enormous areas of infestation (Moran 1980). Similarly, the species richness of phytophagous insects is very low on several species of exotic trees, both in Britain and in South Africa, even though they have been introduced for periods of 100–300 years (Southwood *et al.* 1982 a, b). *Eucalyptus* is attacked by many insect species in Australia, where it is native (Morrow & LaMarche 1978). In California, where it has been widely planted for over 100 years, it suffers very little insect damage and, perhaps as a result, may look oddly vigorous to Australians (Strong 1979). One of the authors (D.R.S.) has extensively examined western US

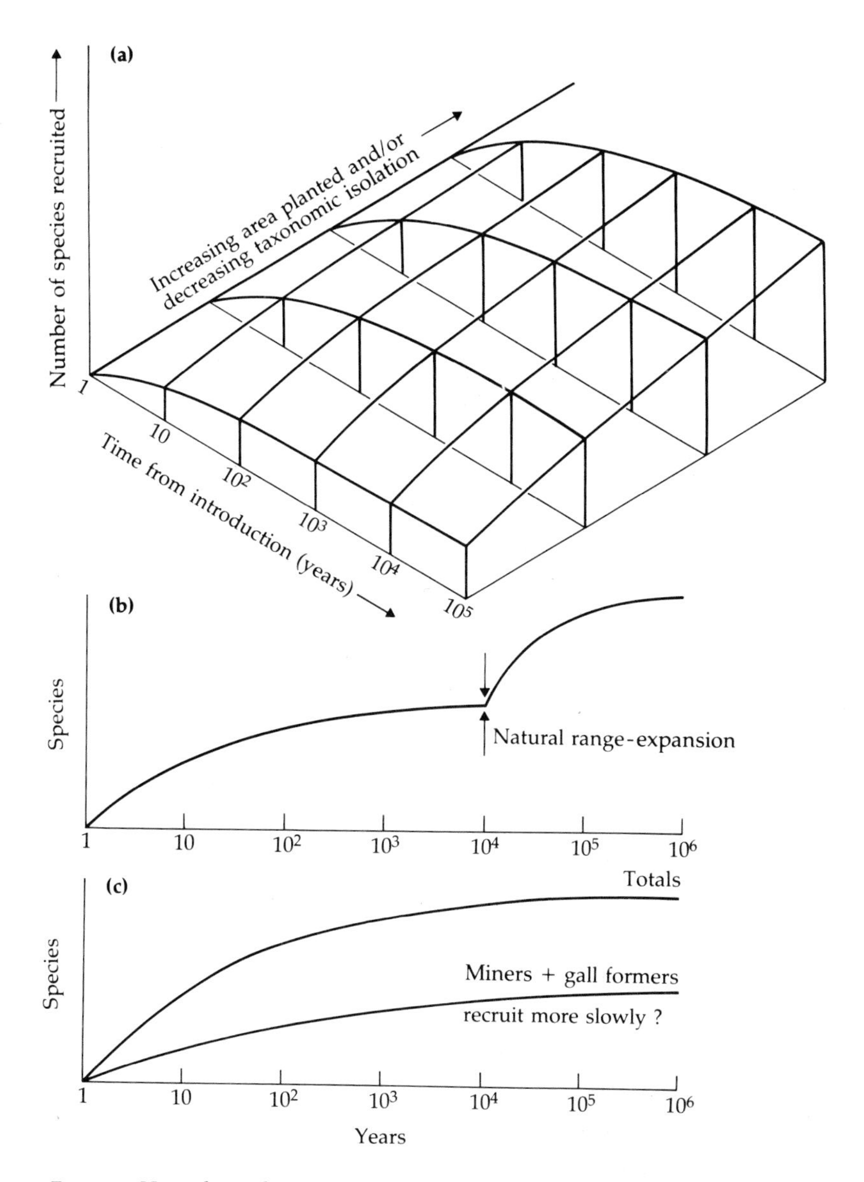

Fig. 4.6. Hypothetical recruitment curves for insects on introduced plants. (**a**) Phytophagous insects accumulate asymptotically with time, reaching higher levels the greater the 'exposure' of the plant (i.e. the bigger the area planted), and reaching higher levels on taxonomically less isolated plants. (**b**) A natural range expansion leads to a new bout of colonization (see Fig. 4.5). (**c**) Endophagous species (miners and gall-formers) may recruit more slowly than external feeders.

Eucalyptus for insects and found only one example of clear phytophagy. On the campus of the University of Arizona at Tucson a species of bagworm (Psycidae, Lepidoptera) was abundant upon ornamental *Eucalyptus* and damaging it heavily. It is interesting that bagworms are among the native eucalypt insects in Australia.

Theoretical Explanations

A hypothesis (Fig. 4.6) to unify these observations is that rates of recruitment of phytophages on to new host plants are affected by two factors (see Section 3.2.2): (i) the area planted (or the size of the geographic range achieved by natural spread) and (ii) the taxonomic, phenological, biochemical and morphological match between introduced plant and native flora. The more unusual a plant, the lower will be the 'predilection' for local insects to colonize it (Southwood 1961b, 1977a). Plants from novel, taxonomically isolated families (novel and isolated in their adopted countries that is) with unusual growth forms and/or phytochemistries, like the Cactaceae transported from the New World to the Old World, or *Eucalyptus* (Myrtaceae) from Australia planted in Europe, may remain immune to attack by resident insects for long periods of time. In contrast, introduced plant species with close or ecologically similar relatives in the native flora may recruit a high proportion of the fauna of such relatives reasonably quickly. For example, in Britain, the North American ornamental cypresses (*Chamaecyparis*, Cupressaceae) have recruited one of the two native species of mirid bug found on the native co-familial juniper, and one of the two native species of shield bug (Southwood 1957) (the second shield bug is probably extinct in Britain). Similarly, *Nothofagus* species (southern beech, Fagaceae) introduced into Britain from South America and Australia have acquired breeding populations of at least six species of Typhlocybinae (leaf-hoppers), all of which normally feed on related native British trees in the same family (*Fagus*, beech, and *Quercus*, oak) (Claridge & Wilson 1976).

A Test of Theory: Insects on British Trees

The interacting effects on insect colonization of host geographic range, time, and taxonomic isolation can best be illustrated by reference once more to data from Britain. The analysis, which has not previously been published, uses the most recent faunal lists and the best historical information available (the former compiled by C.E.J. Kennedy and T.R.E. Southwood and the latter taken from Birks 1980). Previous interest in the insects found on British trees has been great (Southwood 1961a, 1977a; Strong 1974b; Claridge & Wilson 1978; Connor *et al.* 1980; Birks 1980). The new model developed by Kennedy and Southwood may be stated as:

$$\log_{10} (\text{insect species}) = 0.41 \log_{10} (\text{abundance}) + 0.001 (\text{age}) - 0.51 (\text{'evergreenness'}) + 0.013 (\text{taxonomic isolation}) + 0.13 (\text{'coniferous'}) - 0.44$$

($F_{5,22} = 13.9$; multiple $r^2 = 0.77$; $p < 0.005$).

In this equation, 'tree abundance' is measured using estimates of the number of occupied tetrads (2 × 2 km squares) in Britain. 'Age' is the length of time, in years, that the tree has been present in Britain, ranging from 13 000 years for native, immediate post-glacial colonists (*Betula*, *Salix* and *Juniperus*) to 350 years for the introduced *Larix*. 'Evergreenness' and 'coniferous' are dummy variables, and only the latter among the five variables does not enter the stepwise multiple regression in a statistically significant manner (although its inclusion marginally improves the fit of the whole model). 'Taxonomic isolation' is measured as the number of species present in Britain belonging to the same order as the plant.

As expected, present abundance of the tree is the main determinant of insect species richness but, as indicated in Fig. 4.6, both the length of time a tree has been present in Britain and its taxonomic isolation also influence insect species richness. Obviously, since the 'youngest' species used in this analysis (*Larix*) has been in Britain for 350 years, recruitment of insects on to British trees must have continued for much longer periods than is apparent either for cacao or for sugarcane, where recruitment of phytophages is so slow after a few hundred years as to be undetectable. This difference is probably a result of the better data from Britain, but we cannot really know.

Long-term Effects

Although rates of recruitment of new herbivores appear generally to

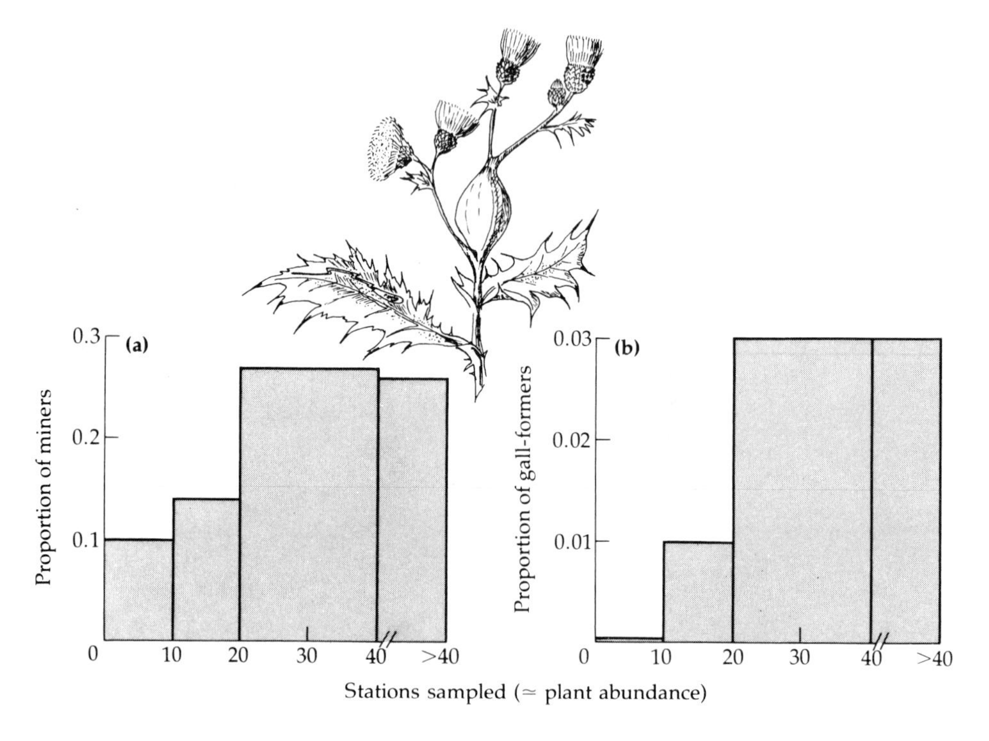

Fig. 4.7. Proportion of (**a**) leaf- and stem-miners and (**b**) gall-formers in the total phytophagous insect faunas of European Cynareae (a tribe within the Compositae), as a function of host abundance (Lawton & Schröder 1978).

slow down with the passage of time, it is not clear whether the ranges of any plants stay constant enough for long enough for recruitment of insects to stop completely. For example, the cynipid gall wasp *Andricus quercuscalicis* entered the rich and long-established British oak fauna as recently as the early 1960s, probably by natural dispersal from the continent (Darlington 1974; Martin 1982; Collins *et al*. 1983). In Illinois the longhorn beetle *Tetraopes tetrophthalmus* has recently added the milkweed *Asclepias verticillata* to its diet (Price & Willson 1976). Habitat disturbance rather than an expansion of host plant range may have facilitated this shift. The scarcity of such reports in the entomological literature strongly reinforces our view that colonization slows to a trickle on plants with broadly constant geographic ranges and patterns of distribution. Over longer time-scales (that ecologists can never hope to study reliably), colonization probably never stops completely.

In particular, we suspect that recruitment curves for endophagous miners and gall-formers may be substantially slower than those for external chewing and sucking species (Lawton & Price 1979; Strong 1979; Kogan 1981; Fig. 4.6). Certainly the data in Table 4.1 and Fig. 4.4 suggest that they are slower. Moreover, the proportion of miners and gall-formers is higher on more widespread native European Cynareae (thistles and their relatives) than it is on local species (Lawton & Schröder 1978; Fig. 4.7). These data are consistent with the view that colonization by miners and gall-formers is a slow process, and is more likely on very widespread plants than on rare ones. Patterns of host colonization by external chewing and sucking insects compared with endophagous miners and gall-formers deserve further study.

4.1.4 *Why Are Recruitment Curves Asymptotic?*

Two Alternatives

We see two ways to generate asymptotic recruitment curves: the 'niche-saturation model' and the 'pool-exhaustion model' (Lawton & Strong 1981). The saturation model is synecological; it stresses species interactions. The pool-exhaustion model is more autecological in terms of the phytophages; it stresses independent colonization events by different phytophage species, without effects of interactions between species.

We encountered the niche-saturation model (or ecological-saturation hypothesis) in a slightly different form and context in Section 2.3.1. The model supposes that species colonization rates are slowed by lack of available niches. As successful colonists pre-empt available resources, invasion of the community by new species becomes progressively more difficult. Under this hypothesis, limits on the number of species exploiting a plant are determined by interspecific competition (see, for example, MacArthur 1972; Hutchinson 1978; May 1981). A major variant of the model proposes that 'enemy-free space' or 'escape space' (Askew 1961; Gilbert & Singer 1975; Lawton 1978, 1983a; Lawton & Strong 1981, Atsatt 1981a), not food, is the resource in critically short supply. A probable (but not absolutely inevitable) consequence of both variants of the niche-saturation model is that there should be few unexploited resources in local communities of long-established plants.

The pool-exhaustion model makes the diametrically opposite prediction, that of many vacant niches on most plants. Under the pool-exhaustion hypothesis, it is the pool of potential colonists that is exhausted, not resources; or to extend the analogy in Section 2.3.1, borrowed from Hutchinson (1965), it is suitable actors that are in short supply, not good parts. A number of species have a high predilection for colonizing a plant spreading or introduced into a new area, either because they are highly polyphagous or because the plant is functionally quite close to their normal host(s). Such insects find and exploit the plant quite quickly. The vast majority of insect species in the regional pool have a vanishingly small probability of ever being able to make the necessary biochemical, physiological and behavioural jump on to the new host. Of course, some species are intermediate between these two extremes. Hence the regional pool of potential colonists is quickly exhausted, and colonization is asymptotic.

Which Mechanism?

We believe, on present evidence, that a standard niche-saturation model based on interspecific competitive feeding interactions can be rejected as the major mechanism generating asymptotic recruitment curves for insects on host plants; interspecific competition for food between phytophagous insects seems to be too feeble or rare to provide a strong limit to the number of colonists exploiting most plants (Lawton & Strong 1981; Chapter 5).

Enemy-free space is a stronger candidate, not least because introductions of several species of phytophagous insects for biological control of weeds have purportedly failed because of the impact of resident natural enemies (Goeden & Louda 1976), and theoretical studies leave little room to doubt that sharing natural enemies can lead to exclusion of one or more phytophage species from a habitat (Williamson 1957; Holt 1977). The hypothesis relies on the notion that different types of enemies (e.g. spiders, predatory insects, parasitoids and birds), and different species of each type, hunt for their victims in characteristically different ways and on different parts of the plant. Hence phytophagous insects vulnerable to one species or guild of enemies may be safe from another, although few refuges are absolute. In consequence, there are limits to the numbers of phytophage species able to coexist under the influence of a given set of natural enemies.

Tests of the role of natural enemies in moulding niches and

determining community structure of phytophagous insects are still very few. Ants feeding on the nectaries of certain species of plants constitute a special case, and their impact on phytophages can be considerable (Chapter 6). Also consistent with the enemy-free space hypothesis are the relatively constant proportions of predator to phytophage species found in insect communities on trees (Moran & Southwood 1982). While these data are encouraging, they are no more than a tentative start to the detailed examination of the role of natural enemies in setting limits to the number of insect species colonizing plants. In contrast, the pool-exhaustion hypothesis has rather more support.

Good evidence of 'empty niches' can be found in phytophage communities. Within groups of closely related, well-established native plants it is easy to identify plant parts that are never utilized by phytophages on one species of plant, but which are utilized on a close relative. For

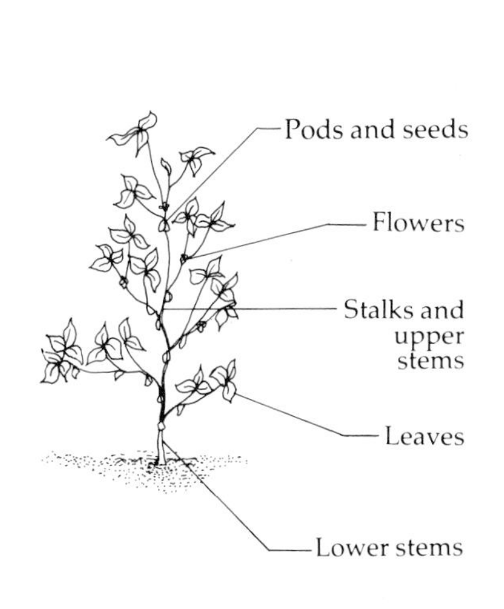

	North America		Central and South America		Orient	
Pods and seeds	6+	2	4+	3	5+	1
Flowers	3+	1	2+	1	1+	2
Stalks and upper stems	0	2	3	1	1	1
Leaves	16+	1	10+	2	11+	3
Lower stems	3	2	2+	3	3	3

Fig. 4.8. Feeding positions of phytophagous insects on soybean in three regions of the world. The upper left-hand portion of each box gives the number of species exploiting each plant part in each region (+ indicates minimum estimate). The lower right-hand portion of each box gives the number of 'vacant niches' — plant parts exploited in a particular way (e.g. stem-mining) in one or both other regions, but lacking an equivalent species in the area in question. (After Turnipseed & Kogan 1976.)

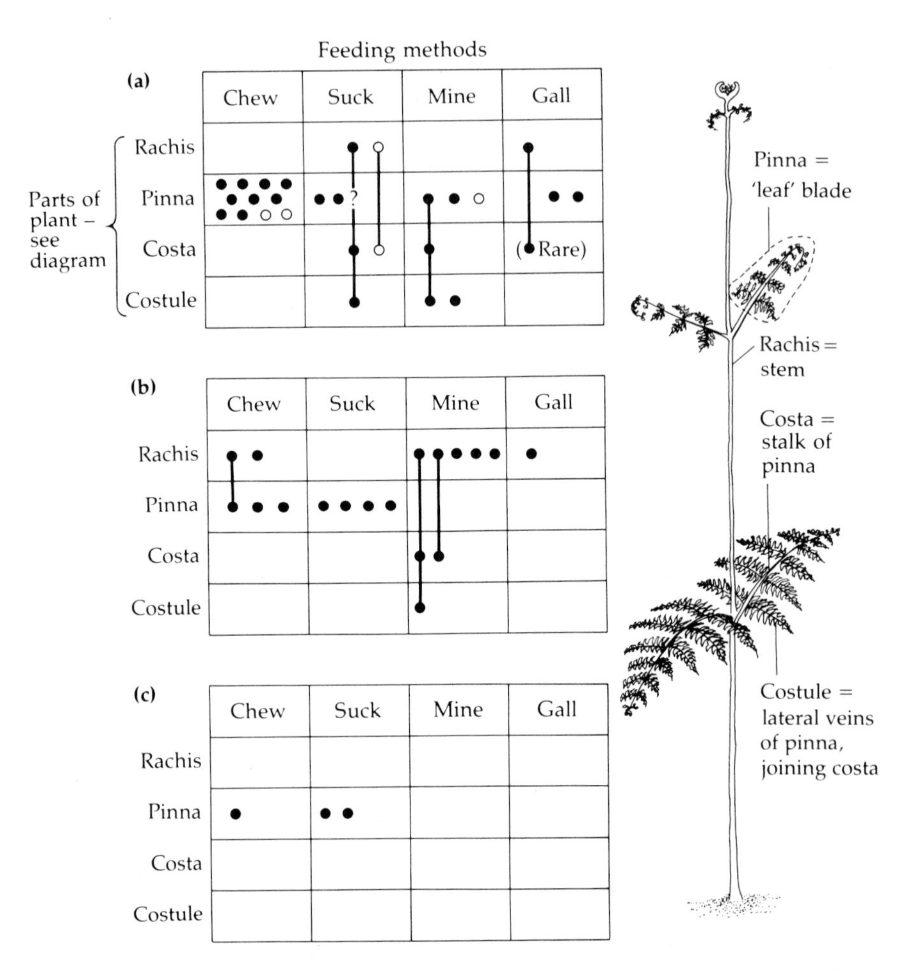

Fig. 4.9. Feeding sites and feeding methods of phytophagous insects attacking bracken fern (*Pteridium aquilinum*) at three intensively studied sites in widely different parts of the world. The same species of bracken is native and long established in all three regions: (**a**) Skipwith Common, Yorkshire, England; (**b**) Hombrom Bluff, Papua New Guinea; (**c**) Sierra Blanca, New Mexico, USA. None of the sites have any species of insects in common. Each dot refers to the feeding site and mode of attack of one species of insect; feeding sites of species exploiting more than one part of the frond are joined by lines. The Skipwith site (a) has been studied for much longer than (b) and (c); open circles in (a) refer to species recorded in less than 50 per cent of the years of study (1972–1981). (a) and (c) are open sites, (b) is bracken under savannah woodland. (From Lawton 1982, 1983b.) These communities differ markedly in the numbers of species they contain, and in the distribution of species across resources. There are many apparently 'vacant niches'.

example, unlike the majority, some species of Umbelliferae in temperate regions may not support any species of swallowtail butterfly (Slansky 1973), others no leaf- and stem-mining agromyzid flies (Lawton & Price 1979). The part of the plant left unutilized varies in an apparently haphazard way from plant species to plant species. Moreover, for a species of plant on which rapid phytophage recruitment is no longer occurring, comparisons between both regional faunas (e.g. soybean, Fig. 4.8) and local communities (e.g. bracken, Fig. 4.9) in different geographic regions reveal very different numbers of phytophagous species using the same resources in different areas, and idiosyncratic gaps in patterns of resource utilization. It would be illogical to attribute asymptomatic recruitment curves in such plants to niche saturation, although different natural enemies on different plant species or on the same plant in different areas may account for some of the gaps and differences in resource use. The simplest explanation is that such data are consistent with the pool-exhaustion hypothesis.

4.1.5 *Where are Colonists Recruited From?*

Native host plants are constantly subjected to potential insect colonists. Large-scale and local movements of insects are common (e.g. Rainey 1976; Schulz & Meijer 1978; Taylor & Taylor 1979), as are mistakes in plant identification by ovipositing females (Chew 1975; Chipperfield 1980; Side 1955; Singer 1971; Straatman 1962), chance mutations of insects (for example the loss of or changes in diapause pattern (Klausner *et al.* 1980; Tauber & Tauber 1981)), and unusual environmental conditions (van Emden 1978), which constantly bring together new insect–plant combinations. A truly astronomical number of individual insects find themselves on the wrong host and either leave or perish. Colonists are the minute fraction that stay and survive (Southwood 1961b, 1977a).

An excellent summary of the sources of colonists moving on to one particular plant — the legume soybean, *Glycine max* — is provided by Turnipseed and Kogan (1976). They write:

> 'Colonization by three main components of the native fauna is generally observed. Complexes of euryphagous (polyphagous) species ... readily move into the new crop. A second group of colonizers is the complex of stenophagous species adapted to wild legumes ..., grain legumes ..., and forage legumes. A third category

of colonizers consists of certain stenophagous (oligophagous) species that seem to have shifted their host preferences, sometimes from other plant families. As an example, a weed borer, *Dectes texanus* (Coleoptera, Cerambycidae), normally associated with composite weeds of the genus *Xanthium*, is now found on soybean. In general, the patterns of colonization of soybean that can be deduced from existing records suggest that phytophagous insect species accumulate asymptotically .... This rapid accumulation is mainly caused by colonizers of the first two categories. The third category contributes a small number of species.'

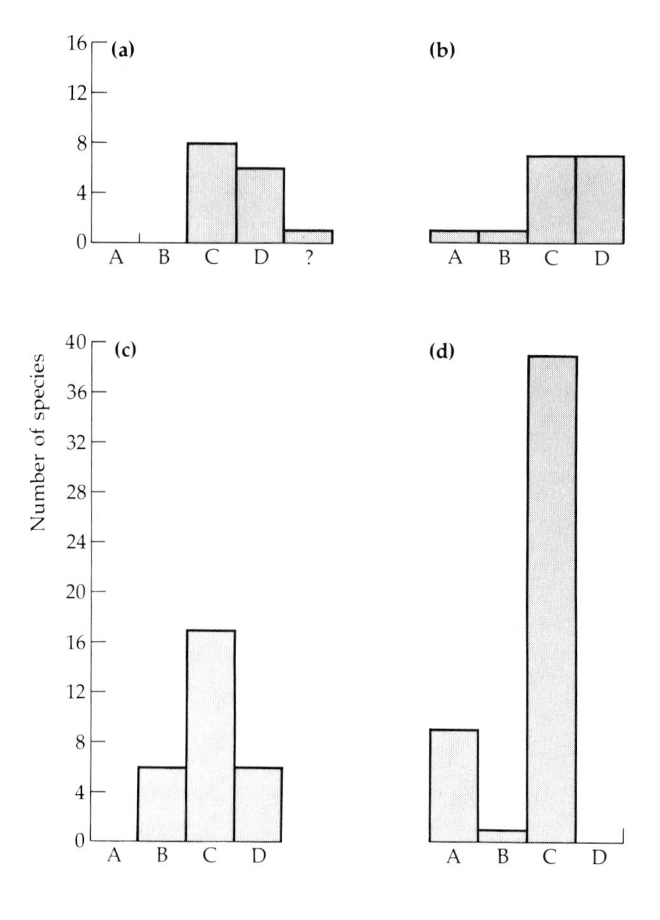

On present evidence this scenario is an excellent description of the colonization of most plant species. Some examples are given in Fig. 4.10.

Chemical Similarity as an Explanation for Major Host Shifts

Colonization by highly polyphagous species, or by species exploiting plants in the same genus or family, is not very surprising. Occasional host shifts* by stenophagous species from one plant family to another are more remarkable and, *a priori*, appear totally unpredictable (Fig. 4.10). Some could involve similarities in chemistry (enhanced predilection) between the original and the new host, unrelated to taxonomic affinity (cf. Chapter 3, Sections 3.2.2 and 3.4), but data to test this hypothesis are scarce. A well-known example is the addition of the garden nasturtium (*Tropaeolum majus*) to the host plants of the white

* 'Host shift' is a widely used term of convenience. It does not imply than an insect population abandons an old host to take up occupation on a new host species. Rather the new host is simply added to the insect's dietary repertoire.

Fig. 4.10. Taxonomic relationships between introduced plants and plants from which insect colonists have been recruited. A–D are insects grouped according to their original host plants:

A — Stenophagous insects, normally feeding on native plants in the *same genus* as the introduced plants.

B — Stenophagous insects, normally feeding on native plants in the *same family* but a *different genus* to the introduced plants.

C — Polyphagous insects, attacking plants in *more than one family*.

D — Stenophagous species attacking a narrow range of hosts in a *different family* to the introduced plants.

The histograms show numbers of species of insects (**a, b, d**) or plants (**c**), as follows.

(**a**) All insects recruited to *Buddleia davidii* in Britain (Owen & Whiteway 1980).

(**b and d**) Lepidoptera recruited to *Pinus contorta* in Britain (**b**) and to *Picea sitchensis* (**d**) (Winter 1974).

(**c**) The number of introduced plant species and hybrid plants exploited by 17 species of phytophagous Miridae in the grounds of the Agricultural College of Norway (plants are trees and shrubs), e.g. nine plants have been colonized by mirids normally feeding on relatives in the same genus as the non-native plant (from Taksdal 1965).

butterflies (*Pieris brassicae* and *P. rapae*) in western Europe. The garden nasturtium was introduced into Britain from Peru almost 300 years ago, and it belongs to a small family allied to the Geraniaceae. Use of nasturtium by *Pieris* is understandable, because nasturtiums contain mustard oils, which are otherwise characteristic of the Cruciferaceae, the 'normal' host plants of these butterflies. (That so simple an explanation for major host shifts cannot always be found, however, is revealed by the other common pest of garden nasturtium, a black aphid. This is apparently the same species that originally lived on a composite, *Arctium*, which lacks mustard oils (Taylor 1959; Robinson 1980).) Another example of chemical similarities between shared, but taxonomically unrelated, hosts is Berenbaum's (1981a) discovery that those few host plants used by North American species of swallowtail butterflies in the *Papilio machaon* complex that are not umbellifers all contain linear furanocoumarins — chemicals characteristic of the normal hosts in the subfamily Apioideae (Umbelliferae).

Some host shifts by phytophages lend support to radical views on plant classification. The Chinese shrub *Buddleia davidii* was introduced into western Europe only at the end of the last century, and according to most plant taxonomists the Buddlejaceae are placed near the Primulaceae. Although apparently taxonomically isolated, particularly from other trees and shrubs, *Buddleia* has accumulated a sizeable fauna in western Europe (Owen & Whiteway 1980; Moran & Southwood 1982; T.R.E. Southwood & C.E.J. Kennedy, unpublished; Figs 4.3 and 4.10). Among its new phytophages are three insects from mullein, *Verbascum* spp. (Scrophulariaceae), the mullein moth (*Cucullia verbasci*), the mullein aphid (*Aphis verbasci*) and the mirid *Campylomma verbasci*, and two species of *Cionus* weevils normally confined to Scrophulariaceae. These host shifts all support an alternative, radical view of the plant's affinities: namely that the Buddlejaceae and Scrophulariaceae are related and hence share other characteristics (Cronquist 1973).

Ecological Opportunity

Such examples aside, a major determinant of novel host plant relationships appears to be *ecological*, rather than phylogenetic or chemical, opportunity — the close proximity of widespread and abundant normal hosts to the new food plant. Thus, most of the major host shifts on to conifers noted by Winter (Fig. 4.10) in Britain are by insects feeding on abundant but unrelated native moorland plants, *Myrica gale* (Myricaceae), *Vaccinium*, *Erica* and *Calluna* (Ericaceae), among which the conifers have been extensively planted. Gould (1979) has reproduced such host shifts experimentally in the laboratory by growing cucumber plants (a toxic, resistant species) among lima beans (the initial, favourable host), resulting in rapid adaptation of the phytophagous mite *Tetranychus urticae* to cucumber. Other laboratory experiments in which, following repeated exposure and initially high mortalities, phytophagous insects adopted new host plants, many radically different from their normal hosts, were carried out by Schroder (1903), Pictet (1911), Harrison (1927), Kozhanchikov (1950) and Brower *et al.* (1967).

Contemporary patterns of host use, particularly those in which congeners feed on taxonomically unrelated plants, suggest that host shifts between unrelated plants have been as common in the past as they are on introduced plants at the present time. Examples in which closely related insects exploit unrelated host plants are found amongst sawflies (Benson 1950), macrolepidoptera (Holloway & Hebert 1979), microlepidoptera (Powell 1980), Lycaenidae (Atsatt 1981), aphids and psyllids (Eastop 1979; Hille Ris Lambers 1979), and indeed amongst phytophages in general (Brues 1946; Jermy 1976). Thus we argue that the undoubted close affinities of certain insect groups with sets of closely related plants, and host switching between close relatives (Brues 1946; Benson *et al.* 1976; Gilbert 1977; Gilbert & Smiley 1978; Cornell & Washburn 1979; Berenbaum 1981a, b), are not the only major determinants of insect–plant relationships (Strong 1979). We explore this idea further in Chapter 7.

Summarizing, plants recruit insects from a variety of hosts, spanning the full gamut from close relatives to those with no obvious structural or biochemical affinities but merely close physical proximity. The result is a fauna on most plants that is a pot-pourri of the coevolved, the preadapted and the opportunistic in varied and at present unpredictable proportions.

4.2 From Colonization to Speciation

4.2.1 The Problem

Many native plants have at least one species-specific phytophagous insect, and many plants have several (Chapter 6). Complete monophagy might be achieved either by the disappearance (extinction) of all but one of the hosts of a formerly oliphagous insect or by insect speciation. Issues about colonization of novel hosts and the evolution of new insect species are vexatious. Data are poor, time spans uncertain, and mechanisms hotly debated (Mayr 1963; Bush 1975a; White 1978; Hammond 1980; Futuyma & Majer 1980; Key 1981; Williamson 1981; Futuyma 1983). In this section, we briefly review the problem of speciation by phytophagous insects; without understanding speciation, we cannot hope to understand how plants acquire specialised species of phytophagous insects, a process that follows logically from colonization.

Circumstantial evidence suggests that the colonization of hosts may occasionally lead relatively quickly to the formation of new species of specialized phytophages. The Polynesians settled the Hawaiian islands about 1000 years ago, bringing with them bananas from Tahiti. Intriguingly, at least five species of Hawaiian pyraustid moths in the genus *Hedylepta* are now obligate feeders on feral Polynesian varieties of banana in Hawaii according to Zimmerman (1958, 1960). Zimmerman maintains that none of these species feed on any other plants, and all five form a closely linked group of endemic species. He believes they evolved from related *Hedylepta* exploiting *Pritchardia* palms. These native palms are now virtually extinct, leaving the five *Hedylepta* species isolated on their new host. *Hedylepta blackburni*, which Zimmerman suspects is ancestral and very closely related to the obligate banana feeders, exploits relict *Pritchardia*, Polynesian wild bananas and modern cultivated bananas, as well as its primary host, introduced coconut palms. Certainly, critical review of this scenario is necessary, but it does provide a possible example of polyphagy preceding the rapid evolution of monophagy.

An analogous but feebly documented situation apparently holds on Rapa, a remote island half-way between northern New Zealand and South America. 'It is strange that the (sole) food plants of several of the endemic species of moth (on Rapa) are 'weed' species which have been introduced' (Clarke 1971). If these moths genuinely have no native hosts, rapid speciation (within the last 1000 years) may again be implicated.

Glimpses of the stages before full speciation may be provided by *Rhagoletis* flies (Tephritidae), whose larvae feed on living fruits and hence are phytophagous in the broad sense of the word. In Oregon,

Washington and British Columbia, *Rhagoletis indifferens* established permanent populations on domestic cherries shortly before 1913, some 89 years after the crop was introduced. Less than 100 years later, according to Bush (1975b), two distinct races of *R. indifferens* coexist in the area, one infesting domestic cherries at low altitude in late June and early July and the other, the original host (wild bitter cherry, *Prunus emarginata*), at higher altitudes in July and August. It is, as Bush points out, arbitrary when a 'race' becomes a 'sub-species' or 'species', but if *Rhagoletis* on domestic and cultivated cherries are indeed genetically isolated and distinct, then such incipient speciation in 100 years (or generations) on cherries is consistent with Zimmerman's evidence for complete speciation in 1000 years on bananas.

Really, there are two classes of problems associated with speciation on a new host. How frequently does it occur as rapidly as these examples suggest? Obviously some sort of distribution of rates is to be expected (Williamson 1981). Of what shape is this distribution and where is the mode? Secondly, what are the mechanisms? A consideration of the second question throws some light upon the first.

4.2.2 *Mechanisms of Speciation*

Theories

The conventional view of speciation is that it takes place allopatrically (Mayr 1963; Bush 1975a; Hammond 1980). A gene pool is split by geographic barriers into two or more demes; genetic drift and different selection pressures operating upon the two isolated gene pools then lead to the evolution of two species where formerly there was one.

The *essential* requirement for speciation is the interruption of gene flow, the tendency for biased (assortative) rather than random mating (Maynard Smith 1966) and different selection pressures. As Bush (1975a, b) has pointed out for animals that live on their food supply, barriers interrupting gene flow may develop and new species arise within the same geographic region. This process is termed sympatric speciation (Mayr 1963).

Intermediate between sympatric and allopatric speciation lie distinctions of various semantic and conceptual shades — 'parapatric', 'stasipatric' and 'quasi-sympatric', etc. — in which gene pools meet and

Table 4.2. Examples of insect species that are relatively restricted (oligophagous) or very restricted (monophagous) in their diet at any one locality, but which feed on different food plants at different localities (regional polyphagy). Such species are obvious candidates for allopatric speciation.

Author(s)	Insect(s)	Notes
Cates 1981	*Estigmene acraea, Hyles lineata, Vanessa cardui* (Lepidoptera)	Individuals from different local populations exhibit 'a high level of local specialization'
Downey & Fuller 1961	*Plebejus icarioides* (Lepidoptera)	16 species of *Lupinus* are used as food plants throughout range, but individual populations are monophagous; in four cases, even when a *Lupinus* species is used elsewhere and is present in an area, it may be ignored by individuals
Dugdale 1975	*Dodonidia* sp. (Lepidoptera)	This New Zealand butterfly 'exists in two populations', one on *Chionochloa* from eastern Nelson southwards, the other on *Gahnia* in northwest Nelson and over North Island
Fox & Morrow 1981	Many *Eucalyptus* feeding insects	Of 80 species collected at one locality, 30 were entirely monophagous at that locality; 51 (including the 30) were oligophagous. 'Available data suggest that ... insects at this site use other ... species elsewhere'
Gilbert 1979	*Heliconius* spp. (Lepidoptera)	Local populations of some species, e.g. *H. cydne, H. melpomene,* are extremely restricted in their choice of food plants at one locality, but feed on many species over a range of localities (8 and 31 respectively for the examples given)
Higgins & Riley (1970)	European butterflies (Lepidoptera)	Numerous examples of varieties and subspecies showing regional differences in food-plant use.
Hsiao 1978	*Leptinotarsa decemlineata* (Coleoptera)	Populations from Arizona are adapted to feed on *Solanum elaeagnifolium,* which is unsuitable as a host for populations from Utah, New Mexico and Texas
Knerer & Atwood 1973	*Neodiprion* spp. (Hymenoptera)	Two species, *N. abietis* and *N. pratti,* exhibit complex regional differences in choice of food plants, in combination with sympatric host-

		race formation (see Table 4.3)
McNeill, S. (pers. comm.)	*Leptoterna dolobrata* (Heteroptera)	Wide variation exists in the number and types of host grasses used by this species in North America, because of variation in flowering time of the grasses and in the hatching of the nymphs
Pielou 1974	*Cachryphora serotinea* (Homoptera)	An aphid using three different but closely related *Solidago* species in different parts of its range
Singer 1971 Rausher 1982	*Euphydryas editha* (Lepidoptera)	Adults in different local populations had markedly different oviposition preferences (Singer); there are also differences in larval growth characteristics and digestive physiology (Rausher)
Wiklund 1973, 1974	*Papilio machaon* (Lepidoptera)	*Papilio* uses different umbellifer hosts in different geographic areas and habitats in Sweden
Wiklund 1977	*Leptidea sinapis* (Lepidoptera)	*Leptidea* selected different principal hosts in woodland (*Lathyrus montanus*) and meadows (*L. pratensis*)

overlap in various ways (Key 1974, 1981; White 1978; Pielou 1979; Hammond 1980). The clearest of these distributions are termed parapatric, in which two gene pools abut along a common boundary, but do not overlap. They have been best documented for Australian grasshoppers in the subfamily Morabinae (Key 1981). In order to keep the present discussion within reasonable bounds, we will focus upon speciation mechanisms at the theoretical extremes, dealing first with allopatric speciation (in which gene pools are geographically isolated) before moving to a consideration of sympatric speciation (where they are not).

Allopatric Speciation

Widespread species of phytophagous insects obviously experience different selection pressures in different parts of their ranges. For example, some 'polyphagous' species actually consist of monophagous or oligophagous local populations (Table 4.2; see also Fox & Morrow 1981) and provide ample raw material for speciation were these populations to become isolated by geographical barriers (the existence of local food-plant races implies that partial barriers to gene flow already exist). An excellent example is the British subspecies of the swallowtail butterfly *Papilio machaon*, geographically isolated from members of the same species on the continent and, unlike them, restricted to one species of Umbelliferae, *Peucedanum palustre* (Ford 1945). If such isolation continued for long enough, the end result could be two species where formerly there was one. How long might such speciation take?

Bush (1975a) recognizes two sorts of allopatric speciation. 'Type 1a' occurs 'slowly', probably over many thousands of generations, as large but isolated gene pools gradually diverge under different selective regimes. 'Type 1b' is 'rapid' and driven by founder effects in tiny, geographically isolated fragments of the gene pool. Type 1b speciation is probably capable of creating new species rather quickly. Indeed, recent experiments with *Drosophila* emphasize the speed and extent to which small, isolated founder populations can become reproductively isolated from parent stock, particularly if the pioneers are subject to repeated catastrophic reductions in numbers before becoming firmly established (Templeton 1979; Jones 1981). The analogy with some colonizing phytophagous insects is surely a close one.

The relative roles of Type 1a and Type 1b allopatric speciation in the evolution of host plant specialists are probably very different. The introduction of a novel food plant into an area is highly unlikely to be a primary or even a measurable cause of Type 1a speciation. A widespread polyphagous species is unlikely to speciate as a consequence of adding a novel food plant to its local diet.

In contrast, Type 1b allopatric speciation is probably a prime mover in the evolution of new host specialists. The necessary small, isolated founder populations may be generated by insect dispersal (Southwood 1978a) or by more complex mechanisms. For example, all five of Zimmerman's recently evolved *Hedylepta* have different, restricted, and non-overlapping distributions, consistent with recent allopatric speciation (Pielou 1979). Each may have evolved from an initially very small number of local colonizers making the required host shift from vanishing lowland *Pritchardia* palms to upland bananas. Of course, this is pure guesswork, and other explanations are possible.

One other comment is in order. For relatively immobile insects, the attenuation of gene flow, leading to allopatric speciation, might occur over very small physical distances. Thus the semantic and practical problems involved in deciding exactly how gene pools are fragmented and when ranges are 'continuous' (sympatric) or 'fragmented' (allopatric) may be prodigious (Bush 1975a; Endler 1977; Southwood 1978a; Pielou 1979; Hammond 1980).

Sympatric Speciation

For phytophagous insects, it has been suggested that relatively rapid differentiation of host specialists can occur *despite* the fact that novel and original hosts often grow very close to one another, ruling out spatial isolation of gene pools for anything but the most immobile insect species. Bush (1975a, b) outlines the necessary conditions for such sympatric speciation to occur. The first, essential step is that the appropriate alleles needed to overcome the resistance of the new host plant, to feed, reproduce and survive, should be present among some of the individuals that reach it by normal dispersal processes (Southwood 1978a). Whether such 'pioneers' merely increase the host range of the species or represent incipient speciation depends on the extent to which their host plant shift has attenuated gene flow — the extent to which there is an enhanced probability that they will mate with each other in the 'pioneer colony' (Maynard Smith 1966). Bush (1975a, b) suggests that for sympatric speciation to occur the following conditions must hold.

(i) Mating should occur on, or near to, the host plant.

(ii) Ovipositing females select the food plant; larvae have no choice in what species they eat.

(iii) The insect should be monophagous or oligophagous prior to making the initial shift to the new host.

These conditions hold for many phytophagous insects, especially members of the Chrysomelidae, Curculionidae, Diprionidae, Agromyzidae and Tephritidae.

Bush's conditions contribute to, but are not themselves, the mechanisms of assortative mating that cause the attenuation of gene flow. The most likely mechanism is the separation in time (as opposed to space) of the periods of mating — allochronic isolation (Tauber & Tauber 1981). There is now good experimental and observational evidence that the development rates of phytophages are different on

Table 4.3. Examples of possible sympatric host-race formation. Sympatric host-races exploit different species of food plants in the same habitat or community and, while retaining the capacity to interbreed, may differ in details of morphology, behaviour, and phenology, implying restricted gene flow between individuals on different host plants (see Jaenike 1981).

Author	Insect	Notes
Brues 1946	*Rhagoletis pomonella* (Diptera)	Different 'strains' exist on different hosts, although individuals may be transferred between hosts and strains are inter-fertile. This example, which was first reviewed by Brues, has subsequently attracted attention from many authors (see Bush 1975a, Jaenike 1981). The status of Brues' 'strains' is still debatable
Baltensweiler *et al.* 1977; Baltensweiler 1971	*Zeiraphera diniana* (Lepidoptera)	Two sympatric morphs are distinguishable as final instar caterpillars: a dark morph on *Larix* and an orange–yellow morph on *Pinus*
Guttman *et al.* 1981; Wood 1980; Wood & Guttman 1981, 1982	*Enchenopa binotata* (Homoptera)	Sympatric host races differ in morphology, phenology, etc., and in genotype. Moreover, when females are given a choice, they select mates from their own host. This study constitutes the best-documented example of host-race formation.
Klausner *et al.* 1980	*Oncopeltus fasciatus* (Heteroptera)	Individuals with different (genetically controlled) diapause characteristics exploit different food plants
Knerer & Atwood 1973	*Neodiprion abietis* (Hymenoptera)	Individuals from three species of host tree are morphologically identical but have different seasonal phenologies, and there is some evidence for genetic isolation
Le Quesne 1965	Various leaf-hoppers (Homoptera)	Different forms and life histories on different food plants and habitats
Menken 1981	*Yponomeuta padellus* (Lepidoptera)	Genetic differences exist between populations on different tree species. Mating is thought to take place exclusively upon the host
Mitter *et al.* 1979	*Alsophila pometaria* (Lepidoptera)	Markedly different genotypes exist upon different hosts; the moth is, however, parthenogenetic

Phillips & Barnes 1975	*Laspeyresia pomonella* (Lepidoptera)	Host races of the coddling moth exist on apples, walnuts and plums. However, the orchards are isolated from one another, so populations are not strictly sympatric
Price & Willson 1976	*Tetraopes tetrophthalmus* (Coleoptera)	Beetles from two different hosts (one recently colonized — see page 90) differ in phenology, size and longevity. Males mate with females from the same host more often than with females from the other host plant
Tavormina 1982	*Liriomyza brassicae* (Diptera)	Insects collected from one species of host produced a greater proportion of their mines on that host than individuals of the same species collected from other hosts. Some evidence that larval growth is slightly faster on parental host

different hosts (Le Quesne 1965; Knerer & Atwood 1973; Ahmad & Abror 1977; Wood 1980; Wood & Guttman 1981, 1982). Therefore, if the period of adult life (more specifically that of mating) is brief, gene flow between the two populations will be greatly reduced. An analogous situation, but with a longer time scale, is found in the periodic cicadas (Alexander & Moore 1962).

The effectiveness of allochronic isolation due to differences in time of development on different host plants will be enhanced (Southwood 1978a) if the following conditions are met.

(i) There is no diapause or other mechanism for ensuring synchrony of *adult* emergence.

(ii) Pre-reproductive movement is minimal. This condition will be met particularly frequently if long-distance migration is rare and where the normal mode of locomotion for local movement is walking (e.g. Hemiptera and Coleoptera in temperate regions) rather than flight (e.g. Lepidoptera) or, if trivial movements are usually by flight, when courtship and mating occur around the host plant, as in Lycaenidae (Gilbert 1979).

This scenario predicts, for example, that most groups of arboreal Lepidoptera, which do not fulfil either condition, will have broad host ranges (a prediction supported by Futuyma (1976)) and will not speciate sympatrically. However, many tree-dwelling Hemiptera do conform to both conditions, and among these the British Miridae usually have narrow host ranges, while there appears to have been explosive speciation in the genera *Lygocoris* (*Neolygus*) and *Lopidea*, mainly on trees (Southwood 1978a). A range of other examples of potential sympatric host-race formation and speciation is given in Table 4.3. That four of these 11 examples are drawn from tree-feeding Lepidoptera reveals the shaky nature of contemporary generalizations in this area.

Without doubt, the best current evidence for sympatric host-race formation and incipient speciation is the *Enchenopa binotata* 'species complex' of membracids in North America (Wood 1980; Guttman *et al.* 1981; Wood & Guttman 1981, 1982). Individuals occurring on different hosts are morphologically and genetically distinct and have different patterns of seasonal development. Moreover, females prefer to mate with males from their own host plants, and movements between hosts are restricted. We know of no other examples so well documented as this one. As Futuyma and Mayer (1980), and Jaenike (1981) point out, few of the other published examples purporting to show sympatric host-race formation and speciation are known to meet all or even most of the necessary criteria for it to occur.

In brief, whether most monophagous or oligophagous species of phytophagous insects have evolved sympatrically, as Bush (1975a, b) believes, remains uncertain and challenging. Under these conditions, speculation about rates of sympatric speciation seems idle. Even for the best-studied example, *Enchenopa binotata*, estimates for the length of time since host *races* (not species) diverged range between approximately 2×10^4 and 2×10^5 years for one race and between a quarter of a million and 3 million years for another, depending upon the method employed (Wood & Guttman 1981)!

4.2.3 *Epilogue*

Allopatric and sympatric speciation, together with their hybrid cousins and relatives of varying semantic and conceptual shades, provide a theoretical basis for the evolution of new host specialists. It is therefore important to remember that virtually none of the vast number of documented colonizations summarized in this chapter has yet led to speciation, although very few have been studied well enough to detect host-race formation. How much time is required for speciation? Short of saying that Type 1a allopatric speciation occurs 'slowly' and Type 1b and sympatric speciation occur 'quickly', we remain abysmally ignorant about what these rates are in an absolute sense for phytophagous insects. That speciation has not been reported for any colonists after 100 years is hardly surprising. It may possibly have occurred within 1000 years in Zimmerman's moths. Hence we guess that the evolution of significant numbers of host specialists on novel food plants will be negligible until plant and insects have been associated for at least 10^4 years, and time periods as long as 10^6 years for significant numbers of species to evolve would not be surprising. Moreover, we expect very different speciation mechanisms to operate at different rates for polyphages, oligophages and monophages, and for species that differ in their dispersal abilities, diapause patterns and breeding systems. Here are rich, but undoubtedly difficult, grounds for future research.

4.3 Summary

Insects often find and colonize newly introduced host plants, or naturally spreading hosts, remarkably quickly. However, there are some interesting exceptions, such as *Opuntia* cacti in the Old World; these plants have apparently recruited no herbivores after several hundred years. We suggest that two factors determine rates of colonization of new hosts by insects: (i) the area planted (or the size of range achieved by natural spread) and (ii) the taxonomic, phenological, biochemical and morphological match between introduced plants and the native flora. Data for native and introduced British trees, in particular, support these suggestions.

Often the first insects to colonize new host plants are polyphagous species. Moreover, rates of colonization by external chewing and sucking herbivores appear to be higher than those shown by endophagous gall-formers and miners. Rates of colonization by all groups decrease with time, being more rapid at first and subsequently (after tens or hundreds of years) slowing to a trickle. Such asymptotic recruitment curves are probably best explained by a 'pool-exhaustion' model — in which the supply of suitable potential colonists is exhausted — rather than by a model based on interspecific competition setting limits to the number of coexisting species.

Colonists are recruited from a variety of sources. Some, as we have noted, are broadly polyphagous species that already attack a wide range

Chapter 5
Species Interactions in Communities: The Animals

5.1 Introduction — Populations as the Basis of Communities

A community is a group of species that share the same habitat. Some, at least, of the component populations will interact with each other. These interactions and the resulting population dynamics* underlie broad patterns in the structure of ecological communities. In this chapter, and the one that follows, we summarize those aspects of population dynamics and population interactions that are most important for an understanding of the structure of phytophagous insect communities.

Our focus is upon the population dynamics of single species of phytophages and their interactions 'horizontally' with species that are potential or actual competitors, and 'vertically' with natural enemies. So we are concerned primarily with interactions involving other animals, as competitors, predators and parasitoids. Microorganisms, diseases and pathogens also enter the story as important natural enemies.

Interactions between phytophagous insects and their host plants are addressed in Chapters 6 and 7.

5.1.1 The Link Between Population Dynamics and Community Structure

Population Dynamics

Three important questions in population dynamics are why and how much populations fluctuate, and how populations are regulated. As a broad generalization, populations are regulated and controlled by *density-dependent* processes, and perturbed and disturbed by *density-independent* processes. The extent to which any particular population

* For general accounts of population dynamics see Clark *et al.* (1967), Williamson (1972), Varley *et al.* (1973), Hassell (1976, 1978), Elseth & Baumgardner (1981), May (1981) and Begon & Mortimer (1981).

stays constant or fluctuates in numbers from one generation to the next depends upon the form and relative magnitudes of both density-dependent and density-independent events in its life history.

An appreciation of population dynamics is central to an understanding of two aspects of community structure. First, the form and relative magnitudes of the density-dependent and density-independent events in each species' life history ultimately determine community predictability, i.e. the constancy of species' relative and absolute abundances from one year (or generation) to the next. Secondly, the nature of the density-dependent control determines which species interact and how they do so, in particular whether interspecific competition is important in structuring the community.

Although it is usual to assume that *density dependence* regulates populations and thus dampens out fluctuations caused by density-independent disturbances, this is certainly not always true. *Delayed density dependence* can be an important reason why populations fluctuate. The precise conditions under which delayed density dependence drives population fluctuations are considered by, amongst others, May (1972, 1973, 1974b, 1975), May *et al.* (1974), Southwood *et al.* (1974), Hassell *et al.* (1976), and Nisbet and Gurney (1982). Obvious biological causes of delayed density dependence are sudden food shortages in scramble competition, or delayed reactions by natural enemies to an increase in the availability of victims. Seasonally discrete life history stages are also an important cause of time delays, and high rates of population increase exacerbate them; both are characteristic of many phytophagous insect populations.

Theoretically Ideal Community Types

Now, depending upon details of their population dynamics, groups of species may behave in one of several ways.

(i) The most extreme possibility is for density dependence to have no effect, or only very feeble sporadic effects, at any density on most of the component populations of a community. The result is a shambles of randomly fluctuating populations. Local extinctions, large changes in abundance, and very high densities should be commonplace in this uncontrolled world; which species are common, rare or absent should be highly variable in space and time.

(ii) A second possibility is for populations to be controlled by density-dependent processes (but not ones that operate with significant time lags), with component populations operating more or less independently of all other populations in the same trophic level. This might happen for two reasons. Each population could be resource limited but the community nowhere near saturated with species, for example because of lack of time for species to evolve to fill available niches (Chapter 2), or because isolation limits the invasion of potential colonists (Chapters 3 and 4), or because host plants and individual species' populations are patchily distributed. Patchy distributions markedly reduce the likelihood of effective interspecific competition (see Shorrocks *et al.* 1979; Atkinson & Shorrocks 1981, and references therein). Alternatively, natural enemies may keep each population well below the point where resource limitation becomes important. In either case, the result is a community that is reasonably constant in composition and species abundances, but not one that is influenced to any significant degree by competitive interactions between species on the same trophic level.

(iii) An important variant on (ii) embraces those populations with fluctations of large amplitude driven by delayed density dependence. The result will be communities that vary markedly in species abundances (although if the fluctuations are regular, population numbers are not unpredictable, in marked contrast to (i)). More importantly, competition for resources may only occur sporadically, at the peaks of each cycle. Such 'occasional competition' may still be significant in structuring the community.

(iv) A final class of ideal communities is one in which interspecific competition for limiting resources is a persistent and important feature of community organization. These communities have a predictable structure (and hence resemble those in (ii)), but are more or less saturated with species; component populations are not kept rare relative to available resources by natural enemies.

Possibilities (i)–(iv) represent no more than high points on a multi-dimensional continuum. Most ecologists now accept that different sorts of organisms and habitat templets (Southwood 1977b) combine to generate real communities throughout this continuum, with the possibility that within any one real community, populations may be drawn from more than one of the ideal sets.

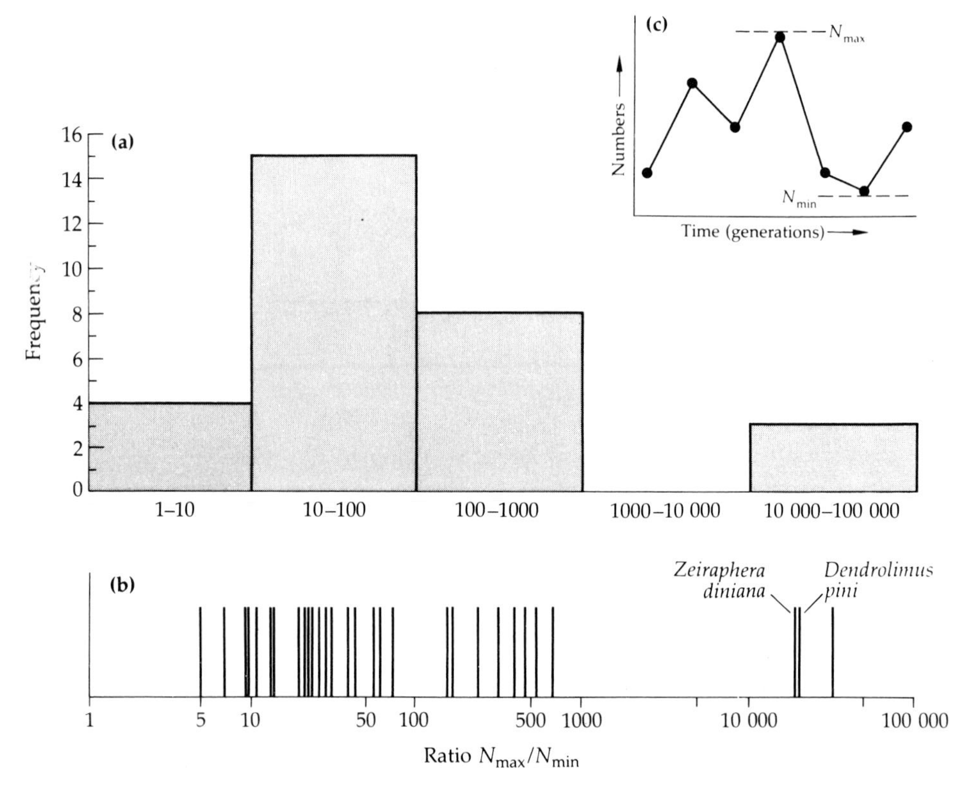

Fig. 5.1. The ratio of population maxima to population minima (N_{max}/N_{min}) in 30 populations of phytophagous insects, comparing the same life history stage for each species in successive generations. The minimum number of generations (or years) that each population was studied for is five, with a maximum of 60 and a mean of 19. The data include 21 Lepidoptera, four Hemiptera, two Diptera, two Coleoptera and one Symphyta. (**a**) Summary of the number of populations falling into order of magnitude classes; (**b**) the individual data; (**c**) how the data are calculated. The species named in **b** are discussed individually in the text (Sections 5.1.2 and 5.3.3). (From data in

It is also important to realise that interspecific competitive interactions (in (iii) and (iv)) may be contemporary and amenable to study by experimental manipulation; or they may be 'ghosts' (Connell 1980), detectable because species' ecologies and morphologies have evolved to minimize competitive effects; or they may lie somewhere between these two extremes. Moreover, competitive interactions need not be for obvious resources like food, or oviposition sites. One interesting possibility, discussed in Chapter 4, is that species might compete for 'enemy-free space'.

The Nature of the Evidence for Phytophagous Insect Communities

How do real phytophagous insect communities match up to these theoretical possibilities? There now exists extensive evidence on which to base an answer to this question. This evidence may be grouped under four headings: the amplitude of population fluctuations (Section 5.1.2); the analysis of life tables to discover the frequency and type of density dependence (Section 5.1.3); the influence of interspecific competition (Section 5.2); and, finally, the role of natural enemies (Section 5.3).

5.1.2 The Amplitude of Population Fluctuations

Some examples of the upper and lower bounds within which populations of phytophagous insects fluctuate over long runs of population data are gathered in Fig. 5.1. A few populations are extremely unstable. For example, counts of larvae of the moth *Dendrolimus pini* hibernating below their host pine trees at Letzlingden in Germany varied 20 000-fold over 60 generations (Varley 1949). However, most populations vary less than this, with extreme maximum and minimum counts of the same life history stage in different generations falling within one or two orders of magnitude of one another. By most sensible criteria these are moderately constrained, but definitely not tightly regulated populations.

The species in Fig. 5.1 are a haphazard selection, gathered by population ecologists over the past 30–40 years. Most come from widely different sorts of communities. Comparable data for 22 species over 11 years, the entire community of phytophagous insects on bracken (*Pteridium aquilinum*) at Skipwith Common in Yorkshire, are given in Fig. 5.2. The picture is much the same as that in Fig. 5.1. Some species vary little in abundance from year to year, most fluctuate within wider but still fairly narrow bounds, and a few populations are extremely unstable.

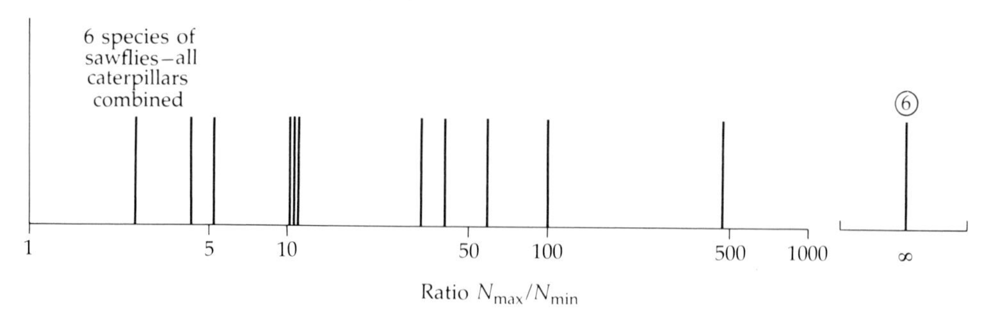

Fig. 5.2. The ratio of population maxima to population minima (N_{max}/N_{min}) for the entire community of phytophagous insects feeding on bracken (*Pteridium aquilinum*) at Skipwith Common, Yorkshire, between 1972 and 1982. Populations were sampled in seven of these eleven years. Data on six species of sawfly caterpillars have been pooled, because early in the study larvae were not identified to species. Three years' data on individual species all show $N_{max}/N_{min} < 5$. Six (non-sawfly) species with $N_{max}/N_{min} = \infty$ apparently became extinct on the site in one or more years. (Lawton 1982, 1983b, 1983c, and unpublished observations.)

The species in Fig. 5.1 are almost entirely from temperate regions. Somewhat unexpectedly, most available evidence indicates that phytophagous insects from the climatically more stable tropics do not have more stable populations (Bigger 1976; Wolda 1978a, b).

These data say simply that many phytophagous insect communities must be rather inconsistent in the abundance of their component species from one year to the next. Exactly how predictable different communities are from generation to generation has not often been examined. Figures 5.3 and 5.4 give two examples.

The community of bracken-feeding insects at Skipwith Common (Fig. 5.2) has a moderately constant structure (Fig. 5.3). Despite fluctuations in numbers, rare species tend to be rare in most years and common species are almost always common. The result is a significant correlation between species' rank abundances comparing each year with all other years for which there are data. But there are exceptions. 1977 was an unusual year, and the three non-significant correlations in Fig. 5.3 all involve comparisons with this year (1977–1981, 1977–1975 and 1977–1972). Species rank abundances changed so much in 1977 that community structure in this year was not significantly correlated with community structure in any other year.

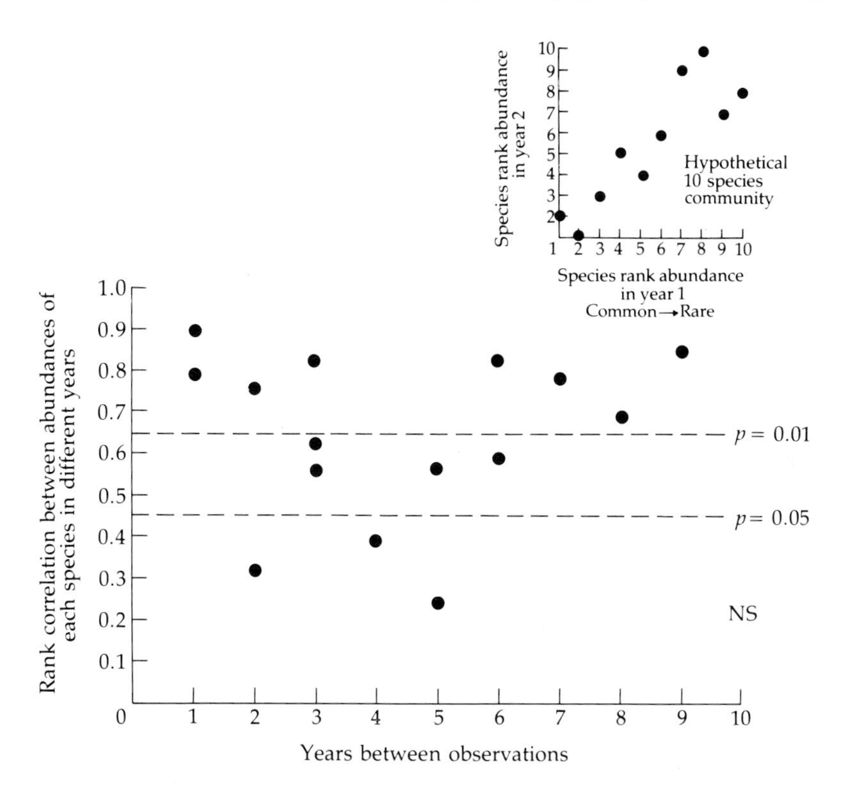

Fig. 5.3. Spearman rank correlation coefficients between species abundances for the community of phytophagous insects feeding on bracken at Skipwith Common, Yorkshire. Each point on the graph is a correlation between two years, as shown in the (hypothetical) inset. All possible pairs of years for which data are available between 1972 and 1981 have been compared. The maximum number of years between observations is therefore nine (1972 with 1981). Three species only recorded in one year (three of the six species where $N_{max}/N_{min} = \infty$ in Fig. 5.2) have been omitted from these calculations to minimize tied ranks (see Siegel 1956). Significant rank correlations between most pairs of years indicate that rare species have remained rare, and common species have remained common throughout the ten years of study; in other words, the community has a reasonably predictable but not absolutely constant structure. (After Lawton 1983c.)

Similar data are presented in rather a different way in Fig. 5.4 for grassland leaf-hoppers at Silwood Park in Berkshire. This is apparently a less stable group of insects, with three otherwise very rare or absent species occasionally becoming very common (e.g. *Macrosteles laevis,*

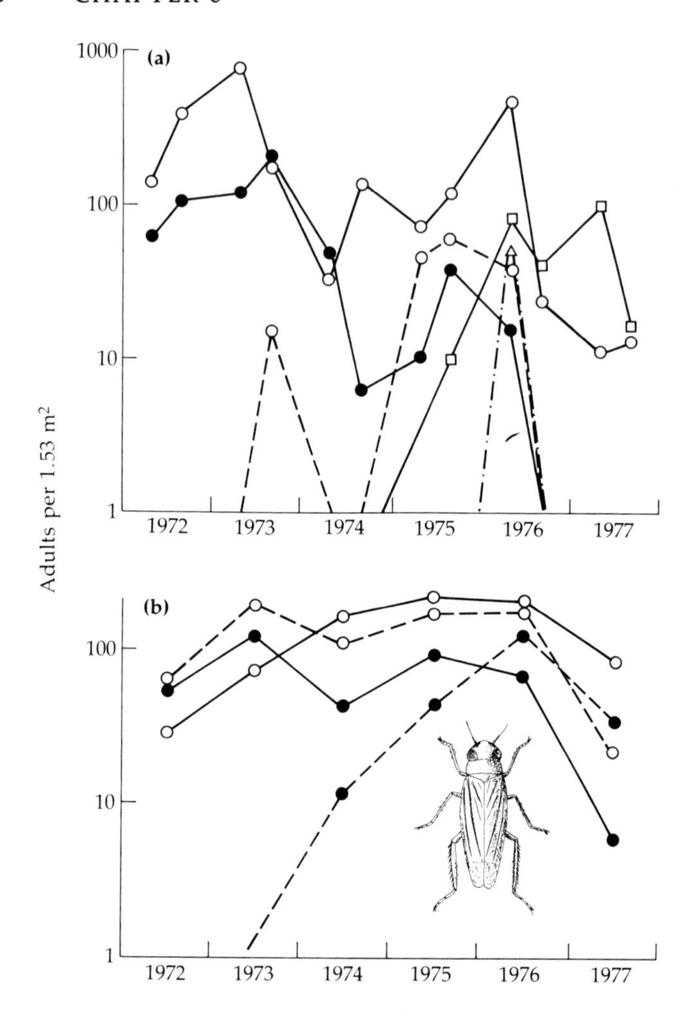

Fig. 5.4. Adult population densities in nine species of grassland leaf-hoppers (Auchenorrhyncha, Homoptera) at Silwood Park, Berkshire. Each different combination of dots and lines represents one species. (**a**) Bivoltine species; (**b**) univoltine species. (After Waloff & Thompson 1980.)

Cicadellidae) and an initially common species (*Dicranotropis hamata*, Delphacidae) that declined to very low levels in 1976 and 1977. Waloff and Thompson (1980), who gathered these data, wrote that 'the leaf-hopper complex was labile, in its population levels and in species

Table 5.1. Spearman rank correlation coefficients, r_s (Siegel 1956), calculated on the abundances of Heteroptera taken in light traps at Rothamsted Experimental Station in the four years 1946–1949. (**a**) Species associated with trees (Acanthosomatidae, Pentatomidae and Miridae); (**b**) Species associated with herbaceous vegetation (all Miridae). Entries in the tables are Spearman rank correlations between species' abundances in year i vs. the same species' abundances in year j, as illustrated in the inset to Fig. 5.3. (Data from Table XVII in Southwood 1960b.) Perfect correlations (the commonest species in year i is the commonest in year j and so on down to both years with the rarest species also the same) have $r_s = 1$; no correlation between rank abundances (species are common one year, rare another, etc.) gives $r_s = 0$.

(**a**) Species associated with trees

	1946	1947	1948
1947	0.48	—	—
1948	0.71	0.48	—
1949	0.51	0.72	0.48

$n = 21$

Critical values for r_s; $P = 0.05$, $r_s = 0.36$
$P = 0.01$, $r_s = 0.52$

(**b**) Species associated with herbs

	1946	1947	1948
1947	0.66	—	—
1948	0.80	0.71	—
1949	0.67	0.92	0.68

$n = 16$

Critical values for r_s; $P = 0.05$, $r_s = 0.43$
$P = 0.01$, $r_s = 0.60$

In each year, three traps were used, situated in mixed (i.e. arable and pasture) farmland, close to orchards, shelter belts and woodland. Two traps were in the same places in all four years; the third trap was in arable and pasture in 1946–1947 and in a garden, closer to trees and bushes, in 1948–1949. Data from all three traps have been pooled in each year. Moving the third trap will undoubtedly have reduced the correlations between species' rank abundances in the first two versus the last two years. Omitted from the analyses are species represented by single individuals taken only in one or two years.

composition. Only four of the nine species [in Fig. 5.4] were sufficiently abundant, throughout 1972–77 to be consistently sampled'. Their study is considered further in Table 5.3B.

Almost certainly the degree of stability in component populations, and hence of predictability in community structure, will vary among plant types and habitats (Southwood 1977b; Price 1980, 1983). An example is presented in Table 5.1 for Heteroptera (Hemiptera) collected in light traps at Rothamsted in southern England (Southwood 1960b). Species rank abundances were significantly correlated within two habitats in all pairs of years, but correlations were better for the Heteroptera associated with herbs than for those associated with trees. Perhaps the opposite result would have been more consistent with *a priori* notions of habitat stability, although many herbaceous plant communities are very persistent. Unfortunately, light-trap data are not ideal for these purposes; for example, light traps draw insects from quite wide areas and mixtures of communities. It would be valuable to have data like those in Figs 5.3 and 5.4 and Table 5.1, from a range of different habitats, specifically gathered to test the constancy of community structure.

5.1.3 Life Tables for Phytophagous Insects

Table 5.2 is a summary of life tables for phytophagous insects, analysed by Varley and Gradwell's *k*-factor analysis (Varley & Gradwell 1960; Varley *et al.* 1973), so assembled that we can use the data to answer two questions.

(a) How frequently has density-dependent or delayed density-dependent population control been detected in phytophagous insect populations?

(b) What form does density-dependent population control most commonly take?

Answers to both questions must be hedged around with qualifications and elaborations.

The Frequency of Density Dependence

There are 31 life tables in Table 5.2. The original authors detected density dependence (or delayed density dependence) in just under half of these (14, 45%). Subsequent studies indicated in the table strongly suggest that density dependence or delayed density dependence is present in another three populations, and it is suspected in two more. So even the most charitable analysis of Table 5.2 leaves 12 populations (38–39%) with no obvious density-dependent control.

These data highlight a well-known problem with *k*-factor analysis. For various reasons, particularly shortage of data, but also for other reasons (e.g. Luck 1971; Benson 1973; Bulmer 1975; Slade 1977, and references therein), the method may fail to detect density-dependent population control even though it is there. Unfortunately, the converse is also true: it may sometimes identify density dependence when none exists (Luck 1971; Royama 1981a, b). In other words, *k*-factor analyses are much less precise than one would like. Hence the conclusion to draw from Table 5.2 is that density dependence is either too feeble, too sporadic, or both, to be detected in somewhere between one-half and one-third of phytophagous insect life tables. However, it could be rather more common than this, or rather less. The 'real' figure is anyway unimportant; so high a frequency of feeble density dependence must often generate communities that are rather unpredictable in species composition and relative abundances through time.

The Form of Density Dependence

Our second question was: where density dependence has been detected, what form does it take? Table 5.2 suggests that natural enemies are of overriding importance; examples include parasitoids, predators and pathogens. Intraspecific competition for food is, in contrast, quite rare, with only three examples, although it is possible, even probable, that some of the other examples of density dependence involving changes in adult fecundity and so on are a response to intraspecific competition for food. Nonetheless, enemies still outnumber intraspecific competition as agents of density-dependent control by a ratio of about 2:1

Conclusions

If competition occurs through the exploitation of a limiting resource like food, then intraspecific competition is a necessary though not sufficient condition for interspecific competition. Furthermore, population suppression by natural enemies tends to reduce or eliminate the depletion of food resources. In other words, the first conclusion to be drawn from analyses of life tables is that the dominant density-dependent influences on phytophagous insect populations and communities usually operate vertically through the food chain, and not horizontally with competitors.

Table 5.2. Summary of main density-dependent controlling factors (discovered by authors or by subsequent analysis) operating on populations of phytophagous insects, revealed by life table studies (Varley *et al.* 1973). Data are from Podoler & Rogers (1975), supplemented by Stubbs (1977), and other sources given in the table.

Species	Density-dependent mortalities
	Parasitism, predation and disease
West Indian cane fly *Saccharosydne saccharivora* (Diptera)	
'2 Dundas'	1. Egg parasitism by *Tetrastichus*
'D-Piece'	2. Parasitism of early nymphs by *Stenocranophilus*
Yew gall midge *Taxomyia taxi* (Diptera)	3. None identified: the life cycle of this herbivore is very complex, and parasitism by *Mesopolobus* may be density dependent in one-year life cycle galls (Redfern & Cameron 1978)
Cabbage root fly *Erioischia brassicae* (Diptera)	4. Pupal parasitism by *Aleochara* and some predation on pupae
Broom beetle *Phytodecta olivacea* (Coleoptera)	5. None identified: mortality in the soil due to predation up to emergence of adults in autumn may be density dependent
Spring usher *Erannis leucophaearia* (Lepidoptera)	6. Parasitism and predation (Stubbs 1977)
Pine looper *Bupalus piniarius* (Lepidoptera)	7. Larval mortality due mainly to parasitism by *Eucarcelia* and *Poecilostictus*. Infectious disease also important (Anderson & May 1980, 1981)
Mottled umber *Erannis decemlineata* (Lepidoptera)	8. Parasitism and predation (Stubbs 1977)
Winter moth *Operophtera brumata* (Lepidoptera)	
England	9. Pupal predation, by carabid and staphylinid beetles and small mammals (Figs 5.5 and 5.8). Infectious disease also important (Anderson & May 1981)
Canada	10. None identified, but parasitism of caterpillars by *Cyzenis* (and *Agrypon*?) is strongly regulatory (Hassell 1980; Beddington *et al.* 1978)
Grey larch moth *Zeiraphera diniana* (Lepidoptera)	11. None identified: parasites or predators on eggs or pupae may be a 'major cause' (delayed density dependence) of cycles (Varley & Gradwell 1970). More likely, the cycles are driven by infectious disease (Anderson & May 1980, 1981)
Black-headed budworm *Acleris variana* (Lepidoptera)	12. None identified: parasites or predators on eggs or pupae may be a 'major cause' (delayed density dependence) of cycles (Varley & Gradwell 1970). More likely, the cycles are driven by infectious disease (Anderson & May 1980, 1981)

Citrus swallowtail *Papilio xuthus* (Lepidoptera)	13. Egg and pupal parasitoids are spatially and temporally density dependent (Hirose *et al.* 1980) (cf. example 23 in a different habitat)
	Competition for Food
Grass mirid *Leptoterna dolobrata* (Heteroptera)	14. Competition for high-nitrogen feeding sites by adults, and possibly also by late-instar nymphs (Fig. 5.5)
Colorado beetle *Leptinotarsa decemlineata* (Coleoptera)	15. Starvation of older larvae through food shortage. Parasitism of pupae by *Doryphorophaga* may also be density dependent (Stubbs 1977)
Cinnabar moth *Tyria jacobaeae* (Lepidoptera)	16. Death of caterpillars from starvation and delayed density-dependent reduction in adult fecundity (Dempster 1975; Dempster & Lakhani 1979; see Table 5.3)
	Other
Large copper *Lycaena dispar* (Lepidoptera)	17. Decrease in adult fecundity (Stubbs 1977)
Thistle lady beetle (Coleoptera) *Henosepilachna pustulosa*	18. Adult dispersal and fecundity were density dependent (Nakamura & Ohgushi (1981)
Frit fly *Oscinella frit* (Diptera)	19. None identified, but adult mortality, seasonal migration and variation in fecundity appear to be density dependent in combination. Cause unknown
Tea moth *Andracea bipunctata* (Lepidoptera)	20. None identified (Banerjee 1979)
Olive scale *Parlotoria oleae* (Homoptera)	
Hills Valley	21. None identified
Herndon	22. None identified
Swallowtail *Papilio xuthus* (Lepidoptera)	23. None identified (Watanabe 1981) (cf. example 13 in a different habitat)
Leafhoppers (Auchenorrhyncha, Homoptera)	24 29. Six species of grassland leafhoppers. None identified (Waloff & Thompson 1980; see Table 5.3)
Sulphur butterfly *Colias alexandra* (Lepidoptera)	30. None identified (Hayes 1981; see Table 5.3)
Viburnum whiteful *Aleurotrachelus jelinekii* (Homoptera)	31. None identified (Southwood & Reader 1976)

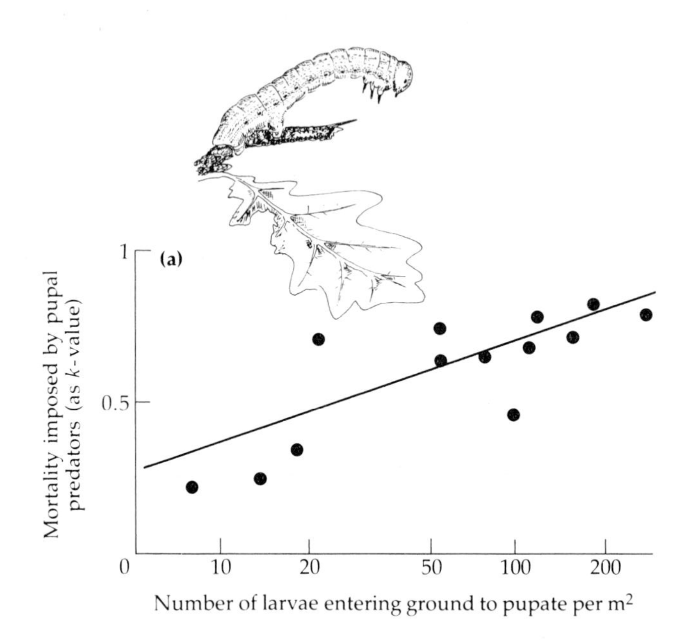

(a)
Mortality imposed by pupal predators (as k-value)
1
0.5
0
10
20
50
100
200
Number of larvae entering ground to pupate per m²

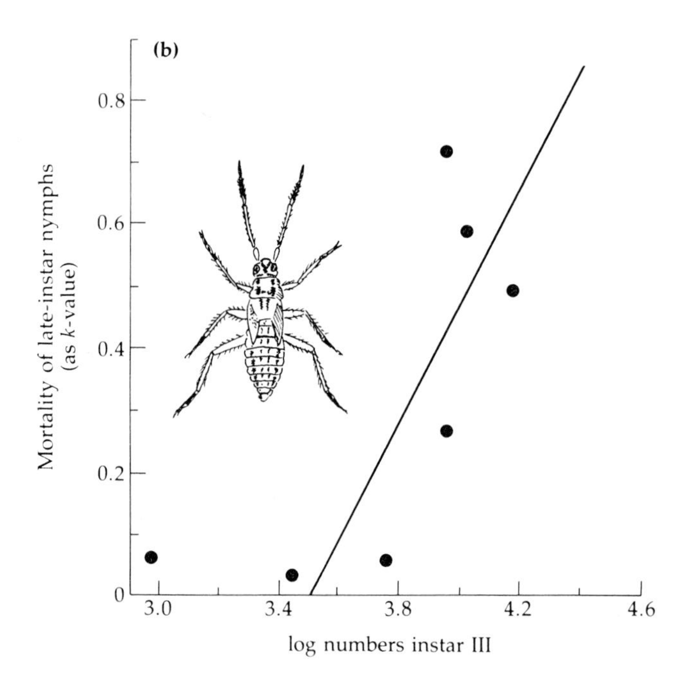

(b)
Mortality of late-instar nymphs (as k-value)
0.8
0.6
0.4
0.2
0
3.0
3.4
3.8
4.2
4.6
log numbers instar III

The second conclusion is that density dependence is often too feeble to be detected in many populations. At best, phytophagous insect communities will only be moderately deterministic in their structure.

5.1.4 *Density-Vague and Density-Independent Responses*

It is useful at this point to digress briefly and consider in more detail the main causes of indeterminism operating on populations of phytophagous insects. There are two.

First, density-dependent control processes are themselves far from perfect. Whilst much theoretical effort during the past decades has focused upon average effects of density-dependent processes in population regulation, attention is now increasingly drawn toward accommodating the abundant natural variation in population processes that was described by Milne (1957) as 'imperfection'. Examples are shown in Fig. 5.5. Many such relationships are well described as 'density vague': on average the relationship is density dependent, but there is much variation round the regression line (Strong 1983a).

Secondly, the great majority of mortalities acting on populations are density independent, and many are highly variable in their impact from year to year, or place to place. Some examples of the many and various things that kill phytophagous insects are summarized for three populations in Table 5.3.

Because of the many factors that normally affect them, it is not difficult to see why populations of phytophagous insects fluctuate as

Fig. 5.5. Examples of density-dependent mortalities acting on two species of phytophagous insect. (**a**) Pupal predation in the winter moth *Operophtera brumata*; the main predators are carabid and staphylinid beetles, and small mammals — moles, mice and particularly shrews (Varley *et al.* 1973; see also Fig. 5.8). (**b**) Death of late-instar nymphs of the bug *Leptoterna dolobrata*. Most of the scatter round this regression is apparently due to the effects of temperature, mortality being lower in years with low June temperatures. Mortality appears to be due to competition between nymphs for a limited number of high-nitrogen (good quality) feeding sites. High June temperatures increase this stress (McNeill 1973).

Mortalities on both graphs are expressed as k-values, i.e. $\log_{10} (N_t/N_s)$ where N_t and N_s are numbers entering and surviving the life history stage in question.

Table 5.3. Three life table studies that illustrate a range of abiotic and biotic factors operating upon sequential stages in the life history of phytophagous insects. (DD = density dependent; DI = density independent.)

Stage in life history	Factor	Mortality range (%) (Low High)	Density relationship
A. Cinnabar moth on tansy ragwort at Wheeting, England. Data for 8 generations. (Dempster 1975)			
Adult	Reduced fecundity	0–75	DD, delayed?
Egg	Infertility	1–5	DI
	Failure to hatch	1–3	DI
	Rabbit grazing and miscellaneous	0–9	DI
Larvae I and II	Starvation	0–25	DD
	Predators and unknown	24–84	DI
Larvae III and IV	Starvation	0–58	DD
	Predators and unknown	0–34	DI
Larvae V	*Apanteles* parasitism	5–36	Inv DD?
	Starvation and failure to pupate	2–83	DI
Pupae	Vertebrate predators and failure to emerge	51–96	DI
B. Leaf-hopper *Errastunus ocellaris* on grasses *Holcus mollis* and *H. lanatus* at Silwood Park, England. Data for 12 generations, 6 years. (Waloff & Thompson 1980)			
Egg	Reduced fecundity	23–55	DI
	Mortality (parasitism, plant phenology, and drought)	38–99	DI
Nymph I, II	Parasitoids (total)	0.5–38	DI
	Pipunculidae	0–37, 0.5–20	DI
Nymph III, IV	Dryinidae	0–8	DI
Nymph V		0–6	DI
Adults	Parasitoids (total)	0–40	DI
	Pipunculidae	5–40	
	Dryinidae	0–5	
C. Sulphur butterfly, *Colias alexandra*, on *Lathyrus leucanthus* in Rocky Mountains, USA. Data for 5 generations (years). (Hayes 1981)			
Egg	Infertility	33–40	DI

	Failure to hatch	5 7	
	Other: dessication, dislodgement then predation by ants and mites, cannibalism	22 30	
Larvae I	Failure to initiate feeding	32 37	DI
Larvae II	Dislodgement from plant,	28 60	DI
Larvae III	followed by predation and desiccation	35 70	Inv DD
Overwinter by Larvae III (Diapause)	Flooding, freezing and some predation	50 95	DI
Larvae IV and V, Pupae (Post Diapause)	Parasitism by *Apanteles*, predation by snakes and rodents	0 60	DI

A. At Wheeting, cinnabar moths give perhaps the clearest known example of food limitation for any phytophagous insect, reflected in the density dependence in larval survivorship and egg production (Table 5.2); larvae starve and the next generation of adults are lean on eggs after early members of the cohort graze the ragwort to the point of defoliation. The plant recovers in the next year by means of its underground reserves to create a boom–bust cycle between the plant, the moth, and the main parasitoid *Apanteles*. The parasitoid does not measurably affect cinnabar moth dynamics. See Chapter 6 for further discussion.

B. This life table is part of an analysis of six species of leaf-hoppers at Silwood (see Fig. 5.4). The mortality patterns were quite similar among the six species and are characterized by Waloff and Thompson (p. 409) as showing an 'absence of any obvious density-dependent process' (Table 5.2). Predation and parasitism account for much of the mortality, but lack of hopper coordination with plant phenology and plant dryness due to the weather kill both eggs and nymphs.

C. Unlike the cinnabar moth at Wheeting, *Colias* suffers little if any starvation from low quality or quantity of host plants; they do not deplete their food supply. The short run of data (five generations) inevitably means that detecting density relationships is difficult. However, all stages had coefficients that indicated decreasing mortality with increasing log density, except for Larvae III in diapause, which had the opposite sign. Hayes allows for the possibility of diffuse competition between caterpillars and other phytophages such as beetles, other Lepidoptera, cattle, horses, deer and elk. Some of these phytophages are also predators on small larvae and eggs.

they do (Figs 5.1 and 5.2), and why communities are but moderately predictable in their composition (Figs 5.3 and 5.4, and Table 5.1).

5.1.5 Aphids

The value of *k*-factor analysis is greatest when discrete generations or cohorts of animals can be followed from birth until death. Where births are effectively continuous and generations blurred, as happens for example in aphid populations, the technique is so awkward to apply as to be virtually useless. Fortunately, aphid population dynamics can be investigated by other methods, and a number of species have now been well studied.

In several species, intraspecific competitive effects at high population densities lead to a decline in reproductive rate and/or to increased emigration of winged individuals (e.g. Dixon 1979; N. Gilbert 1980; Barlow & Dixon 1980). However, detailed studies of the lime aphid (*Eucallipterus tiliae*) host-specific on lime trees (*Tilia*), show that population regulation is complex (Barlow & Dixon 1980). In some years, depending upon the initial numbers of aphids and upon the weather, predation by ladybird beetles (Coccinellidae) is sufficient to prevent population outbreaks. However, too many or too few aphids in spring leads to a failure of predator control. The time of maximum numbers also differs between years, depending upon initial conditions, the abundance of predators, and weather. The net result is that although populations are regulated by density-dependent events, it is difficult to attribute regulation to any one mortality factor; a hierarchy of different processes is involved.

More important in the present context is that regulation in the lime aphid is not tight enough to prevent peak population densities from varying over two orders of magnitude on the same tree in successive years, making this aphid's relative and absolute abundances in the community of insects on lime trees unpredictable. Population sizes in many other species of aphids appear to be equally variable from year to year, and from place to place (Taylor & Taylor 1979; Taylor *et al*. 1980).

On a smaller spatial scale, the fate of individual colonies of aphids is also highly variable, typified by the rise and fall of two small infestations of *Aphis barbarae* on thistle, *Cirsium arvense* (Lamb 1980). Each infestation may have arisen from a single immigrant. The first grew to

122 colonies and produced many alate emigrants during June, and the second grew to only 12 colonies and produced no alates during the same period. Both infestations were totally exterminated in July by a group of predators, including lacewing larvae (Neuroptera), ladybird beetles (Coccinellidae), and hoverflies (Syrphidae). A similar picture emerges for the fate of individual colonies in the aphid *Pterocomma populifoliae* on big-toothed aspen, *Populus grandidentata* (Sanders & Knight 1968), and for four species exploiting rosebay willowherb (fireweed), *Chamerion angustifolium* (Addicott 1978b).

In short, the abundance, presence and absence of particular species of aphids is often a very unpredictable feature of local communities of phytophagous insects.

5.2 Interspecific Competition

Competition between species for limited resources is widely regarded as an important process structuring ecological communities, for example in setting limits to the number of coexisting species and in moulding and constraining what those species do — where and how they feed, their body sizes, seasonal distributions, etc. (MacArthur 1972; Cody & Diamond 1975; Hutchinson 1978; May 1981).

The problem is that many ecological phenomena that might be due, in whole or in substantial part, to the effects of interspecific competition can also be explained by alternative hypotheses not involving competition (Lawton & Strong 1981). We discussed an example in Chapter 4. Colonization curves for insects on introduced plants might be asymptotic for two reasons. Either there are limits to the numbers of species able to coexist on one species of plant, set by interspecific competition for food and space; or alternatively the pool of potential colonists is simply exhausted. As we explained in Chapter 4, we favour the 'pool-exhaustion' hypothesis over the 'competition' hypothesis.

Some of the reasons for rejecting interspecific competition as a major force structuring phytophagous insect communities have already been outlined. Not least of these reasons is the observation that significant *intra*specific competition occurs rather infrequently in phytophage life tables (six of 31 examples, *c.* 20%; Table 5.2). Of course this does not mean that phytophagous insect populations never compete. Good examples exist, and are reviewed in Sections 5.2.3 and 5.2.4. Our point is simply that interspecific competition is not nearly the most important or consistent mechanism structuring communities of insects feeding on plants.

5.2.1 *General Evidence for a Lack of Interspecific Competition*

Negative evidence is often difficult to come by, partly because ecologists understandably are reluctant to study and to write about non-events, and partly because editors are reluctant to publish them! But some evidence has been published.

With stunningly simple logic, Hairston *et al.* (1960) argued:

> 'obvious depletion of green plants by herbivores are exceptions to the general picture, in which the green plants are abundant and largely intact .... The (only possible) remaining general method of herbivore control is predation ... including parasitism .... Herbivores are seldom food-limited, appear most often to be predator limited, and therefore are not likely to compete for common resources.'

As Chapter 2 makes plain, green plants are not universally edible and nutritious, but they do not have to be for the general point made by Hairston *et al.* to be valid. Replying to their critics (Murdoch 1966; Ehrlich & Birch 1967), Slobodkin, Smith and Hairston (1967) wrote:

> 'Within every native environment of every species [of plant] several herbivores (leaf eaters and sap suckers) can be found that are capable of extensive injury. The scarcity of most of these herbivores most of the time results in their food being largely unutilized.'

Many other studies now point to a general infrequency of interspecific competition and lack of competitive organization in phytophagous insect communities. Examples from a wide range of habitats and species are gathered in Table 5.4.

5.2.2 *A Detailed Case Study: Hispines on* Heliconia

Tropical hispine beetles on *Heliconia* are good examples of species-rich communities that appear to operate without competition or resource depletion. Ironically, one of us (Strong 1982 a,b) chose to study these beetles just to demonstrate how competition structures phytophagous insect communities; these beetles seem predisposed to compete. Species

Table 5.4. Studies showing or suggesting an absence of significant interspecific competition in phytophagous insects, arranged alphabetically by author. Examples not included in this table are discussed in detail in the text, Section 5.2.3.

Author(s)	System
Addicott (1978a)	Aphids (Homoptera) on fireweed: there was no evidence that the presence of one species on a plant altered the feeding position of any other species on the same plant.
Askew (1962)	*Neuroterus* species (Hymenoptera; Cynipidae) on oak colonize trees independently of one another.
Chew (1981)	*Pieris oleracea* and *P. rapae* (Lepidoptera) in North America; apparent extermination of *oleracea* by *rapae* does not stand up to critical examination.
Faeth & Simberloff (1981a)	*Cameraria* leaf-miners (Lepidoptera) on oak are maintained at densities far below those where competition is likely.
Faeth & Simberloff (1981b)	No evidence of density compensation with a slightly reduced number of species of leaf-miners (Lepidoptera and Coleoptera) on oaks.
Futuyma & Gould (1979)	'Insect species are certainly not as equitably distributed over potential resources as a theory of species packing based on competition would lead one to expect' (Lepidoptera and Tenthredinoidea on woody plants).
Harrison (1964)	No competition apparent between four species of Lepidoptera on bananas; peak populations were well below those required for competition.
Hopkins & Whittaker (1980)	No competition apparent between *Apion violaceum*, *Apion miniatum* (Coleoptera) and *Norellisoma spenimanum* (Diptera) stem-miners in *Rumex*.
Joern (1979) and Otte & Joern (1977)	Diet in grasshoppers (Orthoptera) is probably not strongly influenced by competition although microhabitat utilization may be.
Lawton (1982, 1983b, c)	Insects from several orders feeding on bracken. No evidence of density compensation in faunally impoverished communities or that competitors influence distributions or feeding ecologies.

Table 5.4. (*cont.*)

Le Quesne (1972)	Three species of *Eupteryx* (Homoptera) never, apparently, became common enough for interspecific competition for food to be significant (but see example 17, Table 5.5b).
McClure (1974)	Little evidence that populations of five species of Hemiptera coexisting on sycamore in Illinois ever became dense enough to produce severe interspecific competition.
Pipkin *et al.* (1966)	No evidence of competition for limited food supply in flower-feeding neotropical *Drosophila* (Diptera); even at peak populations there were many unused flower buds.
Redfern & Cameron (1978)	Competition between gall midge (Diptera) and an eriophyid gall mite on yew was 'unimportant'; no detectable density-dependent effect of interspecific competition.
Risch & Carroll (1982)	Eliminating predatory ants allowed many phytophages to increase, yet no evidence of competition even without ants.
Root (1973)	No evidence that the guild structure of *Brassica*-feeding insects from several orders was constrained by species interactions.
Rothschild (1971) and Yasumatsu (1976)	No evidence of competition between different species of rice stem-borers (Lepidoptera), unless two species share same stem internode, which is rare.
Shapiro (1974)	Two species of Lepidoptera whose larvae exploit lupin flowers may compete for food; but two species of larvae exploiting leaves of the same host almost certainly do not.
Shapiro (1981)	Interspecific competition for oviposition sites among North American Pieridae (Lepidoptera) 'seems rare'.
Varley (1949)	Four species of pine-feeding Lepidoptera 'compete for a limited amount of food only on rare occasions'. The species are generally too rare for direct competition to be significant.
Washburn & Cornell (1981)	No evidence of interspecific competition between populations of gall-formers on *Quercus stellata*; in

Table 5.4. (*cont.*)

	particular extinction of *Xanthoteras politum* (Cynipidae) was not enhanced by competitors.
Watanabe & Omata (1978)	Phytophagous species that might compete with a lycaenid butterfly (Lepidoptera) were all too rare to affect it significantly.
Wise (1981)	A removal experiment with 'darkling beetles' (Coleoptera) shows no evidence of interspecific competition. The adults are partly phytophagous.

are closely related and eat virtually the same parts of their host plants, at the same time. Adults are quite long lived and have low fecundity, and thus conform to a '*K*-selected' life history, which is seen as an adaptation to competition (Pianka 1972).

The first observation that one makes about these beetles is their lack of aggressive behaviour, among both larvae and adults, both within and among species. Neither adults nor larvae show any hostile behaviour. Adults walk slowly around or over each other, rest and feed next to each other, and associate without any antagonism. Larvae are less active, and feed or lie quietly next to each other. These insects live intimately together in nature, and interspecific associations can be maintained in small Petri plates for months, with no mortality. This tolerance is in marked contrast to the behaviour of many phytophagous insects that are intolerant of other individuals, which kill and even cannibalize each other (e.g. Fox 1975; Polis 1981; Stiling & Strong in press, b; see also Table 5.5).

In nature, adults of rolled-leaf hispines spend their lives in the scrolls formed by immature leaves of *Heliconia*. As many as eight species can intermingle in the host scrolls at a single site, and as many as five hispine species can simultaneously occupy a single scroll. Associations of beetles in scrolls are continually reshuffled as rolled leaves mature and unroll. Adults may live for a year or more, and move independently among leaves. Each new group of leaf tenants is drawn haphazardly from a continually reshuffled larger community.

Several sorts of evidence indicate that neither interference nor exploitation competition structures the associations of hispine species in or among host leaves. First, host specificity of hispine species does not change as a function of leaf co-occupancy with other species that might be competitors (Strong 1982a; see Fig. 6.9).

Secondly, interspecific association in leaves of single *Heliconia* populations is not structured in a way consistent with any simple competitive interaction among species. In geographic comparisons, net associations among species range from positive to negative but, taken together, show no pattern of interspecific segregation. Rather, net interspecific association is distributed among communities with a pattern that suggests non-interactive coexistence for these species. Furthermore, number of species in a hispine community does not affect interspecific segregation, and tendencies towards interspecific segregation do not increase with beetle density in leaves, as might be expected were either interference or exploitation competition operating among the insects. Rolled-leaf hispines exist at densities far below those that deplete their resources of food and shelter in rolled leaves (Strong 1982b).

The current hypothesis is that predation upon adults, and parasitism of larvae, eggs, and pupae, keep these tropical populations of insects below levels that critically reduce food or precipitate either intra- or interspecific competition (Strong 1983b). Katydids and assassin bugs can take a high toll of the long-lived adult hispines (Seifert 1982, 1983), and parasitic Hymenoptera kill a large fraction of the immature stages. For example, the eggs of one hispine species suffer 35–50 per cent mortality from chalcidid Hymenoptera (Morrison & Strong 1981). Pupae can suffer up to 75 per cent parasitism from species of Hymenoptera similar to those that attack eggs.

Whether parasitoids and other natural enemies are the crucial factors that maintain hispines at such low population levels can only be learned by future experimentation; perhaps host plant spacing and phenology and weather also play a large role. Another possibility, that harmonious coexistence of hispines results from evolution of interspecific accommodation (Schoener 1974; see Section 5.2.6) is extraordinarily difficult to test in any critical fashion. With natural enemies and other factors keeping densities so low relative to resources, little selection pressure would be generated for niche shifts to accommodate other species.

5.2.3 Occasional Competition

'Occasional competition' means either of two things. The first is that good examples of sustained interspecific competition will occasionally be found between pairs of species in most guilds or communities, as we indicated in the general remarks on page 130. The second meaning is that rather more species might compete occasionally than compete regularly.

Infrequent Cases of Competition between Pairs of Species

Amongst several well-studied groups of phytophagous insects are pairs of species that compete. One of the clearest demonstrations is due to Rathcke (1976). This study revealed that although the larvae of nine of 13 co-occurring species of stem-boring insects (eight Coleoptera, two Diptera and three Lepidoptera) showed greater than 70 per cent overlap in resource exploitation with at least one other member of the guild, only two of these species had any perceptible negative influence on each other. A mordellid beetle larva, designated 'Mordellidae 23', and a microlepidopteran caterpillar (*Epiblema* sp.) were found together in field stems less often than expected by chance. Laboratory observations indicated that the significant difference between observed and expected co-occurrences resulted from interference competition. Whenever larvae of these two species encountered one another within stems, the mordellid would attack, injure and eventually kill *Epiblema*. Rathke was even able to provide an estimate of the competition coefficient ($\alpha = 0.24$) for the impact of the mordellid larvae on *Epiblema* caterpillar populations. The reciprocal influence was negligible. The effect of the mordellid on *Epiblema* was therefore the only significant competitive interaction within the guild (Table 5.5a).

Broadly similar results have been obtained by others. Gibson (1980) and Gibson and Visser (1982) found that only two from seven species of grassland mirids competed significantly (Table 5.5b); with seven species, this system has potential for 21 competitive interactions between pairs of species. In similar vein, Seifert and Seifert (1976, 1979a; see also Seifert 1982) analysed in detail the insects in *Heliconia* bracts in two different areas. Species included not only herbivores, but also detritivores, nectivores and predators. The majority (28 out of 40) of pairwise species interactions within these communities were not significantly different from zero. Several of the significant interaction were mutualistic rather than competitive. There were only five significant competitive interactions involving herbivores, and these were not particularly intense (Table 5.5a).

That species may compete occasionally, although most of the time they do not, is revealed by Kareiva's (1982a) study on three species feeding on collards in cultivation — two flea beetles (*Phyllotreta cruciferae* and *P. striolata*) and the white butterfly *Pieris rapae*. In 21 experiments, Kareiva found only two significant interspecific competitive effects, at the highest densities of planted collards: *Pieris rapae* populations depressed the abundance of *P. cruciferae*, and *P. cruciferae* depressed the abundance of *P. striolata*. Reciprocal effects were negligible (Table 5.5a).

Competition During Outbreaks

Although most phytophagous species seem not to compete for food, there are from time to time 'outbreaks', when an insect becomes so abundant as to partly or completely defoliate its host plant. Voûte (1946) pointed out that these outbreaks occurred principally in cultivated vegetation. It is now recognized that they are generally due to reduced pressure from natural enemies (Varley 1949; Milne 1957; Ito 1961; Clark 1964; Southwood 1975). Possible mechanisms include time delays inherent in predator prey or host parasite interactions (see references, page 112; also Anderson & May 1980, 1981); agricultural sprays that kill predators and parasitoids (Entwistle *et al.* 1959; DeBach 1974; Southwood 1977b; Strong 1982b); weather conditions that militate against natural enemies (Clark 1964; Fig. 5.6; Readshaw 1964); or improved host plant quality or quantity (Sections 2.1.4 and 6.2).

Studies of outbreaks suggest that once the prey population passes a certain density, Voûte's (1946) 'escape point', it rises inexorably until checked by intraspecific competition (Clark 1964; Figs 5.6 and 5.13). The two population levels were termed 'endemic' and 'epidemic' by Clark (1964). Further discussion and models of insect outbreaks are provided by Varley (1949), Southwood (1975), Southwood and Comins (1976), Peterman *et al.* (1979), and Hassell (1981).

When a host plant is defoliated or even killed (Churchill *et al.* 1964) by one species of insect during an outbreak, the other species on it must undoubtedly be affected. However, there is little direct evidence on interspecific competition during outbreaks, probably because most studies have been made by applied entomologists whose principal concern was with the abundant pest species; the fate of the rarer species would repay further study. The role of rare events, such as outbreaks,

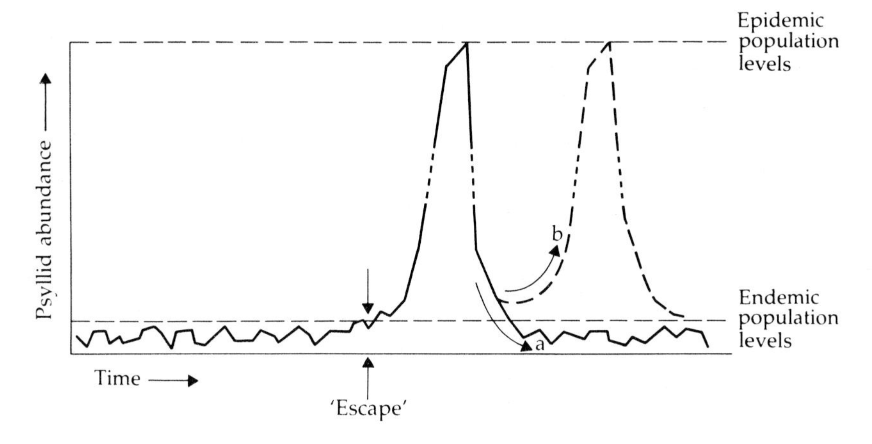

Fig. 5.6. A generalized scheme to explain outbreaks of the psyllid *Cardiaspina albitextura* on *Eucalyptus blakelyi* over the period 1952–63 (After Clark *et al.* 1967). Endemic levels (below the horizontal dotted line) were maintained by predation and parasitism; density-dependent predation by birds on adult psyllids was particularly important in maintaining this lower equilibrium. Nymphs were attacked in a density-independent way by birds, ants, encyrtid parasitoids, etc. 'Escape' from low endemic levels occurred when unusually low temperatures reduced percentage parasitism. Epidemic populations collapsed because of food shortages and intense intraspecific competition. Natural enemies then either regained control until the next outbreak, which did not occur for several years (alternative a) or psyllid numbers were not brought under control of natural enemies and another outbreak followed quickly (alternative b). [In a later study, Geier *et al.* (1983) were unable to find evidence of regulation by birds, etc., at endemic levels, and some populations became locally extinct. This does not alter the importance of intense intraspecific competition during outbreak periods.]

that generate occasional but potentially intense competition, cannot be ignored as a force structuring ecological communities just because they are rare (Lawton & Strong 1981).

5.2.4 *Examples of Interspecific Competition*

In order to present a balanced picture, it is sensible at this stage to summarize the reported cases of interspecific competition between phytophagous insects. These are gathered in Table 5.5.

Examples of interspecific competition are distributed across a wide range of plant types, habitats and insect groups, but two aspects of the data deserve particular comment. There is an indication that competition is especially common among Hemiptera (Homoptera and Heteroptera). If this is a real phenomenon, the reasons for it are unclear.

With more confidence, we can say that highly asymmetrical interactions (amensal interactions — Williamson 1972) outnumber symmetrical 'conventional' competitive interactions by approximately 3:1 if the intensely studied *Erythroneura* leaf-hoppers (example 16) are counted only once, and still outnumber them if all the examples from this genus are included. Amensalism is also a feature of interactions between groups of non-phytophagous insects (Lawton & Hassell 1981, 1983). Amensal interactions imply that many theoretical models of interspecific competition (e.g. May & MacArthur 1972; May 1981) are not appropriate for insects on plants, because they rest on the assumption that competition is broadly symmetrical. This area deserves further study.

5.2.5 *Indirect Competition via Induced Defences*

A growing body of literature attests to the fact that plants sometimes respond to herbivory by turning on inducible defences. Induced defences manifest themselves in a marked reduction in performance, for example in growth rate or survival, in insects feeding on foliage from previously damaged or defoliated branches or whole trees (e.g. Haukioja & Hakala 1975; Haukioja & Niemela 1979; Thielges 1968; Wallner & Walton 1979; Schultz & Baldwin 1982; Fig. 5.7). Similar responses in herbs are also known (Ryan & Green 1974; Carroll & Hoffman 1980; Woodhead 1981). Rhoades (1979) provides a review.

Because this is a problem that involves the interaction of insects with their host plant, we could have chosen to deal with it in Chapter 6. Interest here is on the neglected question of whether defences induced by one species of herbivore might significantly impair the performance of other species on the same host. Theoretically, induced defences open up the possibility of two species competing even though they occur on the host at different times, and/or exploit different parts of that host. There is no hard evidence that this does happen, but it is an interesting possibility. A likely example is Woodhead's (1981) demonstration that phenolic levels in sorghum were significantly and markedly enhanced by feeding of the shoot-fly *Atherigona soccata* and the stem-borer *Chilo partellus*. Since feeding by locusts on sorghum is known to be inhibited (**a**) **Amensalism**

Ref.	Species: Dominant (j)	Species: Subordinate (i)	Mechanism	Taxonomic group, habitat and food
1+	*Danaus plexippus*	*Oncopeltus* spp.	Extreme exploitation of food plant	Lepidoptera and Hemiptera, Heteroptera: milkweed plants
2	*Phyllotreta cruciferae*	*P. striolata*	Exploitation and interference?	Coleoptera: Collards (Section 5.2.3)
2	*Pieris rapae*	*P. cruciferae*	Exploitation and interference?	Lepidoptera and Coleoptera: Collards (Section 5.2.3)
3	*Metrioptera roeselii*	*Platycleis albopunctata*	Acoustic interference	Orthoptera: Probably omnivorous, grass, herbs and partially predatory
4+	*Neacoryphus bicrucis*	*Harmostes reflexulus* and *Lygus pratensis*	Behavioural aggression	Hemiptera, Heteroptera: *Senecio smallii*
5	*Fiorinia externa*	*Tsugaspidiotus tsugae*	Resource exploitation and action of shared parasitoid	Hemiptera, Homoptera: Eastern hemlock foliage
6	*Cromaphis juglandicola*	*Panaphis juglandis*	Habitat modification via excretion	Hemiptera, Homoptera: Walnut foliage
7	'Mordellidae 23'	*Epiblema* sp.	Aggression and death of subordinate	Coleoptera and Lepidoptera: Plant stems in prairie (Section 5.2.3)

Table 5.5(**a**) Examples of amensalism (strongly asymmetrical competition) between phytophagous insect populations in the field. The dominant species, j, has a marked effect upon the subordinate, i, ($\alpha_{ij} \gg 0$) whilst the reciprocal effect is negligible ($\alpha_{ij} \sim 0$); α is the 'competition coefficient' (see, for example, Hassell 1976). Less certain, probably amensal, examples are marked +. There is one case (*) in which the inferior species (i) appears to have a beneficial effect on j. (**b**) Examples of more symmetrical interactions (or interactions of uncertain symmetry) between phytophagous insect populations in the field. NB. Amensalism does *not* imply competitive exclusion, nor symmetrical competition coexistence; both coexistence and exclusion are possible outcomes of both forms of competition (Lawton & Hassell 1981). See footnote

Table 5.5. (*cont.*)

8	*Copestylum* cf *obscurior*	*Quichuana picadoi*	?	Diptera: Water-filled *Heliconia* bracts. Nectar, detritus and plant tissue (Section 5.2.3)
8	*Gillisius* sp.	*Beebeomyia* sp.	?	Coleoptera and Diptera: Water-filled *Heliconia* bracts. Plant tissue (Section 5.2.3)
8	*Gillisius* sp.	*Cephaloleia puncticollis*	?	Coleoptera: Water-filled *Heliconia* bracts. Plant tissue (Section 5.2.3)
8	*Merosargus* sp.	*Gillisius* sp.	?	Diptera and Coleoptera: Water-filled *Heliconia* bracts. Detritus and plant tissue (Section 5.2.3)
8*	*Quichuana picadoi*	*Gillisius* sp.	?	Diptera and Coleoptera: Water-filled *Heliconia* bracts. Detritus and plant tissue (Section 5.2.3)
9	*Copestylum roraima*	*Cephaloleia neglecta*	Disturbance and behavioural avoidance	Diptera and Coleoptera: Water-filled *Heliconia* bracts. Detritus and plant tissue
10	*Cephaloleia neglecta*	*Xenarescus monocerus*	Interference leading to emigration	Coleoptera: *Heliconia* bracts
11+	Several species of mirid	*Orthotylus virescens*	Resource exploitation and direct predation	Hemiptera, Heteroptera: Scotch broom bushes
12	*Urophora solstitialis*	*Rhinocyllus conicus*	Habitat modification via gall formation	Diptera and Coleoptera: Plant tissue in flower-heads of thistle
12	*Larinus sturnus*	*Rhinocyllus conicus*	Probably exploitation of resources	Coleoptera: Plant tissue in flower-heads of thistle
12	*Encosma* and *Homoeosomo*	All other species in flower-heads	Aggression and predation	Lepidoptera and other: Plant tissue in flower-heads of thistle

(b) Symmetrical interactions

Ref.	Species	Mechanism	Taxonomic group, habitat and food
13	*Amplicephalus simplex* *Aphelonema simplex* *Delphacodes detecta*	Resource exploitation. Each species depresses numbers of other two	Hemiptera, Homoptera: *Spartina patens*
14	*Notostira elongata* *Megaloceraea recticornis*	Behavioural interference	Hemiptera, Heteroptera: Grass (Section 5.2.3)
15	*Chrysolina quadrigemina* *C. hyperici* *Zeuxidiplosis giardi*	Intense exploitation competition; this example may more properly belong in Part (a), with *C. q.* as dominant	Coleoptera and Diptera: St John's wort
16	*Erythroneura* spp. 7 species	Resource exploitation leads to significant competition in eight pairs of interactions between the seven species	Hemiptera, Homoptera: American sycamore
17	*Eupteryx cyclops* *E. urticae*	Resource exploitation	Hemiptera, Homoptera: Stinging nettle

Cases within a and b are in alphabetical order, by author, as follows:
1. Blakley & Dingle (1978); 2. Kareiva (1982a); 3. Latimer (1981); 4. McLain (1981); 5. McClure (1980, 1981); 6. Messenger (1975); 7. Rathcke (1976); 8. Seifert & Seifert (1976); 9. Seifert & Seifert (1979a); 10. Seifert & Seifert (1979b); 11. Waloff (1966, 1968a, b), Southwood (1978a); 12. Zwölfer (1979); 13. Denno (1980), Denno *et al.* (1981); 14. Gibson (1980), Gibson & Visser (1982); 15. Huffaker & Kennett (1969); 16. McClure & Price (1975, 1976); 17. Stiling (1980).

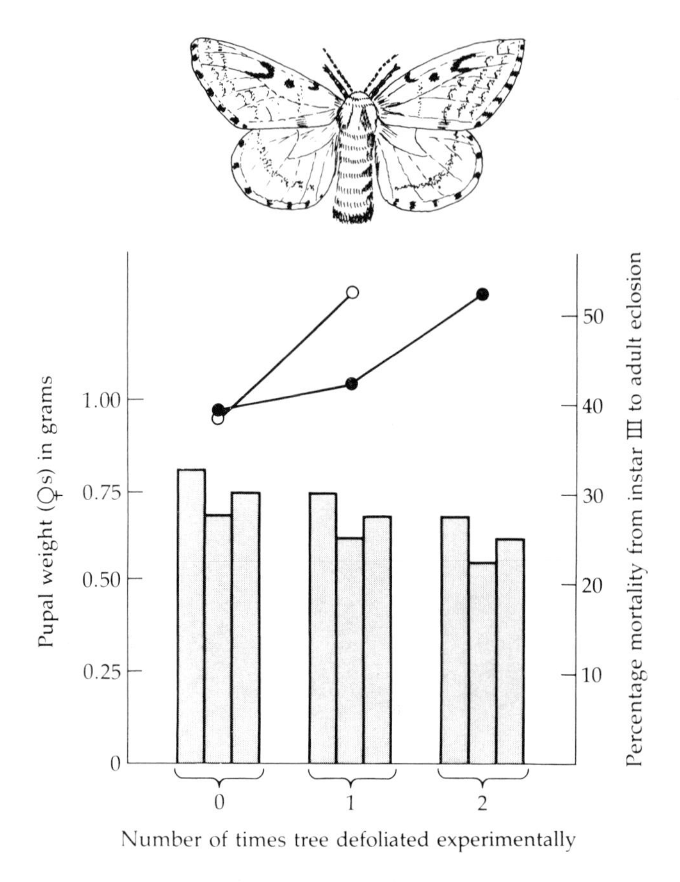

Fig. 5.7. Effects of experimentally defoliating grey birch trees (*Betula populifolia*) on the subsequent growth and survival of gypsy moth (*Lymantria dispar*) with unrestricted access to leaves from control or treated trees (Wallner & Walton 1979). Pupal weights (histograms) were significantly reduced in three populations reared on leaves from defoliated trees: the effect is more marked after two defoliations. Since fecundity in many Lepidoptera is proportional to female size, a herbivore outbreak leading to defoliation may reduce the rate of increase of subsequent generations of moths. Larval survival (filled and open circles) is also lower on previously defoliated trees. (From Caughley & Lawton 1981.)

by phenolics, the potential for indirect competition between stem-borers and foliage-chewers is a real one.

In brief, some competitive interactions between phytophagous insects may be extremely subtle, and while we have no evidence that induced defences are important in interspecific competition, future research could alter this impression.

5.2.6 *The Ghost of Competition Past*

Failure to detect significant interspecific interactions in contemporary communities does not necessarily mean that competition has no part to play in structuring that community. It may instead have left its mark in 'the ghost of competition past' (Connell 1980). That is, species may have evolved to minimize, or eliminate entirely, contemporary competitive effects (e.g. Benson 1978).

Such ideas are tantalizingly plausible but extremely difficult to test rigorously (Connell 1980; Strong 1983b). How much difference ought we to expect between two or more species in such things as where and when they feed, or body size, if these differences have evolved to minimize competition? Are these differences greater than one might expect under some alternative hypothesis? What, for example, is an appropriate null hypothesis?

A weak and partial test of the presence of the ghost of competition past is to ask whether communities of insects independently evolved on the same food plant in different parts of the world show evidence of convergence in community structure. If competition significantly influences species' evolution, then under similar environmental conditions similar communities might be expected to evolve independently in different places (e.g. Cody & Mooney 1978). For one common and widespread plant, bracken (*Pteridium aquilinum*), this is not what has happened. Communities in different parts of the world, even under rather similar climatic regimes, do not conspicuously converge in structure (Lawton 1982, 1983b; Fig. 4.9, particularly communities **a** and **c**).

Perhaps there are better ways of looking for evidence of the ghost of competition past, but until they are developed we are forced to rely heavily on evidence from contemporary species interactions to assess the role of competition in structuring phytophage communities. This evidence says quite simply that although some species of phytophagous insects do compete, most species, most of the time, do not.

5.3 Natural Enemies

Interspecific competition between phytophagous insects is comparatively uncommon because many populations are kept rare, relative to the availability of potentially limiting resources, by the impact of natural enemies (Table 5.2); that is, they are maintained at Clark's (1964) 'endemic level' (Fig. 5.6). This final section of Chapter 5 explores the ecology and natural history of these important natural enemies in more detail.

5.3.1 The Enemies of Winter Moth: a Case Study

One of the best studied populations of phytophagous insects in the world is that of the winter moth *Operophtera brumata* at Wytham Woods, near Oxford in England (Varley *et al*. 1973). The caterpillars feed on the oak *Quercus robur*, as well as on several other species of broad-leaved trees. Successive mortalities acting on the Wytham population are illustrated in Fig. 5.8. As with other phytophagous insects (Tables 5.2 and 5.3), natural enemies take a heavy toll of most life history stages. Figure 5.8 illustrates a number of important points about these natural enemies, as follows.

First, only one group acts in a density-dependent manner. These are predators that kill pupae as they overwinter in the ground underneath the oak trees (k_5 in Figs 5.5 and 5.8). Not only is there significant direct density-dependent mortality at this stage in the life history, but there is also evidence of a delayed component, revealed by joining successive years' data to form an anticlockwise spiral. Two groups of predators are responsible for the heavy mortality of winter moth pupae in the soil; carabid and staphylinid beetles are one and small mammals, particularly shrews, the other. Varley *et al*. (1973) suggest that shrews are responsible

Fig. 5.8. The k-values for the different winter moth mortalities plotted against the population densities on which they acted. k_1 and k_6 are density independent and vary quite a lot; k_2 and k_4 are density independent but are relatively constant; k_3 is weakly inversely density dependent, and k_5 is quite strongly density dependent (see Fig. 5.5), with an important delayed component revealed by the anticlockwise spiral when successive years are joined in sequence. (After Varley *et al*. 1973.)

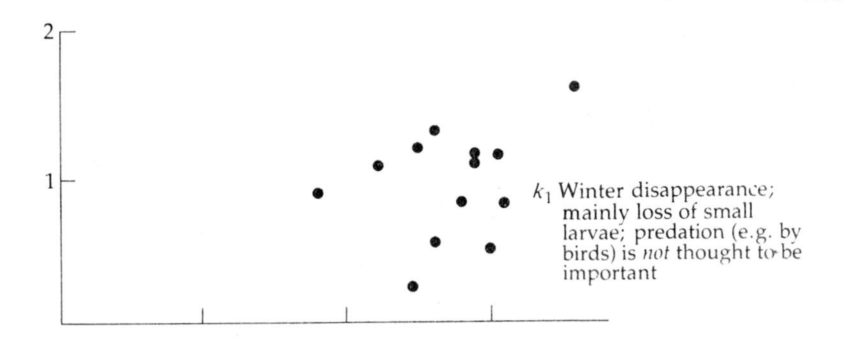
2
1
k_1 Winter disappearance; mainly loss of small larvae; predation (e.g. by birds) is *not* thought to be important

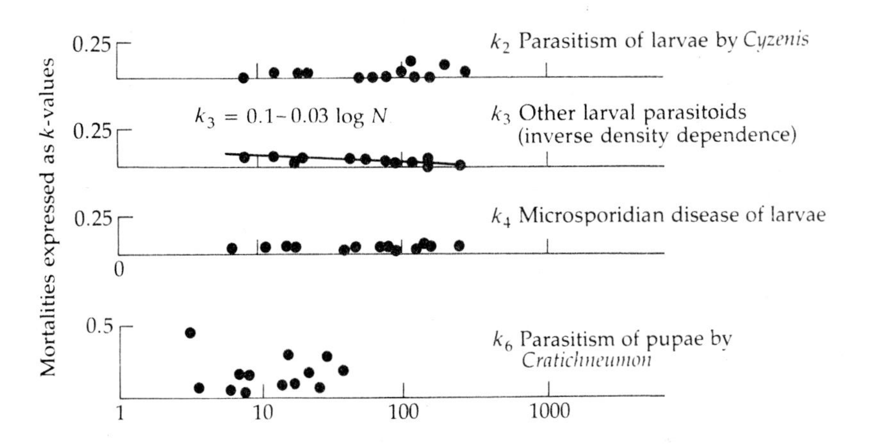
Mortalities expressed as k-values
0.25
k_2 Parasitism of larvae by *Cyzenis*
$k_3 = 0.1 - 0.03 \log N$
0.25
k_3 Other larval parasitoids (inverse density dependence)
0.25
k_4 Microsporidian disease of larvae
0
0.5
k_6 Parasitism of pupae by *Cratichneumon*
1
10
100
1000

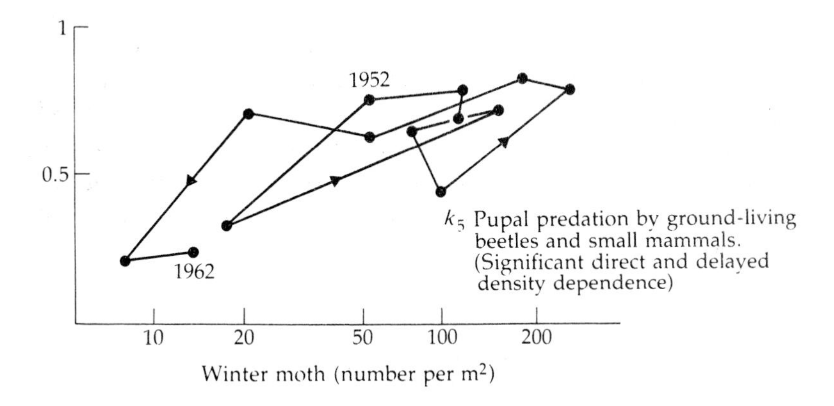
1
0.5
1952
1962
k_5 Pupal predation by ground-living beetles and small mammals. (Significant direct and delayed density dependence)
10
20
50
100
200
Winter moth (number per m²)

for the direct density-dependent component of k_5. The delayed component is probably due to enhanced survival and reproduction of beetles in years of high abundance of the winter moth, leading to enhanced numbers of beetles in the following year (see Kowalski 1977).

Unlike the pupal predators, most natural enemies imposed either density-independent mortality (e.g. k_2 — larval parasitism caused by the parasitic fly *Cyzenis albicans*, and k_6 — pupal parasitism by the ichneumon *Cratichneumon culex*) or *inverse* density-dependent mortality (e.g. that generated by the suite of larval parasitoids that together constitute k_3).

Although not apparent from Fig. 5.8, the impact of particular natural enemies need not be constant from population to population. *Cyzenis albicans* does not regulate winter moth populations in Wytham. But in Canada, where the winter moth was accidentally released in the 1930s and became an important pest of broad-leaved trees, *Cyzenis* proved to be a very effective controlling agent when released in a biocontrol programme (Embree 1971). There is little doubt that in Canada the impact of *Cyzenis* is density dependent and strongly regulatory (Hassell 1980; Table 5.2).

Finally, the variety of natural enemies attacking winter moth at Wytham is impressive. An important group are insect parasitoids, mainly Hymenoptera but also Diptera like *Cyzenis*. However, as we have seen, there are also vertebrate predators — small mammals — and invertebrate predators, particularly ground-living beetles. Last but not least is disease, in the case of winter moth a microsporidian (protozoan) parasite called *Plistophora operophterae* (k_4, Fig. 5.8) and a poorly studied polyhedrosis virus of unknown effect (Anderson & May 1981).

The winter moth has been exceptionally well studied, but there is every reason to believe that its fate is broadly representative of most species of phytophages: the majority of enemies do not have a density-dependent influence; those that regulate differ from place to place and population to population; and enemies take many forms, from microsporidia to mammals and from polyhedrosis viruses to parasitoids.

5.3.2 *Enemy Complexes, Density Independence and Inverse Density Dependence*

If we ignore all vertebrate predators and all microorganisms and

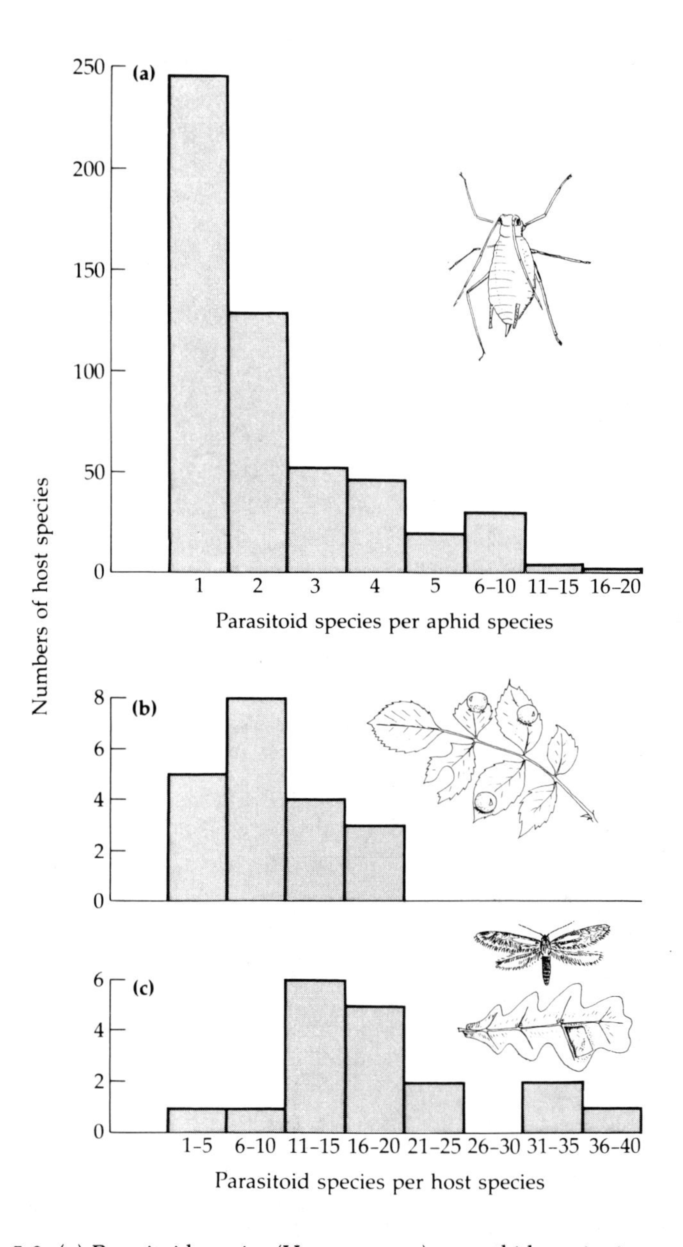

Fig. 5.9. (**a**) Parasitoid species (Hymenoptera) per aphid species in a sample of 528 members of the Aphidoidea from the Palaearctic; only aphids with at least one recorded parasitoid have been included (from Starý & Rejmánek 1981). (**b**) Parasitic Hymenoptera attacking British gall-forming Cynipidae on herbaceous plants, shrubs and trees. (**c**) The same for species attacking leaf-mining Lepidoptera in the genus *Phyllonorycter*. (Data for **b** and **c** from Askew 1980.)

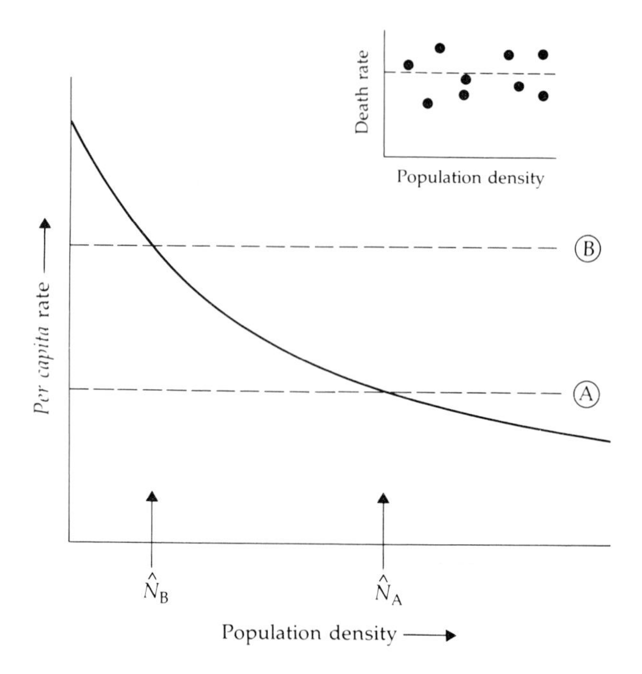

Fig. 5.10. A simple model to show how two different levels of density-independent mortality lead to very different average population sizes. As the population density of the hypothetical phytophagous insect increases, its average fecundity (———) falls in a density-dependent manner. A and B are two different average levels of density-independent predation (— — —). Higher levels of predation (B) lead to lower average population sizes. (Equilibrium populations exist where mean birth rate = mean death rate, i.e. at $\hat{N}_B$ and at $\hat{N}_A$ (Williamson 1972).) In real populations these average rates will inevitably vary (inset), and the population will fluctuate round the equilibrium point.

pathogens, and focus just on insect enemies, the number of species attacking one species of insect herbivore is often large. Figure 5.9 gives three examples; many species are hosts for 10, often more, species of parasitoids. Other examples are given by Cameron (1939), Richards (1940), Thompson (1943–71), Askew (1961), Price (1971), Force (1974), Miller and Renault (1976), Baltensweiler *et al.* (1977) and Pschorn-Walcher (1977). Most insect herbivore species are also attacked by several species of insect predators (e.g. Clausen 1940; Pschorn-Walcher & Zwölfer 1956; Thompson 1943–71; Sunderland & Vickerman 1980).

Even though deaths caused by most natural enemies are not density dependent, moderate or large density-independent mortalities significantly depress the size of victim populations (Williamson 1972) and hence are important (Fig. 5.10). This undoubtedly explains why many species of phytophagous insects are rather rare.

Lastly, some parasitoids and predators may potentially have a *destabilizing* effect on host populations, because they cause inverse density-dependent mortality. A growing number of examples analogous to k_3 in the winter moth life table (Fig. 5.8) now exist in the literature (Morrison & Strong 1980, 1981; Hassell 1982; Stiling & strong 1982), although many of these are 'spatially', not 'temporally', inversely density-dependent, and their effects on population stability may not be the same.

A major problem in understanding communities of phytophagous insects is to tease apart the impact of this battery of enemies, distinguishing those that regulate from those that do not, and amongst those that do not regulate to find which destabilize, which cause fluctuations, which significantly depress prey numbers and which are essentially neutral in their impact.

5.3.3 *Enemy Effects: Some Examples other than those Revealed by Life Table Studies*

From amongst a large number of possibilities, we have selected for further consideration three groups of enemies — parasitoids released for biological control of insect pests, pathogens, and birds.

Biological Control

Biological control is most commonly practised against insects accidentally introduced by man into a new country. Here, relieved of natural enemies, the insects may become very serious pests. Biological control programmes seek to reduce or eliminate the damage caused, by releasing natural enemies from the pest's native country. Most commonly, if the pest is a phytophagous insect, the enemies released are specific parasitoids, usually Hymenoptera, but other sorts of enemies are also used, including predators and pathogens (DeBach 1964; Huffaker 1971; Varley *et al.* 1973; Clausen 1978; Papavizas 1981).

Biological control programmes constitute experiments on a grand scale, and illustrate both the 'escape' of pest species relieved of natural enemies and their demise when enemies are restored to the system. In some cases they provide good estimates of the extent to which parasitoids suppress phytophagous insect populations below the limits set by food supply (Fig. 5.11).

In accord with the conclusions of Sections 5.3.1 and 5.3.2, not all releases of biological control agents are as successful as the examples in

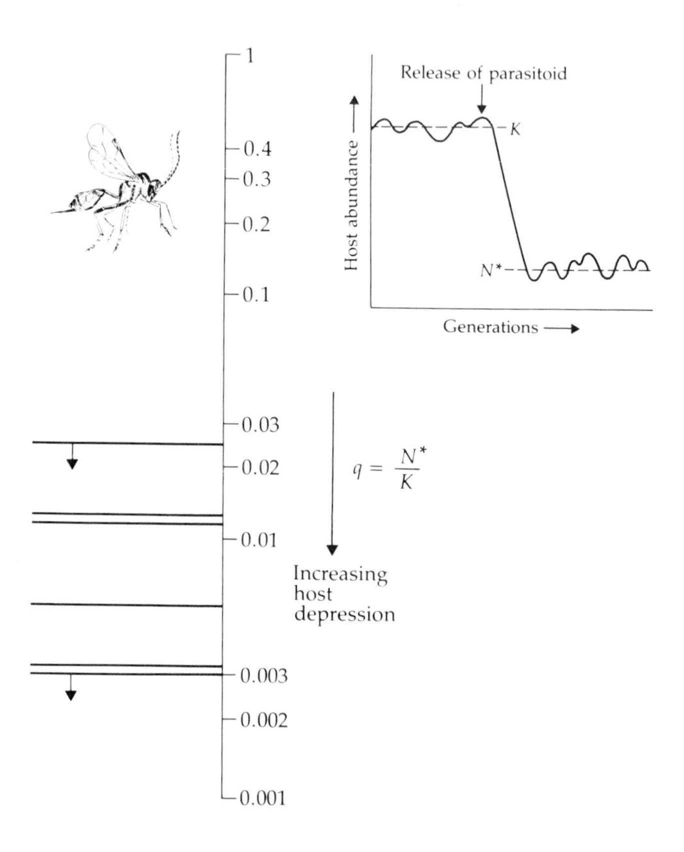

Fig. 5.11. Degree of host depression achieved by the release of single species of parasitoids in six cases of successful biological control; q measures this depression and is defined as the average abundance of the host in the presence of the parasitoid (post-control $= N^*$) divided by the average abundance of the host in the absence of the parasitoid (pre-control $= K$). These q-values are of order of 0.01, i.e. the host populations have been depressed to about 1/100th their former abundance. Arrows imply minimum estimates of the degree of depression. (Data from Beddington *et al.* 1978.)

Fig. 5.11. Indeed, many fail completely, even when the parasitoids become established (Fig. 5.12). A major challenge for future research is to understand better what makes a successful biological control agent (Hassell 1978, 1980; Beddington *et al.* 1978; Heads & Lawton 1983). Obviously, this research ought also to tell us why many species of parasitoids fail to regulate hosts in more natural environments.

The successful biological control programmes speak for themselves (Fig. 5.11). With populations depressed to 1/100th of their abundance in the absence of natural enemies, it is not difficult to see why competition for food is unimportant among many phytophage populations.

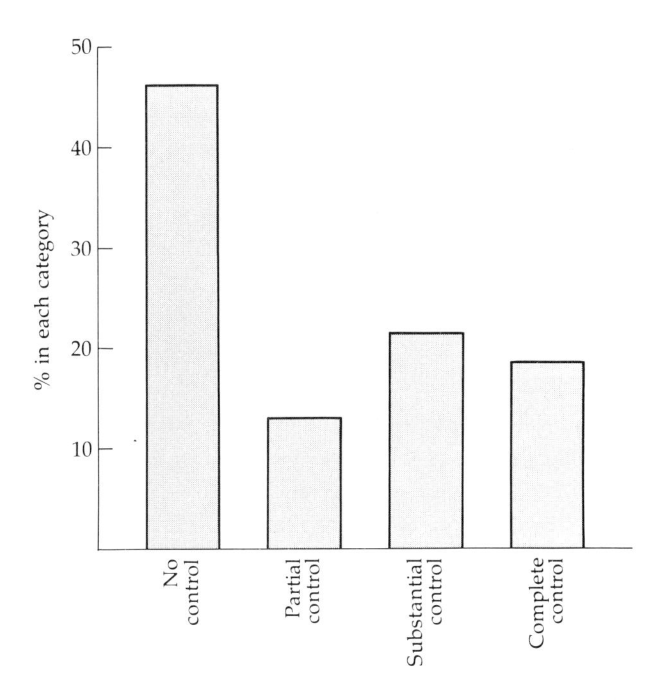

Fig. 5.12. Proportion of successes and failures in programmes of biological control of pest insects by importation of natural enemies; the data are based on a world survey by DeBach (1971). The largest category — no control — includes both cases where natural enemies failed to become established or, if established, have proved to be ineffective. 'Partial', 'Substantial' and 'Complete' control are subjective assessments, but broadly measure the degree of suppression of pest numbers (i.e. 'control' is not being used to imply density-dependent regulation, or lack of it).

Pathogens

Diseases as regulators of phytophagous insect populations have been very poorly studied in comparison with insect parasitoids or predators. However, recent theoretical studies (Anderson & May 1980, 1981) leave little room for doubt that microparasites, broadly defined to include viruses, bacteria, protozoa and fungi, significantly affect natural populations of phytophagous insects. In particular, Anderson and May's studies suggest that virus and microsporidian infections of many temperate forest phytophages drive wide-amplitude stable limit cycles in host abundance with periods in the range 5–12 years. Some of the most extreme population fluctuations in Fig. 5.1 may fall into this category, including those of the larch budmoth (*Zeiraphera diniana*). Populations of the larch budmoth in the European Alps cycle regularly with an amplitude from peak (N_{max}) to trough (N_{min}) of four orders of magnitude and more (Baltensweiler *et al.* 1977). These cycles are probably driven by the long time-delays built into the infection and transmission processes. Fluctuations of the larch budmoth may also be a product of time-delays in the response of the host plant to insect grazing (Fischlin & Baltensweiler 1979).

When diseases are an important source of large-scale oscillations in the abundance of phytophagous insects, there are clear implications for predictability (or lack of constancy) in community structure from one year to the next, and for 'occasional competition' at the high points of the cycle (Sections 5.1.1 and 5.2.3)

Birds

Camouflage, and warning colours combined with distastefulness, bear strong witness to the importance of visually hunting predators in the ecology and evolution of insect herbivores. Tinbergen's (1974) account makes compelling reading. Birds are probably the most important of these predators, although relatively few studies have considered the influence of birds upon insect communities (Dickson *et al.* 1979). In some situations, phytophagous insects can suffer heavily from predation by birds. Buckner and Turnock (1965) found that 43 of 54 species of birds in tamarack bogs of Manitoba, Canada, fed upon the larch sawfly, the tamarack tree's principle defoliator. Of these, 22 species ate sawflies more commonly in areas of high sawfly density, and virtually all the bird species were more abundant in these areas. However, despite more birds, each eating more sawflies where sawflies were common, bird predation was inversely density dependent. At the lower densities of sawflies, the birds took about 65 per cent of adults and six per cent of larvae. At the higher densities only six per cent of adults and half a per cent of the larvae were taken.

Other studies have found that birds have substantial effects upon phytophagous insect populations at relatively low, endemic, densities

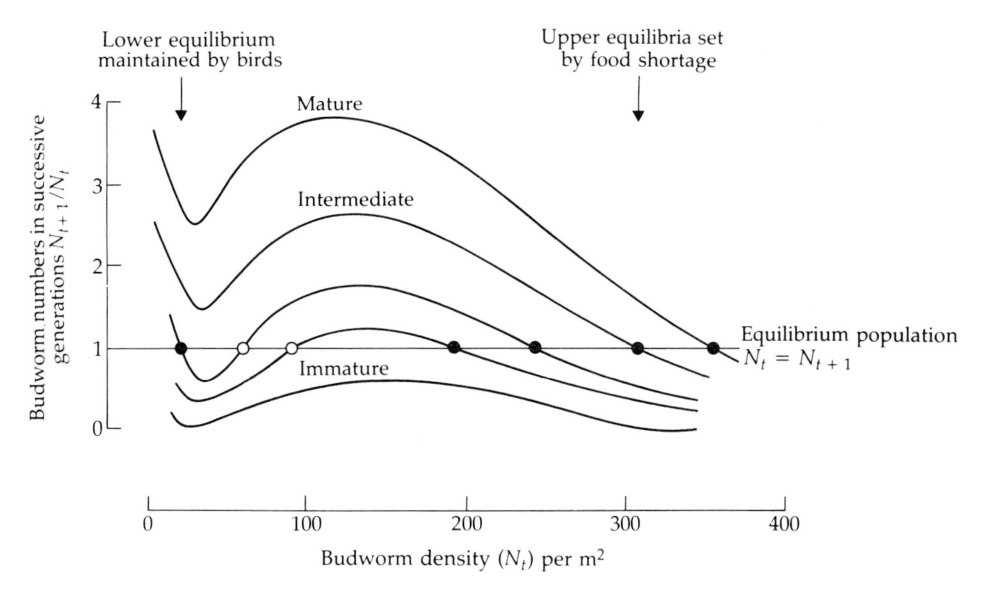

Fig. 5.13. A family of spruce budworm 'recruitment curves' at different levels of forest maturity. Each line shows the net rate of change of the budworm population from generation t to generation $t + 1$. The population shows no net change when $N_{t+1}/N_t = 1$. Stable equilibria are illustrated by solid circles, unstable equilibria by open circles. As the forest matures, the net rate of change of the budworm population increases. The lower equilibrium in forests of intermediate age is apparently maintained by bird predation. As the forest matures, budworms escape from regulation by birds and break out to reach high levels (After Peterman *et al.* 1979.)

(e.g. Solomon & Glen 1979; Pollard 1979) but suggest that the potential for phytophagous insect populations at higher densities to escape from control by avian predators is great. Birds do not usually congregate quickly enough or reproduce fast enough to outstrip a rapidly increasing insect population (Morris *et al.* 1958; Mook 1963; Readshaw 1964; Holling *et al.* 1976; Peterman *et al.* 1979). An example is shown in Fig. 5.13. Low (endemic) densities of the spruce budworm *Choristoneura fumiferana* are apparently maintained by the density-dependent effects of polyphagous predators, particularly birds (Peterman *et al.* 1979). According to Holling and his co-workers, changes in the forest as it matures are responsible for the breakdown of this control and for budworm outbreaks. Unfortunately, the critical importance of bird predation in this system has never properly been tested by experiment. Whilst such experiments are difficult to do, particularly in a spruce forest, they are not impossible.

Holmes *et al.* (1979) kept birds out of parts of the understory of the Hubbard Brook forest in New Hampshire, using netting exclosures. This experiment showed (Fig. 5.14) that densities of externally feeding Lepidoptera larvae were reduced by avian predation, while densities of arachnids, beetles, Homoptera and Hemiptera were unaffected. Lepidoptera may be more apparent than other arthropods to visually orientating vertebrate predators. The experiment does not tell us whether bird predation on Lepidoptera was regulatory or whether numbers of caterpillars increased simply because a density-independent death rate decreased (Fig. 5.10). More experiments along these lines would be valuable.

The general importance of bird predation is hinted at by consistent differences in average levels of defoliation by phytophagous insects in European and North American forests, where defoliation is usually relatively low, and in some Australian *Eucalyptus* forests, where it is high. In Australian *Eucalyptus* forests persistently high populations of phytophagous insects can cause high levels of defoliation of their host trees; from 20 to 50 per cent of leaf area is estimated to be removed consistently by chewing insects (Morrow 1977; Morrow & LaMarche 1978). Comparable European and North American figures are in the range 5–10 per cent (May 1981, page 201). Despite evidence that bird predation may be important in maintaining *Eucalyptus* psyllid populations at low levels (psyllids are sucking herbivores) (Fig. 5.6), P.A. Morrow (pers. comm.)

speculates that the Australian eucalypts may suffer much higher population levels of, and damage from, chewing herbivores because of relatively low populations of insectivorous birds in these forests, which have no nesting migratory birds and relatively few native species. This possibility deserves further investigation.

Chronic persistent defoliation of eucalypts also suggests that parasitoids and other enemies are curiously ineffective (Lawton & McNeil 1979) and that intra- and interspecific competition between phytophagous insects should be more frequent on *Eucalyptus* than in the other systems discussed in this chapter.

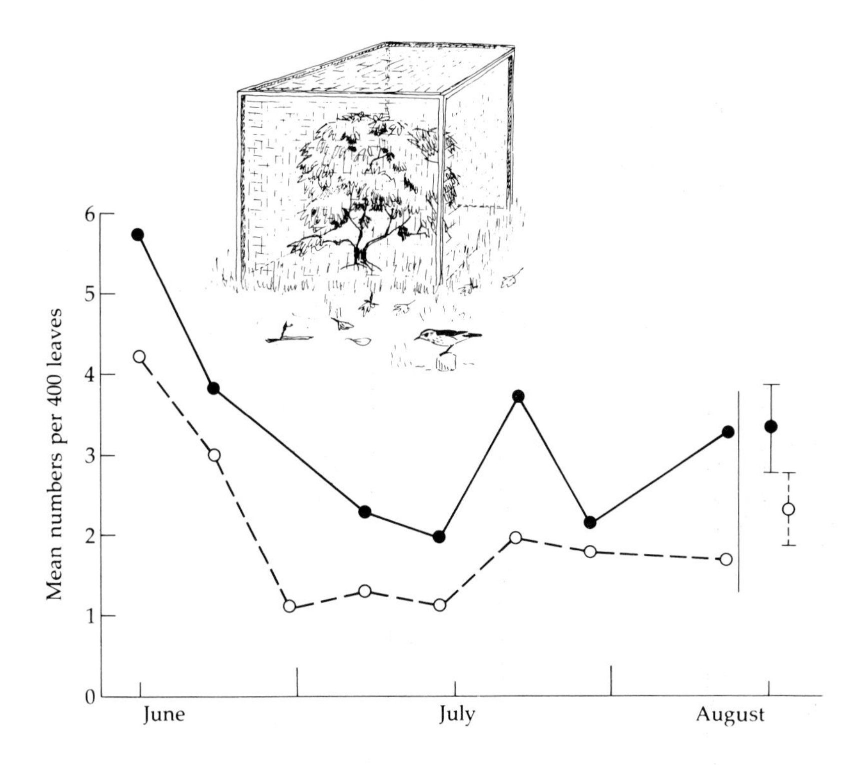

Fig. 5.14. Numbers of caterpillars of Lepidoptera outside (---o---) and inside (—•—) experimental cages, 6m × 6m × 2m tall, covering understory shrubs of striped maple *Acer pensylvanicum*. The 2.2cm mesh of the netting kept birds out, but did not hinder passage of insects. Individual means are averages of 10 cages, 400 leaves per cage, and 10 control areas. Overall means and 95 per cent confidence intervals are shown on the right. Excluding birds increased the numbers of Lepidoptera significantly (After Holmes *et al.* 1979.)

5.4 Conclusions

In Section 5.1.1 we described a set of hypothetical 'ideal' communities. Looking back over the evidence gathered in this chapter we conclude that phytophagous insect communities do not conform closely to example (iv). Interspecific competition is too feeble and sporadic in its effects. Nor are the limited number of available studies a shambles of randomly fluctuating populations (example (i)). In the main, existing examples conform most closely to a mix of examples (ii) and (iii). These communities are moderately deterministic in structure, but with density-independent events, 'density-vague' regulation and delayed density-dependent responses all contributing to fluctuations in the abundance of component populations.

The impact of natural enemies is particularly important in suppressing many populations below levels where food shortages are significant. In other words, key population interactions affecting phytophagous insects are most commonly those acting vertically in the food chain. This theme is developed further in the next chapter, which explores the effects of host plants on the population dynamics of phytophagous insects.

5.5 Summary

Communities are made up of sets of individual species populations that have the potential to interact one with another. An understanding of population dynamics underpins all community ecology.

Depending upon the form and relative magnitude of the density-dependent and density-independent stages in each species' life history, a series of theoretically ideal communities can be defined. They range from essentially random assemblages of species in which density dependence is feeble or non-existent, to highly deterministic systems structured by strong interspecific competition.

Populations of phytophagous insects often fluctuate markedly from generation to generation; hence, at best, communities of phytophagous insects are only moderately constant and predictable in their structure.

An analysis of 31 life tables for phytophagous insects reveals no sign of significant density dependence in twelve of them (38–39 per cent). Where density dependence (or delayed density dependence) has been detected, the effects of natural enemies are much more important than intraspecific competition by a ratio of at least 2:1.

Interspecific competition is unlikely without intraspecific competition, and in accord with the results from life table studies on intraspecific competition, detailed studies of phytophagous insect populations frequently fail to find evidence of significant interspecific competition. The examples that have been reported are reviewed; most are found to be highly asymmetrical (i.e. amensal) interactions. Such examples aside, we conclude that interspecific competition is generally too feeble and sporadic to be the most important or consistent mechanism structuring communities of phytophagous insects.

Interspecific competition is comparatively uncommon because many populations are kept rare, relative to the availability of potentially limiting resources, by the impact of natural enemies — insect parasitoids, insect predators, birds, pathogens, etc. Hence the major processes acting in many communities work vertically through the food chain, not horizontally with other species in the same trophic level.

Table 6.1. Effect of 'island' size on number of insect species found on discrete patches of host plants

Plant	Nature of 'island'	Insect group	Form of species–area relationship S = number of species A = patch or island area	Reference
Collards, *Brassica oleracea*	Experimental patches of 1, 10 and 100 plants	All herbivores	Not specified in detail. Mean number of species per plant increases with increasing patch size	Cromartie (1975)
Rosebay willow-herb (Fireweed) (*Chamerion angustifolium*)	Natural patches surrounded by other vegetation; size range $< 1m^2$ to $10\,000m^2$	All herbivores	$\ln S = 0.08 \ln A + 1.26$ (1979) $\ln S = 0.07 \ln A + 1.60$ (1980)	MacGarvin (1982)
Cord grass (*Spartina alterniflora*)	Real islands surrounded by intertidal mud; size range $< 100m^2$ to $2100m^2$	All arthropods 'associated' with the islands; this includes many herbivorous insects, but also adult Diptera, predatory insects and spiders, etc.	Species number increases from 5 on smallest islands to 15–26 on largest. Linear regression not given, but of form $S = cA + b$ (see Fig. 6.2)	Rey (1981)

Bracken fern (*Pteridium aquilinum*)	Natural patches surrounded by other vegetation; size range $5m^2$ to $1200m^2$	All herbivores	$S = 1.84 \log_{10}A + 5.77$ or $\log_{10}S = 0.09 \log_{10}A + 0.79$	Rigby & Lawton (1981)
Red mangrove (*Rhizophora mangle*)	Real islands in shallow, inshore marine environments; size range $< 50m^2$ to $> 1000m^2$	All arthropods, including phytophagous insects	Experimental reduction in island areas led to fall in S. Species–area relationships not calculated, because successive points from same islands not independent (see Fig. 6.4)	Simberloff (1976, 1978)
Juniper (*Juniperus communis*)	Groups of bushes; size range from single isolated bush to > 3000 bushes	All herbivores	$S = 5.26 \log_{10}N + 0.47$ $S = 3.11 \log_{10}N + 2.76$ Two different areas. N = number of juniper bushes in group	Ward & Lakhani (1977)

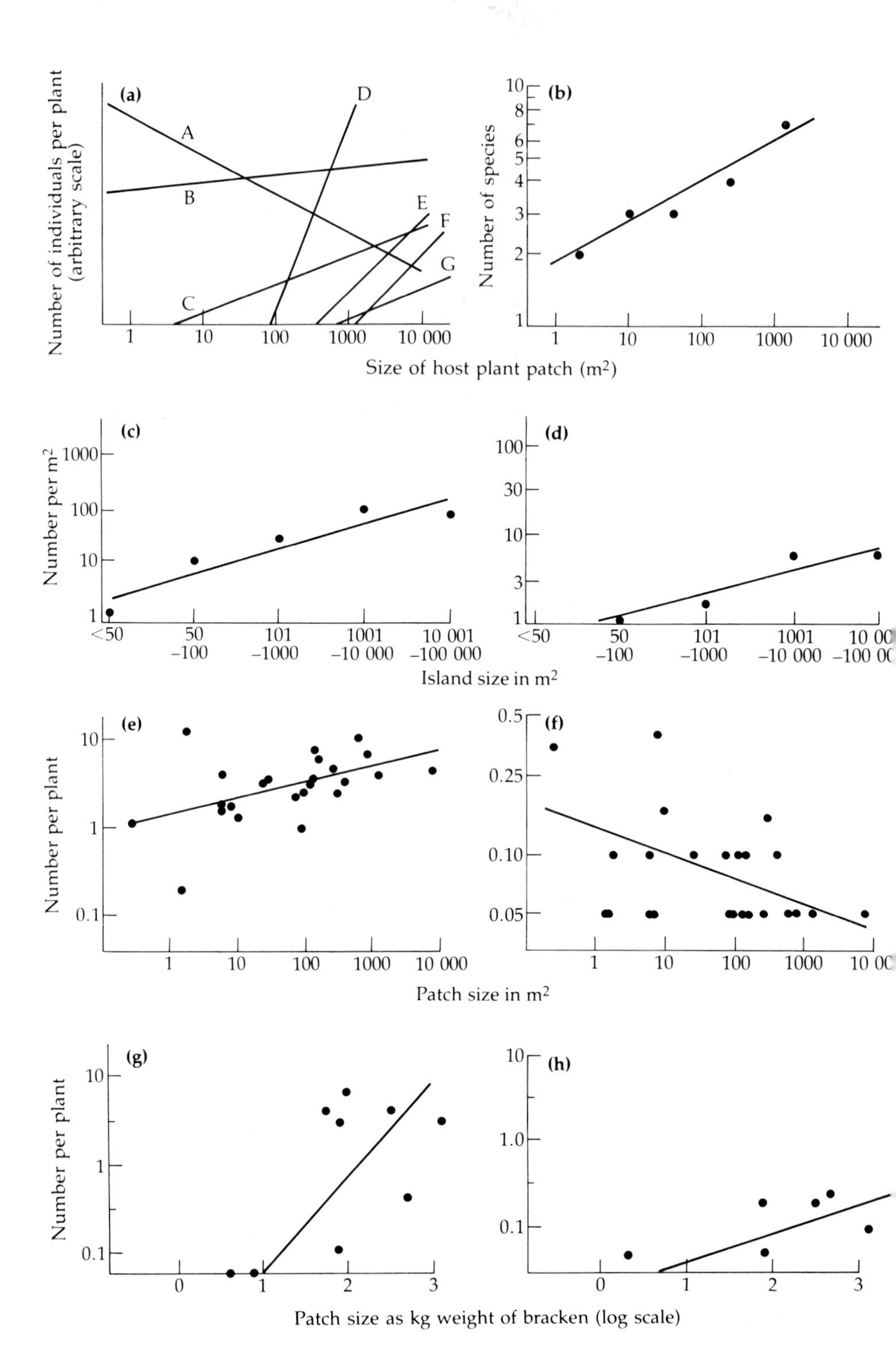
(a)
Number of individuals per plant (arbitrary scale)
A
B
C
D
E
F
G
1
10
100
1000
10 000
(b)
Number of species
1
2
3
4
5
6
8
10
1
10
100
1000
10 000
Size of host plant patch (m²)
(c)
Number per m²
1
10
100
1000
<50
50 -100
101 -1000
1001 -10 000
10 001 -100 000
(d)
1
3
10
30
100
<50
50 -100
101 -1000
1001 -10 000
Island size in m²
(e)
Number per plant
0.1
1
10
1
10
100
1000
10 000
(f)
0.05
0.10
0.25
0.5
1
10
100
1000
Patch size in m²
(g)
Number per plant
0.1
1
10
0
1
2
3
(h)
0.1
1.0
10
0
1
2
3
Patch size as kg weight of bracken (log scale)

rates declined (Fig. 6.3) and immigration rates increased with island area.

That such effects are attributable directly to patch size, and not to habitat variables correlated with patch size, is suggested by Simberloff's (1978) experiments on red mangrove islands, *Rhizophora mangle*, also in the Gulf of Mexico. Simberloff experimentally reduced the sizes of some of these islands (Fig. 6.4), presumably leaving other habitat characteristics more or less unchanged. The number of species decreased on manipulated islands.

However, natural large and small clumps of a particular host plant may differ not only in size, but also in a number of important habitat characteristics; for example, individual plants are sometimes smaller in small patches, there may be less leaf-litter on the ground underneath small patches and so on (Rigby & Lawton 1981). Habitat heterogeneity could then be responsible for generating species–area relationships via species-abundance effects (Fig. 6.1). For example, if death rates are higher on small islands because safe overwintering sites are in short supply in sparse leaf-litter, some species may be absent from small islands (see Rigby & Lawton 1981; Denno *et al.* 1981; and MacGarvin 1982 for further discussion).

Fig. 6.1. (**a**) Hypothetical abundances per plant for seven phytophagous insect species (A–G) on host plant patches of different sizes; species A becomes rarer per plant on larger patches, all the others are commoner per plant on larger patches, and species C–G disappear entirely from patches below a characteristic size.
(**b**) The species–area relationship generated by **a** (see Table 6.1 for some real examples).
(**c**–**h**) Actual species-abundance plots for phytophagous insects on host plant patches of various sizes, for comparison with the hypothetical example in **a**; (**c**) the delphacid *Tumidagena minuta* on the saltmarsh grass *Spartina patens*; (**d**) the cicadellid *Amplicephalus simplex* also on *Spartina*; (**e**) total number of all herbivore species on rosebay willowherb (*Chamerion angustifolium*), and (**f**) eggs of the sawfly *Tenthredo colon* on the same plant; (**g**) the delphacid *Ditropis pteridis* and (**h**) caterpillars of the gelechiid moth *Paltodora cytisella* on patches of bracken (*Pteridium aquilinum*). **c**, **d** after Denno *et al.* (1981), **e**, **f** after MacGarvin (1982), **g**, **h** after Lawton (1978). (Note that the species-abundance measures in **c**–**h** are all on logarithmic scales, so that the exact points at which species disappear from patches cannot be determined by extrapolating the fitted regression lines to zero.)

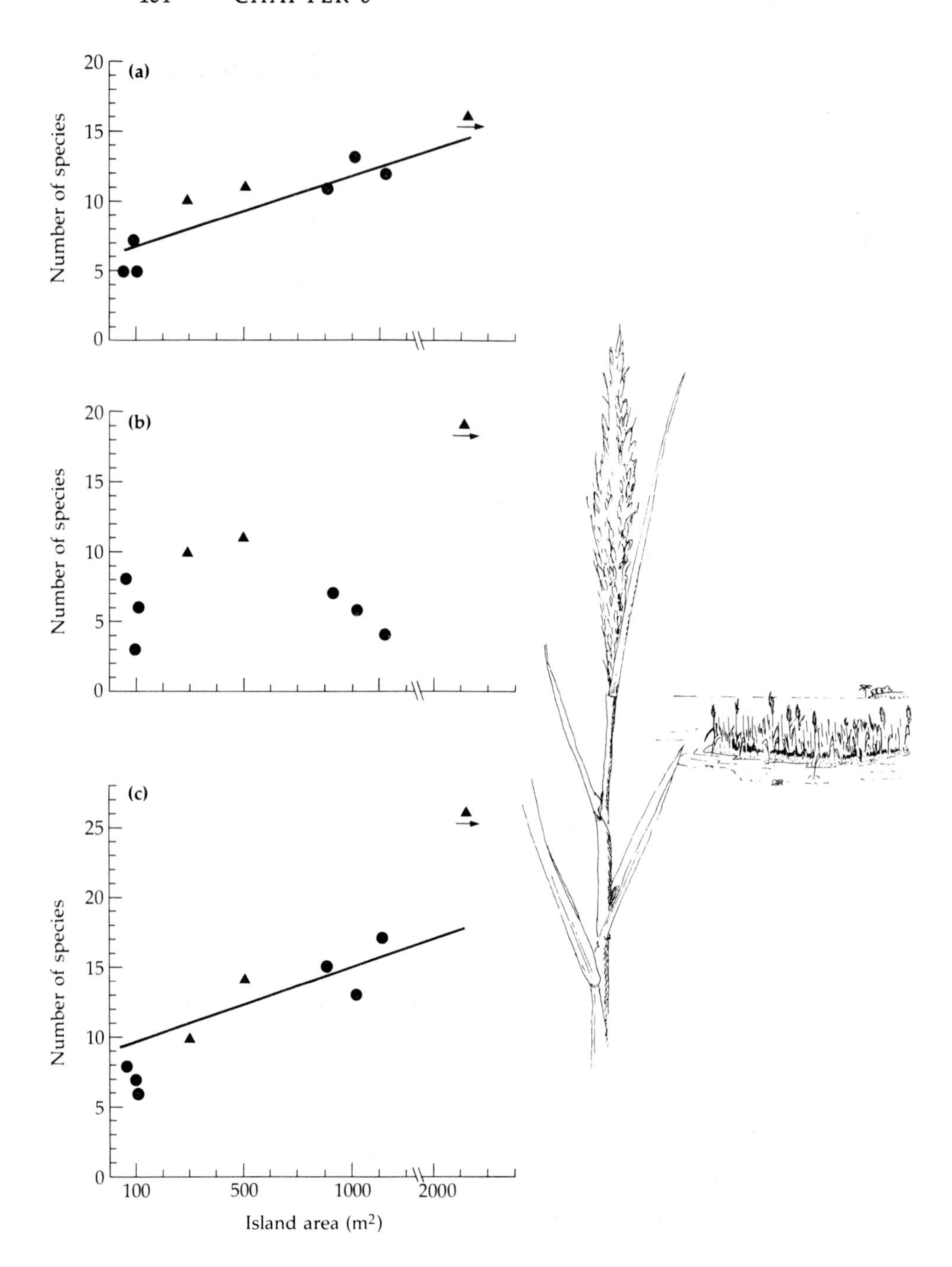
(a)
Number of species
20
15
10
5
0
(b)
Number of species
20
15
10
5
0
(c)
Number of species
25
20
15
10
5
0
100
500
1000
2000
Island area (m2)

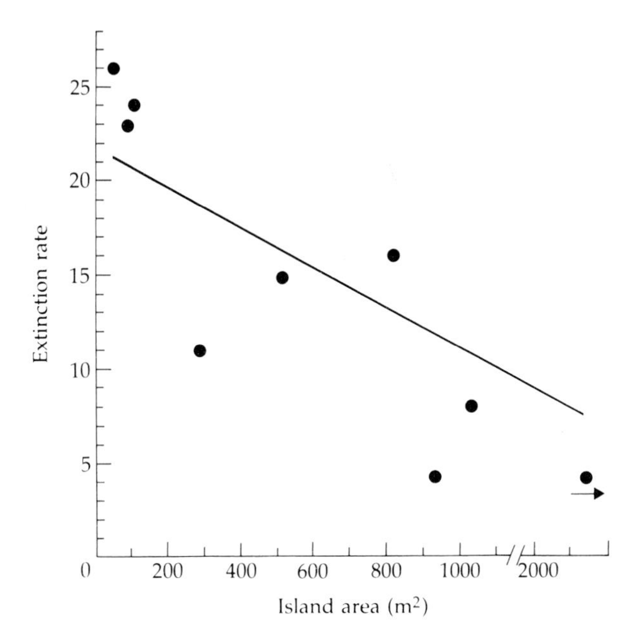

Fig. 6.3. Relationship between extinction rates and area for insects on *Spartina alterniflora* (after Rey 1981) for the islands shown in Fig. 6.2. Extinction was defined as the failure to find a species (previously known to be resident on the island) for two consecutive weeks. Rey started to measure extinction rates twenty weeks after defaunation, once the species area relationship was re-established (Fig. 6.2c). There are some statistical problems in interpreting data on extinction rates as a function of island area (see Rey 1981 for details); these data appear to overcome such difficulties. [Extinction rates on *y*-axis are expressed as number of extinctions per week per average number of species on island × 10^3]

Fig.6.2. Species area relationships for arthropods on islands of *Spartina alterniflora* in Oyster Bay, Florida (Rey 1981). Species are not all phytophagous insects, but also include adult Diptera, parasitic Hymenoptera and mites, as well as predatory insects. (**a**) Pre-defaunation; (**b**) 10 weeks after defaunation with methyl bromide gas; (**c**) 20 weeks after defaunation.
● = defaunated islands, ▲ = control islands. Fitted regression lines are statistically highly significant.

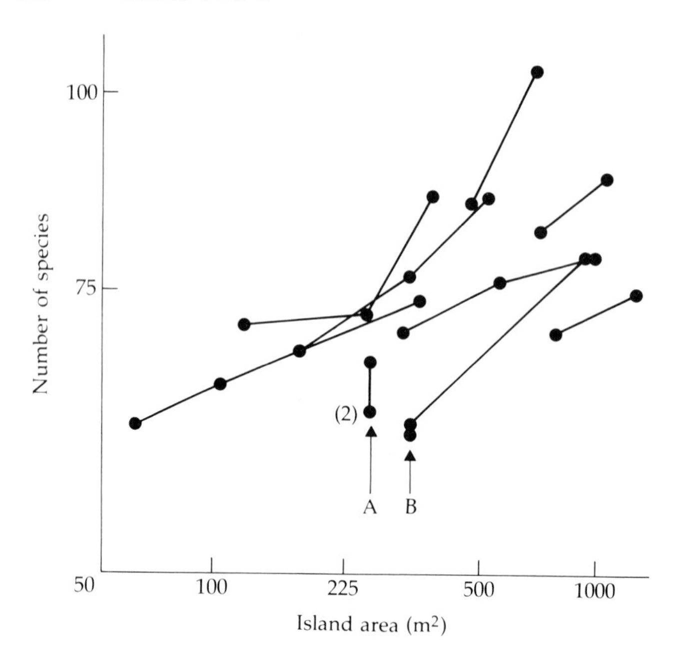

Fig. 6.4. Changes in insect species number with area on islands of red mangrove (*Rhizophora mangle*), experimentally reduced in size (Simberloff 1978). Data from islands successively reduced in size are joined by lines. One island (A) served as a control and held the same number of species on two consecutive censuses, with a slight increase in species number at the third census. Island B was reduced in size between the first census and the second, and was left unmanipulated between the second and third. Every island that was reduced in size showed a fall in the number of species found on it between the first and second censuses (a period of seven months), and between the second and third (a period of twelve months). As in Rey's experiments (Figs 6.2 and 6.3), Simberloff's species involved more than phytophagous insects, but there are no reasons for believing that phytophagous species behaved differently from predators or from non-insect arthropods.

Isolation as an Additional Variable

Although large clumps of a plant support more species of phytophagous insects than small clumps, parallels with traditional island biogeography theory break down when the effects of island isolation are examined (see Chapter 3). Within 'archipelagoes' of local clumps of a single species of plant, isolated patches often have no fewer species than patches close to and surrounded by other patches (Tepedino & Stanton 1976; Ward & Lakhani 1977; Simberloff 1978; Rey 1981; Rigby & Lawton 1981). Lack of significant isolation effects may be due to the difficulty of defining source areas for potential colonists. This difficulty was overcome by experimentally creating extremely isolated patches of the nettle *Urtica dioica* (Davis 1975). The very isolated patches were then found to have impoverished faunas compared with similar-sized groups of plants placed close to established nettle beds.

In passing, it is worth noting that although isolation is an important component of MacArthur and Wilson's theory of island biogeography (Chapter 3), an effect of isolation on species richness is often very difficult to demonstrate, even for real islands (Williamson 1981). Hence, host plant islands are not particularly unusual in this respect.

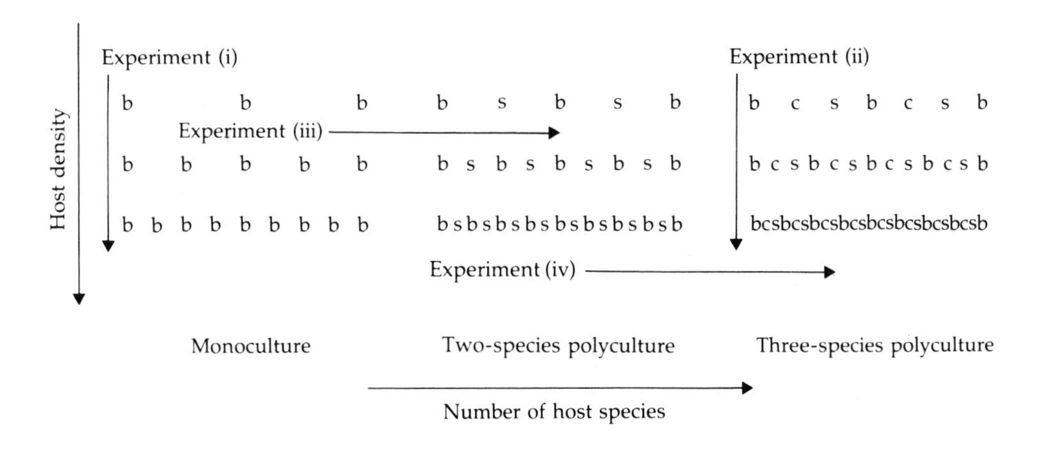

Fig. 6.5. A typical layout for resource concentration studies. In the study, b, s and c are three species of plants, laid out in experimental plots as indicated. Experiments (i) and (ii) vary plant density, but hold plant species diversity constant in monoculture (i) or polyculture (ii). Experiments (iii) and (iv) hold the density of particular species of host plants constant, but vary the number of species with which each is intermingled. In (iii) species b is the host of interest, and species s is added at low density; in (iv) b is again the species of interest, with alternative hosts s and c added at high densities.

6.2.2 *Resource Concentration*

The *resource-concentration hypothesis* states that 'herbivores are more likely to find and remain on hosts that are growing in dense or nearly pure stands; (and) that the most specialised species frequently attain higher relative densities in simple environments' (Root 1973).

Resource concentration is usually studied in terms of two variables, absolute density of a host plant and density relative to other plant species. The other plant species can reduce availability of the target host for the principal phytophages by hiding the plant in a confusing mixture of unsuitable hosts (Atsatt & O'Dowd 1976). These two variables are illustrated in Fig. 6.5.

A growing number of studies have tested the influence of density and plant species mixtures upon the abundance of specific phytophage species (for reviews, see Bach 1980 and Kareiva 1982b). Typical studies

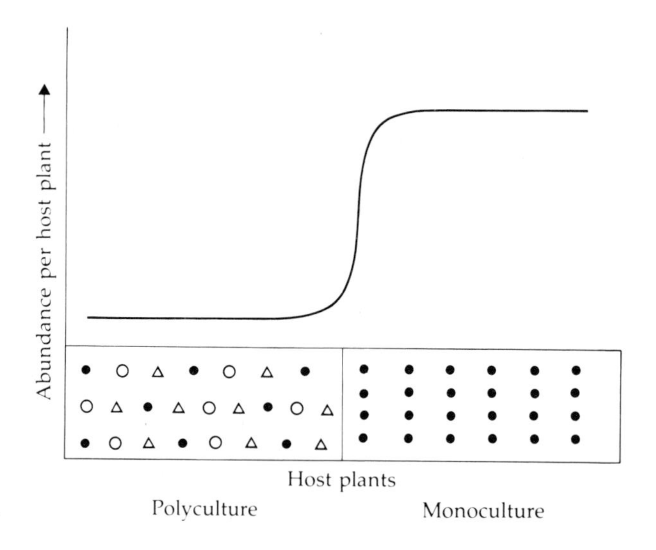

Fig.6.6. A typical result from resource concentration studies in which host plants are grown in monoculture or mixed with others in polyculture (experiment (iii), Fig. 6.5). Specialized species usually increase in abundance in monoculture (see text), although Risch (1981) found that in a polyphagous chrysomelid beetle the reverse was true. Unlike host-specific chrysomelids, which behaved as above, the polyphagous species was more abundant in polyculture.

vary host plant density in monoculture or polyculture (experiments (i) and (ii), Fig. 6.5: e.g. Ralph 1977; Bach 1980; Rausher 1981; Solomon 1981) or vary the number of species of other plants (experiments (iii) and (iv), Fig. 6.5: e.g. Pimentel 1961; Tahvanainen & Root 1972; Root 1973; Cromartie 1975; Bach 1980; Rausher 1981; Risch 1981; Solomon 1981).

Polycultures and Monocultures

Mixing other plant species with the primary host of a specialized phytophagous insect gives a fairly consistent result (Fig. 6.6). Most phytophages can find their hosts more efficiently when no other plant species are present to interfere. This is especially true for insects that have but a single host species, monophages (Tahvanainen & Root 1972; Atsatt & O'Dowd 1976; Perrin 1980; Rausher 1981). Insects are also less inclined to stay on their hosts in polycultures (Bach 1980). Hence many phytophagous insects are more common per plant in monocultures than in polycultures. The biomass of specialist phytophages in collards (*Brassica oleracea*) was consistently higher in pure stands than in perimeter rows with mixtures of plants growing immediately adjacent on either side (Root 1973). Similarly, striped cucumber beetles (*Acalymma vittata*) reached densities 10–30 times greater in monocultures of cucumber (*Cucumis sativis*) than in polycultures of cucumber, corn and broccoli (Bach 1980).

Arguing largely by analogy with Fig. 6.1a, Fig. 6.6 suggests that the species richness of specialist phytophages should be higher in monocultures than when the same number and density of host plants are growing amongst others, in polyculture. In practice, this prediction is refuted by Root's (1973) data on collards. Root found more species of herbivores associated with collards in polyculture than in comparable samples taken from pure stands. Similar studies of other plants would be valuable. Of course, it remains true that the *total* number of phytophagous insect species will usually be higher in mixed vegetation, because there are more species of host plants each with their own associated insects (Murdoch *et al.* 1972; Southwood *et al.* 1979; Fig. 6.7).

Host Density

In contrast to the reasonably consistent effects of additional plant

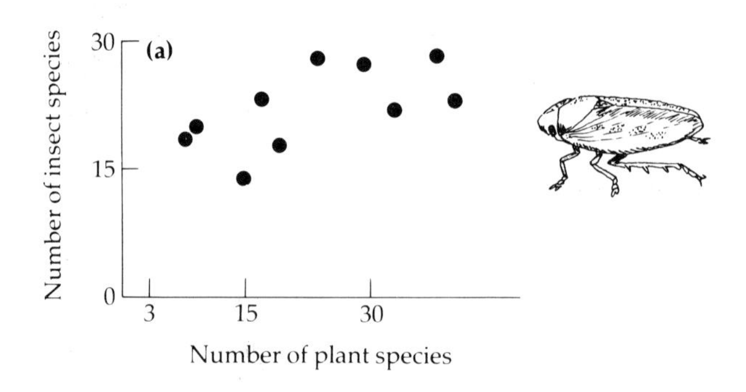
(a)
Number of insect species
30
15
0
3
15
30
Number of plant species

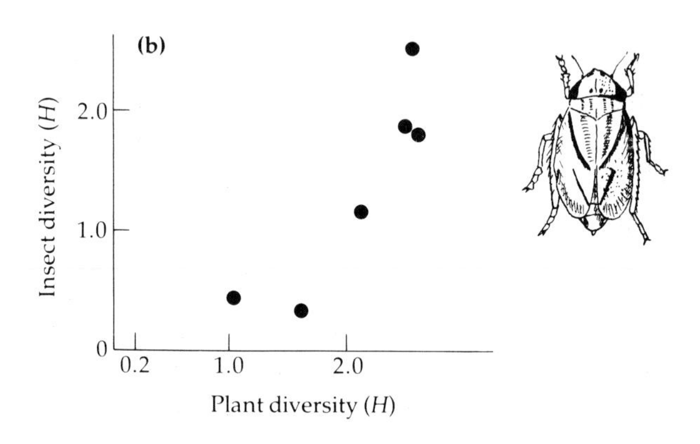
(b)
Insect diversity (H)
2.0
1.0
0
0.2
1.0
2.0
Plant diversity (H)

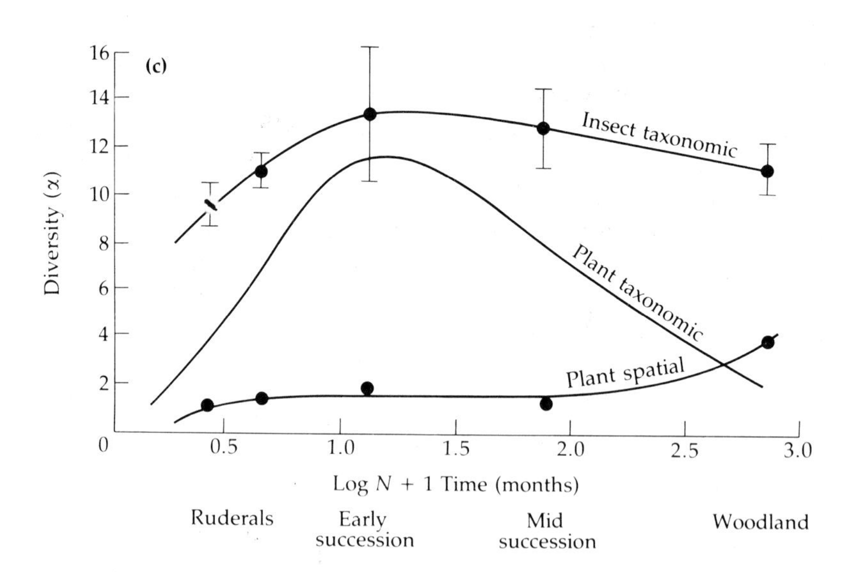
(c)
Diversity (α)
16
14
12
10
8
6
4
2
0
Insect taxonomic
Plant taxonomic
Plant spatial
0.5
1.0
1.5
2.0
2.5
3.0
Log N + 1 Time (months)
Ruderals
Early succession
Mid succession
Woodland

species, the effects of host plant density on phytophage abundance vary widely. Some phytophagous insect species are attracted in greater abundance to, and remain longer on, high-density plots of their hosts (Orians & Solbrig 1977, p. 178; Ralph 1977; van der Meijden 1979; Lemen 1981). Other insects do not behave in this way. Many Lepidoptera oviposit differentially on isolated plants or in low-density patches of their hosts (Jones 1977; Thompson & Price 1977; Shapiro 1981; Solomon 1981; Rausher *et al.* 1981; Mackay & Singer 1982; Courtney & Courtney 1982). Other species are apparently indifferent to host density (McLain 1981 for the bug *Lygus pratensis*; see also Section 6.2.4; Rausher & Feeny 1980). One case is known in which response to host density depends upon the matrix of plant species. The striped cucumber beetle was consistently more abundant per plant in low-density polycultures than in high-density polycultures, but it was indifferent to changes in host density in monocultures (Bach 1980). In another study, laboratory and field data for the same insect and host plant were not consistent. Solomon (1981) found that in the laboratory the gelechiid moth *Frumenta nundinella* was ten times commoner per plant in monoculture than when its host, *Solanum carolinense*, was mixed with another species and hence was relatively less abundant. In the field, no such effects were found.

In brief, because individual species have such different responses to changes in host density, it is impossible to generalize about the effects of host density on the abundances of phytophagous insects, and hence to draw any firm conclusions about how species diversity will be affected. Community structure will almost certainly change with host density, as some insect species decline and become locally extinct and others enter the community. But whether there is anything systematic in these changes remains to be discovered.

Fig. 6.7. Number of species of Homoptera (**a**) and homopteran species diversity measured as the Brillouin Index H (**b**) as a function of plant species richness and diversity in old fields in Michigan (Murdoch *et al.* 1972). More diverse plant communities support more diverse assemblages of plant-feeding Homoptera. (**c**) Insect species diversity (measured as William's α; mean $\pm$SE) tends to increase with plant species diversity during the early stages of succession, but then remains high as plant diversity falls in later stages, because of the enhanced structural diversity of woodland plant communities (Southwood *et al.* 1979). (Indices H and α take into account both the number of species and their relative abundances (Southwood 1978b).)

6.2.3 *Habitat Topography*

It is well known that the local abundance, presence and absence of many species of phytophagous insects are influenced by habitat topography. 'Barriers to air flow such as hills and ridges, buildings, trees, hedges, and the edges or uneven canopies of crops, create sheltered zones in which insects accumulate, and zones where faster winds blow insects away' (Lewis 1969). Small species, aphids for example, are particularly affected in this way. Although poorly studied as a phenomenon influencing the local community structure of phytophagous insects, habitat topography is undoubtably important, and may sometimes blurr or override species–area and resource concentration effects.

6.2.4 *Concluding Remarks: Some Consequences for Interspecific Competition*

Obviously, much remains to be discovered about the ways in which changes in host plant distribution and abundance influence the diversity of phytophagous insect communities. The regional list of species able to feed on a plant (Chapter 3) sets an upper bound to local diversity. However, as we saw in Chapter 3, many of these insects will be absent from parts of larger geographic areas (e.g. Fig. 3.5). The local pool of insects will usually therefore be smaller than the regional pool. The identities of the insect species that actually occur together to make up individual communities depend upon the size of local patches of plants, their density and location, and the presence of intermingled plants of other species.

In the main, each species of insect seems to respond to changes in the distribution and abundance of its host plant quite independently of the other phytophagous insect species in the same community. This is tantamount to saying that some of the most obvious changes in the structure of phytophagous insect communities need not, and often do not, involve interactions between different insect species. In two instances, however, it is known that interspecific competition between phytophagous insects is modified by changes in host density and relative abundance. Both examples were discussed briefly in Chapter 5 (Section 5.2.3 and Table 5.5), but it is worth elaborating on them here.

McLain (1981) studied three species of Hemiptera feeding upon the developing ovules of *Senecio smallii*. *Neacoryphus bicrucis* fed only on this species; *Harmostes reflexulus* fed on *Senecio* and some other flowers in the same meadow (i.e. it was oligophagous or stenophagous); and *Lygus pratensis* was an extreme polyphage. The monophagous *N. bicrucis* was most abundant in large host patches (as expected from Fig. 6.1) and in the parts of these patches with high densities of flowers. The other two bug species were more abundant in smaller patches of *Senecio smallii*, probably because they were excluded from the large patches by the pugnacious monophage. In small patches, *H. reflexulus* was more abundant in places with high bud densities, but it failed to show a similar response to resource concentration on large patches, probably again because it was excluded by *N. bicrucis*.

The second example is Kareiva's (1982a) study of collards, two species of flea beetles and *Pieris* butterflies, in which he varied the density of the host plants and studied how host density influenced interspecific interactions between the insects. As we discussed in Section 5.2.3, most of Kareiva's experiments failed to show many competitive effects. The two significant examples of interspecific competition that he did find were both at the highest density of collards.

It remains to be seen how resource concentration affects interspecific competition in other groups of phytophagous insects. Theoretically, we expect interspecific competition to be more important when population levels are high, and for the outcome of competition to change with variation in the 'carrying capacity' of a habitat (Williamson 1972; Hassell 1976; Begon & Mortimer 1981). Hence we might expect resource concentration to affect interspecific competition in other phytophagous insects.

6.3 Phytochemistry, Morphology and Phenology

This section includes the most important remaining plant influences upon the population dynamics and community ecology of phytophagous insects.

6.3.1 Diet Breadth of Phytophagous Insects

The Degree of Monophagy and Polyphagy

One consequence of the tremendous chemical and physical diversity

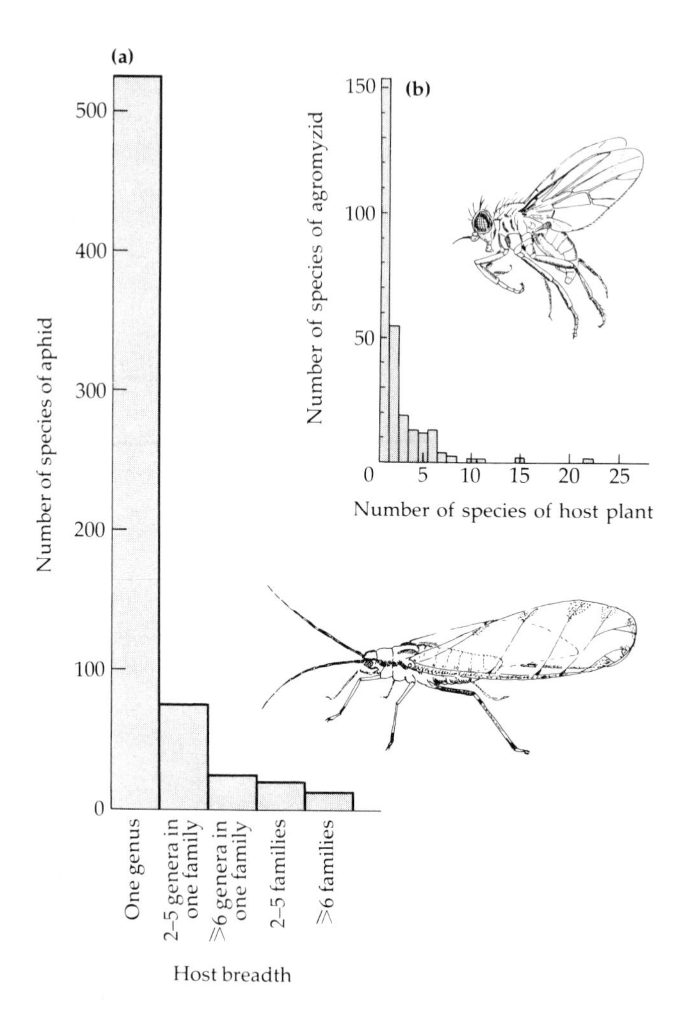

Fig. 6.8. (**a**) Host plant specificity of British Aphididae (Eastop 1973). Most species (514) feed on only one genus of plants. Very few species (13) are broadly polyphagous, feeding on plants in more than six families. (**b**) Host plant specificity of British Agromyzidae (Spencer 1972). Most species are monophagous or oligophagous.

among plants is the narrow diet tolerances of phytophagous insects. Unlike carnivores, many of which can eat a taxonomically broad array of flesh, any phytophage species can eat only a miniscule fraction of plant taxa. Diet breadth typically varies from a single plant taxon for a large proportion of insect species to several plant taxa for a smaller proportion of any group of insects. Two examples are given in Fig. 6.8.

Thompson (1982) suggests that effectively 'parasitic' species, completing all their development on one individual plant, should be more specialized in diet than insects encountering and grazing numerous individual hosts during their lives. Agromyzid flies and aphids (Fig. 6.8) are good examples of plant parasites in Thompson's (and Price's (1980)) sense of the word; grasshoppers are good examples of grazers, and they are often polyphagous (e.g. Bernays & Chapman 1975; Joern 1979b). However, in stick insects, *Carausius morosus*, older individuals are more selective in what they will eat than young individuals (Cassidy 1978), a phenomenon that might occur in other long-lived 'grazing' insects, and not all grazers are polyphagous at any age, so Thompson's views cannot be accepted without further investigation.

Each oligophagous and polyphagous insect species associated with one particular species of host plant will usually exploit a different set of additional hosts. Hence most plants do not have highly distinct, specialized insect faunas, but have rather a collection of insect species with varying alternative hosts (Pimm & Lawton 1980). This pattern can be seen in the great variety of host affiliations among the insects of a single plant taxon. For example, monocots in the genus *Heliconia* are eaten by a variety of insects, from the strictly monophagous hispine beetles to polyphagous Lepidoptera and Homoptera. The polyphages include insects with very different alternative hosts, which range from palms to dicot trees (Auerbach & Strong 1981). The lack of distinct, specialized faunas is also shown by Futuyma and Gould's (1979) study of the Lepidoptera and sawflies in an upland forest of New York, where taxonomically similar insects do not usually attack a similar array of plant species. Much the same point was made in Section 4.1.5.

Definitions

A word of warning is in order here about the terminology used in studies of diet breadth. Different authors use the same word to mean different things. Monophagy, for example, may mean feeding on only one plant species, one genus, or one family (e.g. Dugdale 1975; Holloway & Hebert 1979; Slansky 1976), and care must be taken when comparing studies to accommodate these different usages. Moreover, an insect can be considered polyphagous because it attacks several plant species, but at the same time monophagous because it feeds upon only one specific type of tissue, such as hairy leaves, or plants containing a particular alkaloid (Slansky 1976; Van Emden 1978; Eastop 1979). Finally, diet breadth of insects is relative in a geographical sense. An insect that uses several host species over its range but only one host locally can be considered either monophagous or polyphagous, depending upon the scale of the comparison (Table 4.2; Brues 1946; Downey & Fuller 1961; Singer 1971; Knerer & Atwood 1973; Slansky 1976; Gilbert 1979; Fox & Morrow 1981).

Ecological and Evolutionary Influences

The selective forces leading to greater food plant specialization have been the subject of much speculation in the literature, because evolution of insect diet underlies a number of interesting ecological phenomena (Levins & MacArthur 1969; Joern 1979b; Benson 1978; Jaenike 1978a, b; Smiley 1978; Rhoades 1979). Several theories of diet breadth consider the means of balancing risks of mistakes in host choice with the likelihood of finding a new suitable host in an unpredictable environment. Phylogenetically, monophagy is neither primitive nor derived, because all the available evidence suggests that diet breadth can either contract or expand within an insect lineage (Benson 1950; Holloway & Hebert 1979).

Such speculation takes us firmly beyond the bounds of ecological time and into the realm of evolutionary time. Ecological effects on diet choice, particularly the impact of competition from closely related species have been considered by Strong (1982a) for hispine beetles on *Heliconia* (Section 5.2.2). The presence or absence of potential competitors had no influence on choice of food plants in this group of insects (Fig. 6.9). That this need not always be the case was shown by Fox and Morrow (1981), who found that feeding by several species of insects upon the Australian *Eucalyptus stellulata* was so heavy that the plant produced leaves too small for oviposition by the sawfly *Pseudoperga guerinii*. Insecticides eliminated this diffuse competition (i.e. competition resulting from the cumulative effects of several species) and allowed the plant to produce larger leaves, acceptable to ovipositing sawfly females.

Finally, an interesting correlation between the diet breadth of some desert Lepidoptera and the age and type of plant foliage that they consume will serve as a transition to our next section, on seasonal phenology. Cates (1980, 1981) found that caterpillars of polyphagous species (which he defined as feeding on plants in more than one family) tended to prefer older foliage, while monophagous species (with food plants in only one family) preferred young foliage. Polyphagous species also preferred the *least* abundant plant species in these deserts, whilst annual and herbaceous desert perennials supported proportionately more species of polyphages than did the woody perennials. Cates interpreted these patterns in terms of the theory of 'plant predictability' (Rhoades & Cates 1976) or 'apparency' (Feeny 1976; see Section 3.4.2). These theories suggest that 'unpredictable' plant species should commit

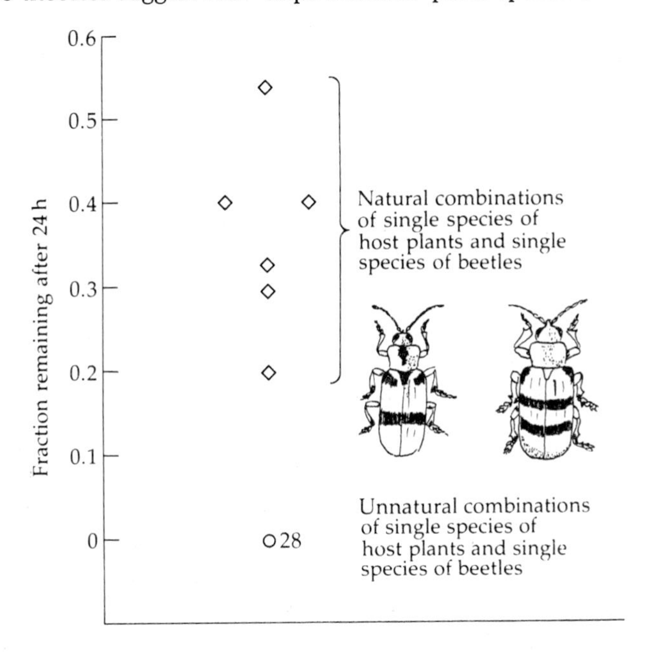

Fig. 6.9. Proportion of adult hispine beetles remaining in rolled leaves of *Heliconia* plants 24 h after being introduced. At least 20 beetles were used in each host–beetle combination. There were six natural combinations (diamonds) and 28 unnatural ones involving host–beetle combinations never recorded naturally in the field. Potential competitors were excluded from all the experimental leaves by bagging. Despite a total absence of competitors, no adult hispines remained on the wrong host after 24 h (Strong 1982a).

less energy to defensive chemistry, because they are more able to escape from herbivores in time and space. However, lower investment in defences may mean that plants are more acceptable to polyphagous species (see, for example, Berenbaum 1981a, b). Cates's data suggest that the less predictable, less easily discovered plant species in his deserts were the seasonally ephemeral annuals, the smaller herbs and rarer plants of all kinds. If so, the observed distribution of polyphagous species was in broad agreement with predictions from the theory.

However, others have found diametrically opposite results, namely proportionately more polyphagous species on highly 'apparent' woody bushes and trees (e.g. Futuyma 1976). A major problem with work in this field is defining host plant apparency and predictability from the insects' point of view. Until our understanding of these problems improves, studies on the proportions of monophagous, oligophagous and polyphagous species in different communities are unlikely to be very enlightening.

6.3.2 *Seasonal Phenology*

A list of species attacking one species of plant throughout its range, or locally within a community, gives a highly misleading impression of the richness of that assemblage at any instant.

Seasons and their weather profoundly affect the growth and development of plants. Seasons also, by entraining the growth of plants, affect growth of insects, and their numbers. These effects are seen as marked changes in abundance, identities, and numbers of insect species on plants through the growing season. Three examples are the phenological development of wild parsnip (*Pastinaca sativa*) and attack by one of its phytophages, shown in Fig. 6.10; bracken and several species of phytophagous insects, shown in Fig. 6.11; and oak and Lepidoptera, shown in Fig. 6.12. Leaves and other parts of plants are not just one resource for phytophages, but a continual series of temporally overlapping resources, from tender new buds, through young blades, finally to senescent leaves, shoots and stems. It follows that many phytophages are seasonal specialists, because by specializing on a type of tissue they specialize upon a period in the phenology of the plant (e.g. Tanton 1962; Slansky 1974; Price 1976; Ikeda *et al.* 1977; Cates & Rhoades 1977; Addicott 1978a; Haukioja *et al.* 1978; Coley 1980; Berenbaum 1981b; Scriber & Slansky 1981; Niemela & Haukioja 1982).

One manifestation of phenology is that insect species less specialized to one season, that is 'seasonal generalists', experience marked changes in population behaviour through the year on a single host plant. Sycamore aphids (*Drepanosiphum platanoides*) on the European sycamore (*Acer pseudoplatanus*) are temporal generalists, in that they occur on trees throughout the growing season (Dixon 1970), but they specialize upon high amino-nitrogen levels in their host trees. Reproduction in this aphid falls to very low levels by late June, after being quite high during the spring. Reproduction rises again in the autumn, when phloem nitrogen again increases in the host plant (Fig. 2.8c).

Being confined to a particular temporal window has a profound in-

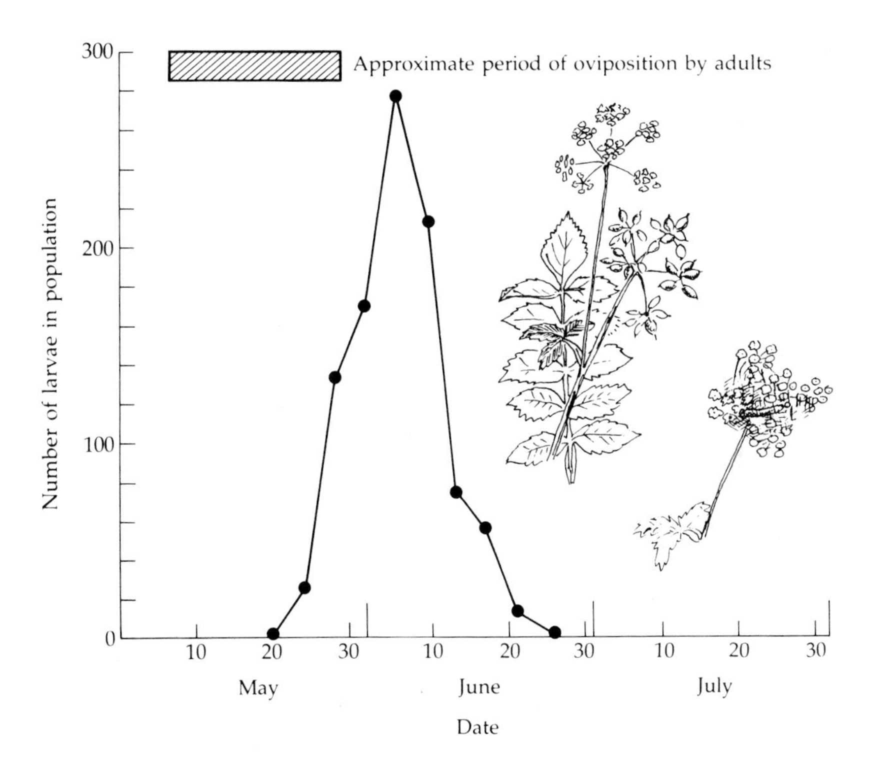

Fig. 6.10. Seasonal phenology of caterpillars of the parsnip webworm, *Depressaria pastinacella* (Lepidoptera: Oecophoridae), on wild parsnip, *Pastinaca sativa* (Umbelliferae). Adults can only oviposit successfully in unopened umbels (flower-heads); caterpillars feed only on the flowers and developing seeds. (Data from Thompson & Price 1977, for larvae on isolated host plants.)

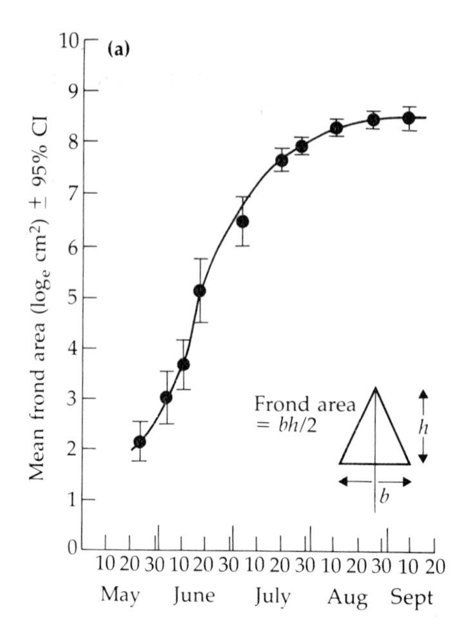
(a)
Mean frond area ($\log_e$ cm²) ± 95% CI
Frond area
= $bh/2$
h
b
0
1
2
3
4
5
6
7
8
9
10
10 20 30 10 20 30 10 20 30 10 20 30 10 20
May June July Aug Sept

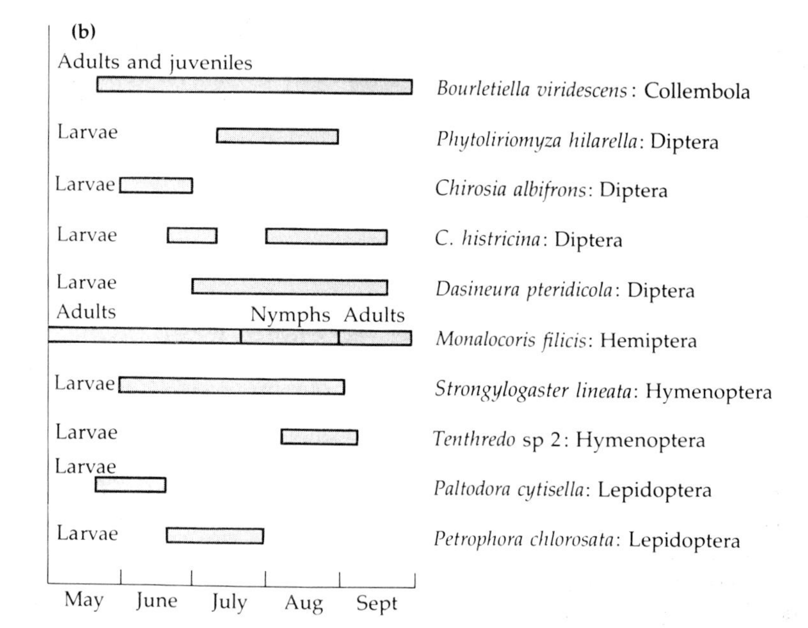
(b)
Adults and juveniles
Bourletiella viridescens: Collembola
Larvae
Phytoliriomyza hilarella: Diptera
Larvae
Chirosia albifrons: Diptera
Larvae
C. histricina: Diptera
Larvae
Dasineura pteridicola: Diptera
Adults
Nymphs
Adults
Monalocoris filicis: Hemiptera
Larvae
Strongylogaster lineata: Hymenoptera
Larvae
Tenthredo sp 2: Hymenoptera
Larvae
Paltodora cytisella: Lepidoptera
Larvae
Petrophora chlorosata: Lepidoptera
May June July Aug Sept

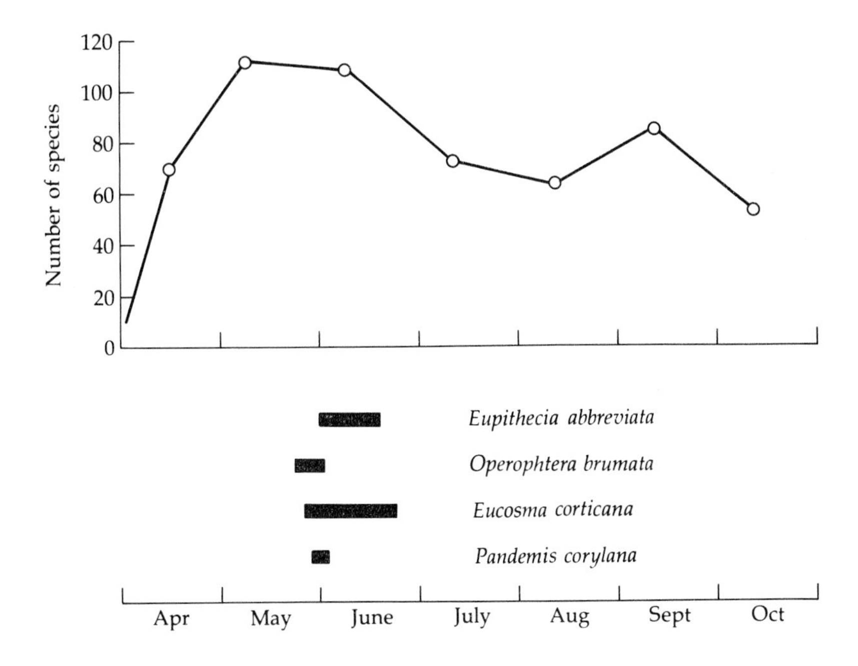

Fig. 6.12. Total number of species of lepidopteran caterpillars exploiting the foliage of oak (*Quercus robur*), together with an indication of the seasonal phenologies of four of these species near Oxford, given by the spread of dates over which they drop from the trees to pupate. Species feed during the weeks to the left of these black bars, and are absent from the phytophage community at all times to the right. There is obviously a marked turnover of species with season. (Data from Feeny 1970.) Changes in the foliage chemistry of oak are summarized in Fig. 2.10a. These changes are probably an important reason why many species are seasonal specialists.

Fig. 6.11. (**a**) Seasonal growth of bracken, and (**b**) the seasonal occurrence of phytophagous insects exploiting the fronds at a site in northern England (Lawton 1978, 1982). Some of the chemical changes that accompany the growth of bracken fronds are shown in Fig. 2.10b. These changes are probably an important reason why many species are seasonal specialists.

fluence upon the dynamics of phytophagous insects. An example is the rigid sequence imposed by the calabash tree (*Crescentia alata*) upon a flea beetle (*Oedionychus* sp.) in seasonally dry areas of Costa Rica (Rockwood 1974). New leaves are produced mainly at the beginning of the spring rains, which bring to an end the long annual dry season in this part of the American tropics. Trees hold leaves for about 11 months, but the flea beetle larvae develop only upon the youngest leaves, which are naturally available only during a few weeks following the first spring rains. Early leaves are not heavily eaten, because the adults surviving the dry season are few, and early egg loads on trees are light. After ageing for a few weeks, leaves toughen and become undesirable to beetles and larvae (Fig. 6.13). Indeed, the rapid phenology of the calabash tree probably causes starvation and prevents reproduction for a majority of beetles in this population.

It is not uncommon to find similar rigid constraints placed upon insect populations by quick phenological sequences in their host plants. The timing of insect development is often crucial, with immatures stalled before metamorphosis or young adults performing very poorly if synchronization with the host slips even one or two weeks out of phase. Examples include the winter moth *Operophtera brumata* on oak, *Quercus robur* (Feeny 1970; see Figs 6.12 and 2.10a); the parsnip webworm, *Depressaria pastinacella*, on wild parsnip, *Pastinaca sativa* (Thompson & Price 1977; Fig. 6.10); the leaf-galling aphid *Pemphigus betae* on cottonwood, *Populus angustifolia* (Whitham 1978); the psyllid *Phytolyma lata* on iroko trees, *Chlorophora* (White 1966); and the aphid *Acyrthosiphon spartii* on broom, *Sarothamnus scoparius* (Waloff 1968a).

Effects on Community Diversity

The effects of seasonal changes in the host plant are marked seasonal changes in the diversity and structure of phytophagous insect communities, as species adapted to use their host at different times and stages of its development enter and leave the community (Price 1976; Lawton 1978, 1983b). Some examples are shown in Figs 6.11, 6.12 and 6.14.

Fig. 6.13. The flea beetle *Oedionychus* sp. and feeding damage caused by the beetle to its host tree, *Crescentia alata* (photographs by L. Rockwood).

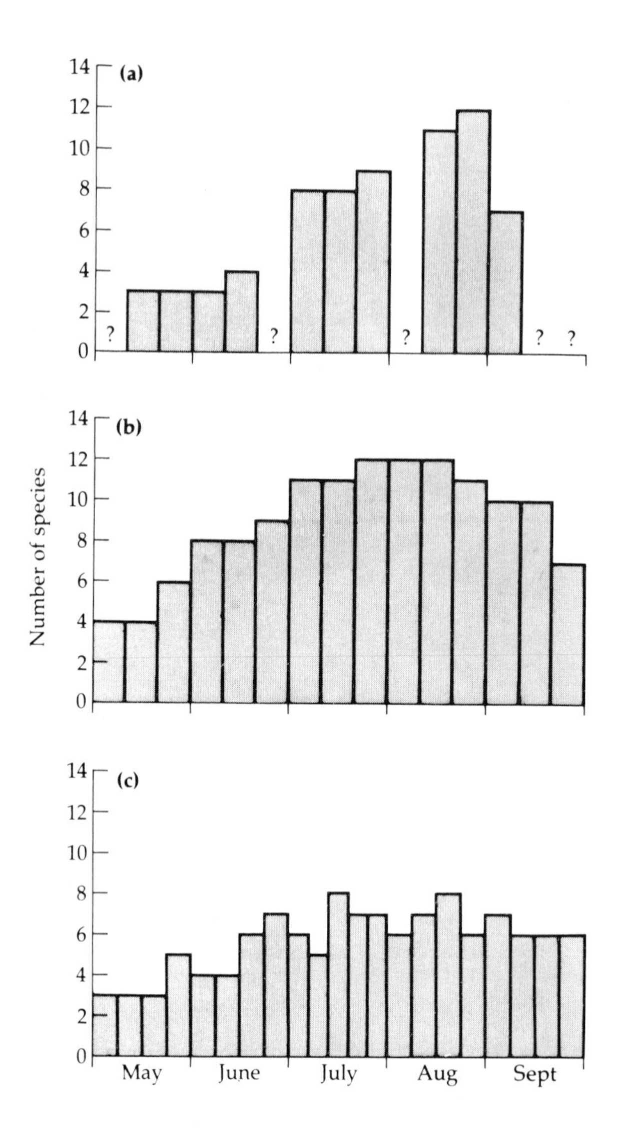

Fig. 6.14. Seasonal changes in the number of phytophagous insect species in local communities of bracken (*Pteridium aquilinum*) in (**a**) woodland and (**b**) in the open at a site in Northern England (Lawton 1982). (**c**) Similar data for nettles *Urtica dioica* in East Anglia (from Davis 1973).

In general, herbaceous plants in temperate regions support peak diversities of phytophagous insects later in the growing season than do trees and woody shrubs (Lawton 1978; Niemela *et al.* 1982). This is probably because the architecture (Chapter 3) of trees (i.e. the variety of potential feeding sites, hiding places and so on) is much richer earlier in the season than that of herbs. Trees and bushes have most of their architecture in place at the start of the growing season; herbaceous plants, in contrast, increase in size and structural complexity as the season progresses (e.g. Fig. 6.11; Lawton 1983a). Some interesting variations on this theme for trees have recently been reported by Niemela and Haukioja (1982). Trees that continue to grow late into the season (e.g. birch, *Betula*, and alder, *Alnus*) have late peak diversities of Lepidoptera. Trees like oak, *Quercus*, that normally finish growing quite early in the summer have peak diversities of lepidopteran caterpillars early in the growing season.

Thus, a major determinant of the diversity of phytophagous insect communities is the seasonal development of the host plant. Once more, key effects are seen to operate vertically through the food chain.

6.3.3 *Resistant and Susceptible Varieties*

Population levels of insects vary greatly between years; even at those times of the year when an insect is normally found on a plant, conditions are often far from ideal, and populations rarely attain their highest possible densities on any host plant (Chapter 5). We have emphasized the role of natural enemies in keeping phytophages rare in time. In space, very similar host plants can vary in ways that greatly affect insect populations. A useful perspective on this subject comes from the difference between 'resistant' and 'susceptible' varieties of crops.

Resistant varieties of crops are bred to reduce the effects of one or more pests. The factors that determine resistance are many (Painter 1951; Day 1972; Jones 1977; Gallun & Khush 1980; Sogawa 1982) and often involve more than nutritional value (e.g. Onuf 1978) or toxicity of the plant to insects (Way & Murdie 1965; Hedin *et al.* 1976). Morphology, anatomy, phenology and autecology may often be involved. The important point for our discussion is that 'resistant' varieties are rarely totally immune from attack. Rather, they support lower average population sizes than the susceptible varieties; target pest

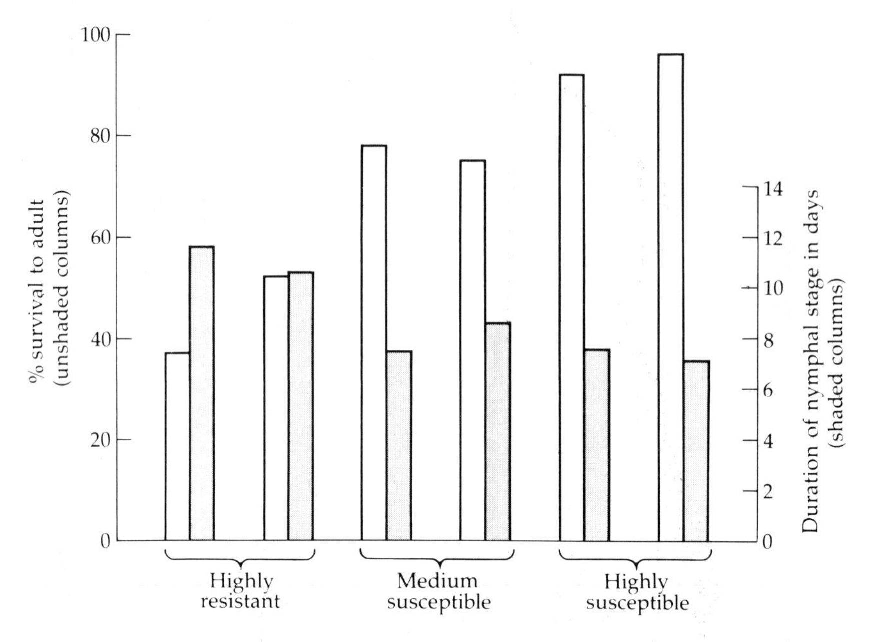

Fig. 6.15. Survival and development times in the cotton jassid *Amrasca devastans* on two resistant varieties of the host plant, two intermediate varieties and two susceptible varieties (from data in Agarwal & Krishnananda 1976). *Amrasca* survives less well and grows more slowly on the resistant varieties.

insects have impaired life history performance on resistant varieties (Figs 6.15 and 6.20).

Differences in phytophage performance that result from plant phenotype, presumably with a genetic basis, are not confined to agricultural plants. Some individuals in wild plant populations, or some populations within a species, support much larger densities of phytophages than others. Examples include the lycaenid butterfly *Glaucopsyche lygdamus* on populations of lupins, which is influenced by differences in alkaloid

Fig. 6.16. Interactions between phytophagous insects, ants and host plants to illustrate three possible modes of interaction. The effects of species A on species B are designated by arrows and signs; for example in (**c**), the plants benefit ants by providing food and shelter, and the ants benefit plants by defending them against herbivores.

+ Positive (beneficial effect)
o Neutral (no effect)
– Negative (harmful effect)
? Doubtful or unknown effects

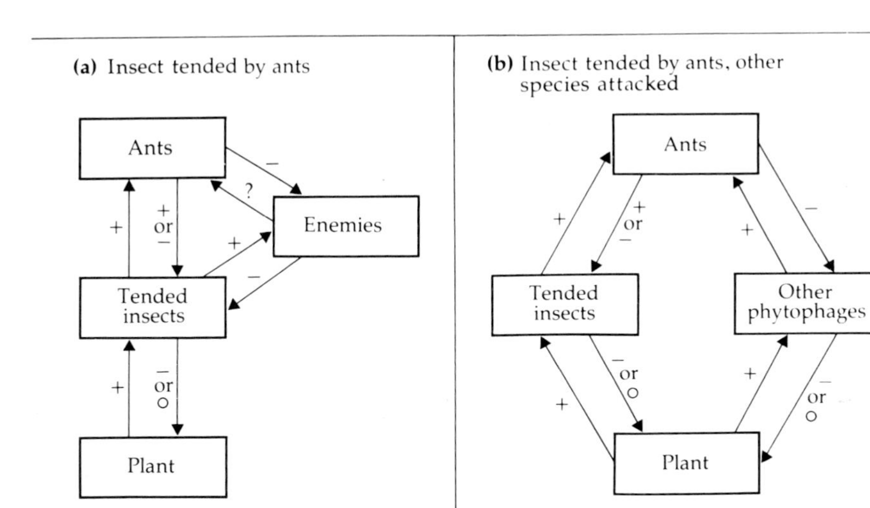

Mutualisms between phytophagous insects and ants

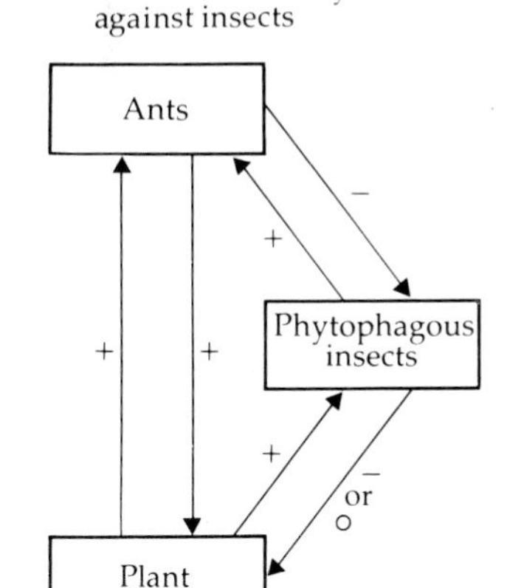

Ant–plant mutualisms

concentrations (Dolinger *et al.* 1973); several insect species among populations of wild parsnip, influenced by furanocoumarin levels (Berenbaum 1981c); differences in damage among bracken fronds, correlated with tannin levels (Tempel 1981); population differences in psyllids of the genus *Glycaspis* on individual eucalyptus trees (Journet 1980), and in the whitefly *Aleurotrachelus jelinekii* on *Viburnum* bushes (Southwood & Reader 1976); and densities of the aphid *Uroleucon caligatum* among clones of goldenrod (Moran 1981).

One of the most intensely studied examples of natural variation in plant susceptibility is that of the ponderosa pine (*Pinus ponderosa*) and the pine leaf scale (*Nuculaspis californica*). Trees in the same population vary enormously in their susceptibility to scale attack. Scale-free trees 'frequently stand for years beside trees infested with as many as ten insects per centimeter of needle, often with intertwining branches .... When plots of trees are sprayed with insecticide to control scales, trees are reinfested after 2–5 years to approximately their original level of infestation' (Edmunds & Alstad 1981). Monoterpene phytochemicals probably determine differences between trees, and terpene levels in trees are probably under genetic control.

Recently, Whitham and Slobodchikoff (1981; see also Whitham 1981) have gone as far as to suggest that different branches on the same tree, or different branches of a rhizomatous plant, may differ in their susceptibility to insect attack, through somatic mutations in the meristems. Rigorous tests of this idea will involve both mutation rate and the rate at which mutations produce protective features.

Each of these various examples reinforces the point that the abundance, presence, and absence of particular species of insects in local communities may be strongly influenced by genetic and phenotypic differences between individual plants, clones of plants, or even parts of the same plant. Add to these influences phenotypic differences due to variation in host plant nitrogen levels (Chapter 2), and also the possibility that high levels of previous damage may induce defences in plants (Chapter 5), and it is clear that variation in the structure of phytophagous insect communities from year to year and from site to site on the same species of host might be due to a confusingly large number of host plant influences.

What is not yet clear is the extent to which the genetic and phenotypic differences discussed in this section can destroy or greatly modify the

Fig. 6.17. Caterpillar of the lycaenid butterfly *Glaucopsyche lygdamus* being tended and defended by an ant (see Pierce & Mead 1981; Photograph by N. Pierce). The interaction is an example of (**a**) in Fig. 6.16. The ant is *Formica fusca,* and has seized a braconid parasitoid that posed a threat to the *Glaucopsyche* caterpillar.

influences of local abundance and distribution of host plants, discussed in Section 6.2. That species–area relationships can be detected in small clumps of plants (Table 6.1), without deliberately controlling for genetic differences between plants in different clumps, suggests that the effects discussed in the present section are of lesser importance than those in Section 6.2. But this is simply a guess.

6.4 Beyond Pairwise Interactions

Here we explore some of the more complex interactions that can be found in some phytophagous insect communities, namely ant–plant and ant–insect mutualisms, and interactions over three trophic levels, from plants through phytophages to natural enemies.

6.4.1 Interactions and Mutualisms with Ants

Ants are often extremely important predators of phytophagous insects (e.g. Jeanne 1979; Faeth 1980; Laine & Niemela 1980; Skinner 1980; Skinner & Whittaker 1981; Risch & Carroll 1982). Their impact on the insect pests of two important tropical crops has been the subject of studies by O'Connor (1950), Way (1953) and Brown (1959) for coconuts, and by Leston (1971, 1973), Greenslade (1971), Room (1971) and Majer (1976a,b) for cocoa.

By simply acting as important predators ants would probably have qualified for brief mention in Section 5.3.3. Ants, however, also play more complex and subtle roles in phytophagous insect communities. We can group these more complex interactions under three headings (Fig. 6.16; see Keeler 1981; Thompson 1982).

Tending by Ants of Phytophagous Insects

This is the least complicated relationship (Fig. 6.16a). Ants gain resources from insects that eat the plant, and these insects are tended by the ants. The influence of the tended insects on the plant range from inconsequential, if densities of the tended insects are low, to highly detrimental if densities are high. The ants' influence upon the tended species may range from positive when tending places the insects on the most nutritious or benign parts of the plant, or when ants drive away

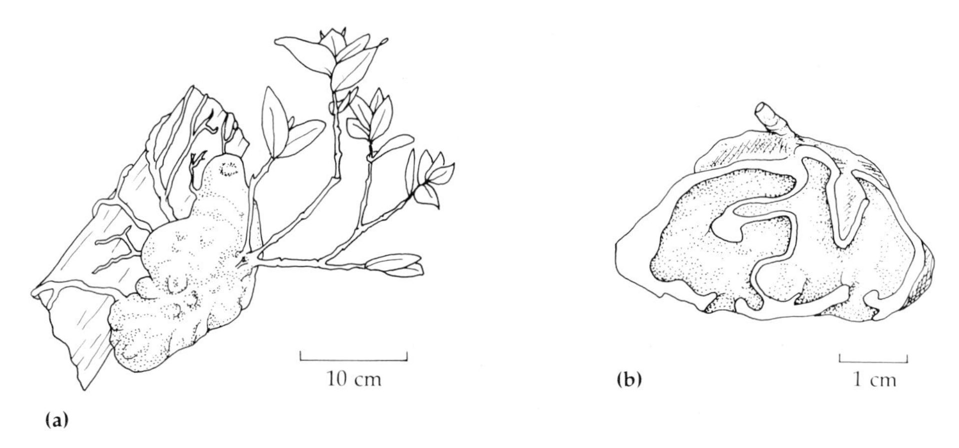

Fig. 6.18. Ant cavities in the epiphyte *Hydnophytum* (Rubiaceae); (**a**) young plant, (**b**) tuber cut open to show cavities. (After Huxley 1980.)

enemies, to negative when the tended insects are harvested by the ants.

The phytophages tended by ants are usually either Homoptera (Way 1963) or caterpillars of the family Lycaenidae (Atsatt 1981a). Some caterpillars of the Riodinidae are also tended (Ross 1966). Ants tend Homoptera and Lepidoptera much as man tends domestic stock animals, guiding them to vigorously growing and nutritious plant tissues, protecting them from predators and parasites, encouraging their reproduction, and harvesting honeydew (an excreted, or in some cases secreted, sugary liquid) or, at times, the whole insect for protein. With lycaenids, the food web can have an additional link, in the form of a parasitic mistletoe plant upon which the ants and caterpillars live. This elaboration is actually quite common in dry parts of Australia (Atsatt 1981b). Lycaenids are also remarkable in having glands which produce substances that mollify and attract ants (Fig. 6.17). Some species of lycaenids may have evolved from mutualists into behavioural parasites of ants, by ceasing to give honeydew while retaining the ants' protection with mollifying and attracting chemicals (Pierce & Mead 1981; Fig. 6.17). Others, like the large blue, *Maculinea arion* (Section 3.2.3), actually spend part of their lives underground in ant nests, as predators of ant larvae.

For most ant-tended species the association is more casual, and of variable benefit. Survival of the membracid *Vanduzea arquata* was enhanced when it was tended by *Formica subsericea* (Fritz 1982). However, tending did not affect the rate of colony growth in *Aphis solicariae* on rosebay willowherb. Two other aphids on the same plant (*A. varians* and *A. helianthi*) benefited from tending when their colonies were small, but large, dense colonies were harvested by the ants for protein (Addicott 1979). Occasionally ant-tending has dramatic beneficial effects on the tended species; the aphid *Periphyllus testudinaceus* was up to 800 per cent more abundant in late spring when tended by *Formica rufa* than when not tended (Skinner & Whittaker 1981).

Tending Ants Attack Other Species

Tending by ants begins to have important consequences for the structure of communities when tending ants attack other, non-tended, species of phytophagous insects (Fig. 6.16b). Because of predation by ants, non-tended species may be much rarer than they would otherwise be, or excluded altogether from certain communities. Ants tending membracids on goldenrods attacked two species of chrysomelid beetles in the genus *Trirhabda*, and prevented the beetles from defoliating their food plants (Messina 1981). Skinner and Whittaker (1981) provide similar examples.

Ant–Plant Mutualisms

Some plants house ants and others feed them, often with important consequences for phytophagous insects (Fig. 6.16c). On plants that house and feed them, ants usually attack phytophagous insects. But they may also tend a small number of species (combining elements of both (a) and (b) in Fig. 6.16), on ant-trees (Thompson 1982) and on plants with extrafloral nectaries. (Ant-trees and extrafloral nectaries are described in detail below.)

The functional diversity of mutualisms between ants and plants is amazing. Even ant-fed plants exist, and have evolved independently at least six times. Usually epiphytes, these plants are found among Bromeliadaceae, Asclepiadaceae, Nepenthaceae, Piperaceae, Polypodiaceae and Rubiaceae. They attract and house ants (Fig. 6.18), which in turn bring organic matter to fertilize their living domiciles (Janzen 1974; Huxley 1980; Thompson 1981). In such cases, the ants gain a place to live, but little or no food from their hosts. Reciprocal effects on

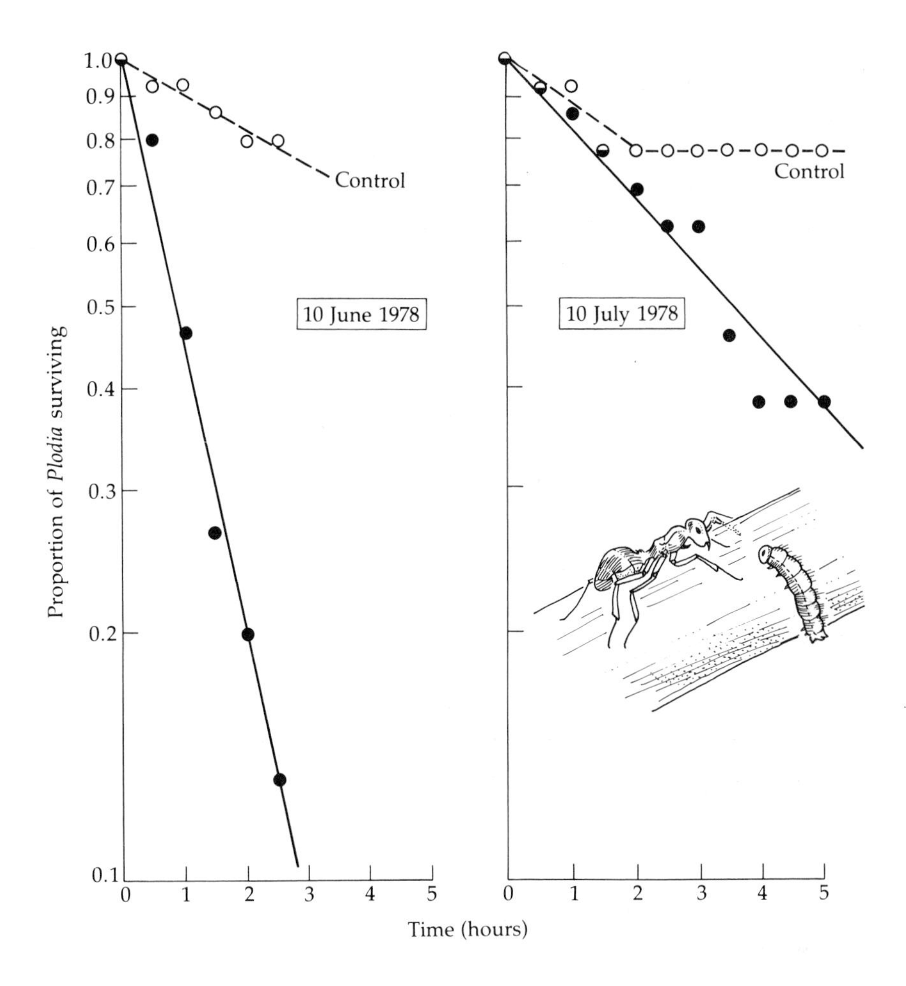

Fig. 6.19. Rate of removal of caterpillars of *Plodia interpunctella* from bracken by ants at Skipwith Common, Yorkshire (J.H. Lawton, unpublished data). A small number of *Plodia* caterpillars disappear from control fronds from which ants are excluded by tree-banding grease; ants rapidly attack and remove *Plodia* from fronds to which they have access. The ants are attracted to bracken fronds by extrafloral nectaries. *Plodia* normally feeds on wheat-bran and is not herbivorous. Caterpillars were experimentally introduced on to bracken fronds to show the effect of ants on a non-adapted species. There is no question of *Plodia* ever naturally colonizing a bracken plant, but these experiments illustrate the potential vulnerability of external, free-living caterpillars to predation by ants.

phytophagous insects exploiting the epiphytes do not appear to have been studied in detail, although Janzen (1974) suggests that they are negligible.

In contrast to these epiphytes, trees in the genera *Acacia, Barteria* and *Cecropia,* collectively known as ant-trees, produce food-bodies on their foliage which attract ants to visit, and the ants have an important role in protecting the plants. Food for the ants may be in the form of protein in sacs or other bodies that grow from stems or leaves (e.g. Risch & Rickson 1981), or of carbohydrates from extrafloral nectaries. In *Barteria,* however, the ants gain most of their food from scale insects and fungi within hollow stems. In all ant-trees, the tree also provides shelter for the ants in specially modified hollow stems and thorns (Janzen 1966, 1967, 1968b; Hocking 1970; Huxley 1980). In return for food and shelter, the ants provide protection against herbivores. *Pachysima* ants protect *Barteria* trees from leaf-eating monkeys (McKey 1974) and even elephants (Janzen 1972)! Obviously the effects of these tight ant–plant mutualisms on the insect communities of ant-trees must be profound. Phytophagous insects are often attacked and removed in large numbers.

Less specific associations between ants and plants often depend upon nectar from floral or extrafloral organs, attractive to ants. Such associations again frequently result in ants benefiting the plants by attacking and removing phytophagous insects, even though the ants are less specific than those on ant-trees and have their nests in other places (e.g. Bentley 1977; Tilman 1978; Inouye & Taylor 1979; Keeler 1979; Koptur 1979; Pickett & Clark 1979; Schemske 1980, 1982; Koptur *et al.* 1982). However, exclusion of ants from the extrafloral nectaries of bracken (*Pteridium aquilinum*) had no measurable effect on herbivorous insects; the resident insects of bracken seem well adapted to avoiding ants (Lawton 1983c and unpublished). Ants do remove experimentally introduced caterpillars of species not normally found on bracken (Fig. 6.19). This finding raises the possibility that ants involved in mutualistic interactions with plants have two sorts of effects. The first is a direct effect on herbivore abundance by depressing population sizes of non-attended species, and would be easily demonstrated. The second is a limit to the first effect and would be less easily detected; certain insect species are excluded entirely from otherwise suitable hosts. That is, species are excluded because of a shortage of 'enemy-free space' (see Section 4.1.4), with ants acting as the important enemies.

6.4.2 *Three Trophic Levels*

This final section brings together the effects of natural enemies and the effects of the host plant on the population dynamics of phytophagous insects. Although it is unusual to treat host plant effects and natural enemies together, the evidence suggests to us that the intersection of these two forces is often of particular importance, and would repay further study (Lawton & McNeill 1979; Price *et al.* 1980).

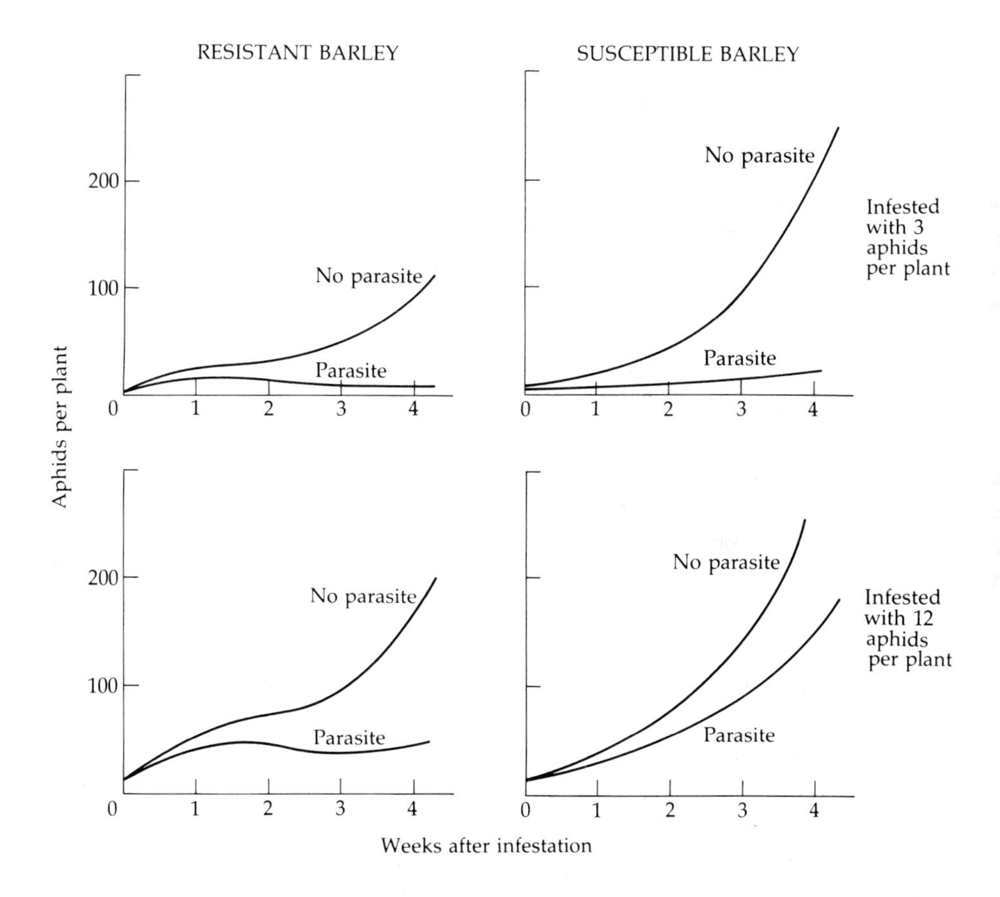

Fig. 6.20. Increase in the aphid *Schizaphis graminum* on resistant and susceptible varieties of barley, in the presence and absence of the parasitoid *Lysiphlebus testaceipes* (Starks *et al.* 1972).

Host Plant Enhancement of Natural Enemy Impact

One approach to integrating host plant influences with those of natural enemies is via deterministic mathematical models of host parasitoid interactions, which show that effective control by natural enemies becomes less likely when the instantaneous rate of increase, *r*, of the phytophage population is greater (Southwood & Comins 1976; Beddington *et al.* 1978; Hassell 1978; Lawton & McNeill 1979). Many of the host plant characteristics discussed in Sections 6.2.2 and 6.3.3 (see also Sections 2.1.2 and 2.1.4) influence the net population growth rate of phytophages, via their effects on development times, number of generations per year, fecundity, and survivorship (e.g. Fig. 6.15). In consequence, a natural enemy can have a greater depressive effect when the plant makes things worse for the insect (Fig. 6.20). Indeed, it is possible that many 'resistant varieties' (Section 6.3.3) would not be

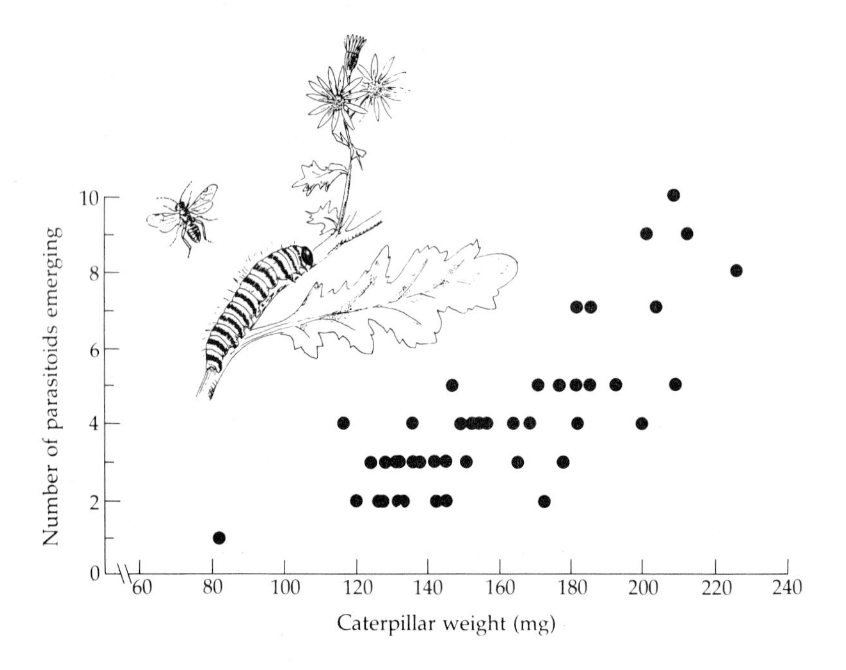

Fig. 6.21. The relation between larval weight of host *Tyria jacobaeae* and the number of larvae of *Apanteles popularis* emerging from that host (after van der Meijden 1980). Small caterpillars give rise to fewer adult parasitoids.

resistant at all if it were not for the presence of natural enemies. For example, one so-called resistant variety of soybean actually suffered *heavier* damage from the Mexican bean beetle *Epilachna varivestis* than did a susceptible variety in the absence of the natural enemies of *Epilachna*. This was because each beetle had to eat more of the resistant variety to complete its development successfully. However, because beetle larvae grew more slowly on the resistant variety, in the presence of a predatory pentatomid bug, *Podisus maculiventris*, they were exposed to predation for longer and suffered much heavier mortality. In the presence of *Podisus*, damage levels on the resistant variety were only 30 per cent of damage levels on the susceptible variety (Price *et al.* 1980; Thompson 1982).

Although this is an artificial system, there are no reasons for thinking that similar host plant influences on rates of individual and population growth of phytophagous insects do not occur in nature (see, for example, Figs 2.4 and 2.8), making many species more vulnerable to natural enemies. Hence, characteristic levels of abundance of phytophagous insects in many communities may be determined by host plant influences from below, and natural enemy effects from above; that is, by interactions involving all three trophic levels (Stiling *et al.* 1982; Vince *et al.* 1981).

Host Plant Impairment of Natural Enemies

Sometimes, host plants impair rather than enhance the impact of natural enemies. For example, the performance of some parasitoids is badly affected by allelochemicals sequestered by phytophagous insects from their host plants (Flanders 1942; Doutt 1959; Smith 1978; Campbell & Duffy 1979; Greenblatt & Barbosa 1981). Invertebrate predators may similarly be adversely affected by aphid prey that are apparently distasteful or toxic, probably because they again sequester allelochemicals from their hosts (e.g. Hodek 1966; Blackman 1967).

The influence of allelochemicals upon vertebrate predators is strikingly emphasized by aposematic (warningly coloured) insects (e.g. Brower & Brower 1968; Rothschild 1973; Tinbergen 1974; Brower & Glazier 1975). Aposematic insects render themselves toxic to visually hunting avian and mammalian predators by sequestering and storing compounds like alkaloids from their food plants. Other examples of

Chapter 7
Coevolution

7.1 Introduction

In Chapter 2 we started our study of insects on plants by looking at problems confronting phytophages, problems of host plant defences, inadequate nutrition, desiccation and attachment. We also explored the general evolutionary history of phytophagous insects. In this closing chapter we return full circle to the opening theme of evolution, focusing particularly on coevolution.

Coevolution is *reciprocal* evolutionary change in interacting species. Species A evolves in response to selection imposed by species B; species B then evolves in response to the change in A. Coevolution could in theory happen in several sorts of pairwise interactions, hosts and parasites, predators and prey, sets of competitors and so on (Thompson 1982). Higher plants might coevolve with pollinators, mycorrhizal fungi, pathogens or phytophages of all kinds. Here we are concerned only with coevolution between phytophagous insects and higher plants.

Coevolution is a familiar theme in studies of insects on plants. Indeed, it has generally been accepted that the great diversity of allelochemicals, or 'secondary' plant substances (Chapter 2), evolved in response to attack by herbivores, particularly insects, and that many insects evolved in response to changes in their host plants. More recently it has become apparent that such close reciprocal evolution is only likely under fairly restricted conditions (Thompson 1982), and a number of workers have increasingly questioned the role of coevolution as a general or even a common mechanism structuring phytophagous insect communities (Janzen 1980; Fox 1981; Futuyma 1983).

Here we explore the evidence for and against the role of coevolution in structuring communities of insects on plants.

7.1.1 Ehrlich and Raven's Model

Most of the interest in coevolution can be traced to Ehrlich and Raven's

Table 7.1. Model of plant–phytophage coevolution, elaborated from Ehrlich and Raven (1964) and Berenbaum (1983).

1 Many plant taxa manufacture a prototypical phytochemical that is mildly noxious to phytophages and that may have an autecological or physiological function in the plant (Harbourne 1982; Seigler & Price 1976; Robinson 1974; see also Section 2.1.4).
2 Some insect taxa feed upon plants with only this and other, similarly mild, phytochemicals, thus reducing plant fitness.
3 Plant mutation and recombination cause novel, more noxious phytochemicals to appear in the plants. The same chemical can appear independently in distantly related plant groups.
4 Insect feeding is reduced because of toxic or repellent properties of the novel phytochemical; thus plants with increasingly noxious chemicals are selected for by the pressure of insect herbivory.
5 The plant, 'protected from the attacks of phytophagous animals, would in a sense have entered a new adaptive zone. Evolutionary radiation of the plants might follow, . . .' (Ehrlich & Raven 1964, p. 602).
6 Insects evolve tolerance of, or even attraction to and utilization of, the novel compound and the plant producing it. An insect can specialize in feeding upon plants with the novel compound; 'here it would be free to diversify largely in the absence of competition from other phytophagous animals' (Ehrlich & Raven 1964, p. 602).
7 The cycle may be repeated, resulting in more phytochemicals and further specialization of insects.

(1964) model, summarized in Table 7.1. This model is obviously plausible; but is it true?

Unfortunately, precisely because it is so plausible, many workers have been willing to accept it rather uncritically. For example, close associations between particular taxonomic groups of plants and certain insects, mediated by plant chemistry, are well known and are sometimes taken as 'proof' of coevolution. One example will suffice.

Glucosinolates (mustard-oil glyconates) in Cruciferae play a major role in the interactions between insects and this group of plants (Feeny 1976; Etten & Tookey 1979; Chew 1979). Glucosinolates are toxic to many herbivores, but provide feeding stimulants for, and are used in host plant location by, a number of crucifer-adapted species (e.g. *Pieris* butterflies, and cabbage aphids *Brevicoryne brassicae*) and by at least one of their parasitoids (*Diaeretilla rapae*) (Van Emden 1978). Such adaptations by herbivores are anticipated by Ehrlich and Raven's model (step 6, Table 7.1); but they are also expected under simple, standard, evolutionary arguments, for example the evolution of resistance to an insecticide, and the acquisition of new herbivores by an introduced plant, discussed in Chapters 3 and 4. The term 'coevolution' should be restricted to situations where it can be shown that there is a *reciprocal* evolutionary effect of adapted insects upon their host plants.

What evidence is there for coevolution? Two examples appear particularly compelling. But as we shall see, this does not mean that they are necessarily typical of other systems of phytophagous insect and host plant.

7.1.2 *Two Coevolved Systems*

Coumarins in Plants, Swallowtails and Other Insects

Simple coumarins are widespread in the plant kingdom (Berenbaum 1978, 1980, 1981a, b, c, 1983; Berenbaum & Feeny 1981). The first elaboration of this basic chemical prototype is the hydroxycoumarin, which occurs in approximately 30 plant families. Next in complexity are the linear furanocoumarins, which occur in eight plant families, but in the largest fraction of genera and species in the Umbelliferae and Rutaceae. Finally, there are the angular furanocoumarins, which occur only in two genera of the Leguminaceae and 11 genera of the Umbelliferae. Consistent with step 5 in the scenario of Table 7.1, the diversity of Umbelliferae is correlated with the stages in this sequence (Fig. 7.1). Genera of Umbelliferae with angular furanocoumarins are much more diverse than those with only linear furanocoumarins, which, in turn, are more diverse than genera without furanocoumarins of either kind. Fascinatingly, and consistent with step 6 in the scenario, the two groups of insects adapted to eating Umbelliferae are most diverse in association with food plants that contain angular and linear furanocoumarins (Fig. 7.2). These two insect groups are the *Papilio* butterflies and oecophorid moths in the tribe Depressarini. Many more *Papilio* species feed upon hosts with angular furanocoumarins than upon host species with either linear furanocoumarins or hydroxycoumarins. More species of Depressarini moths associate with plants with either linear or angular furanocoumarins than with host species with just hydroxycoumarins or

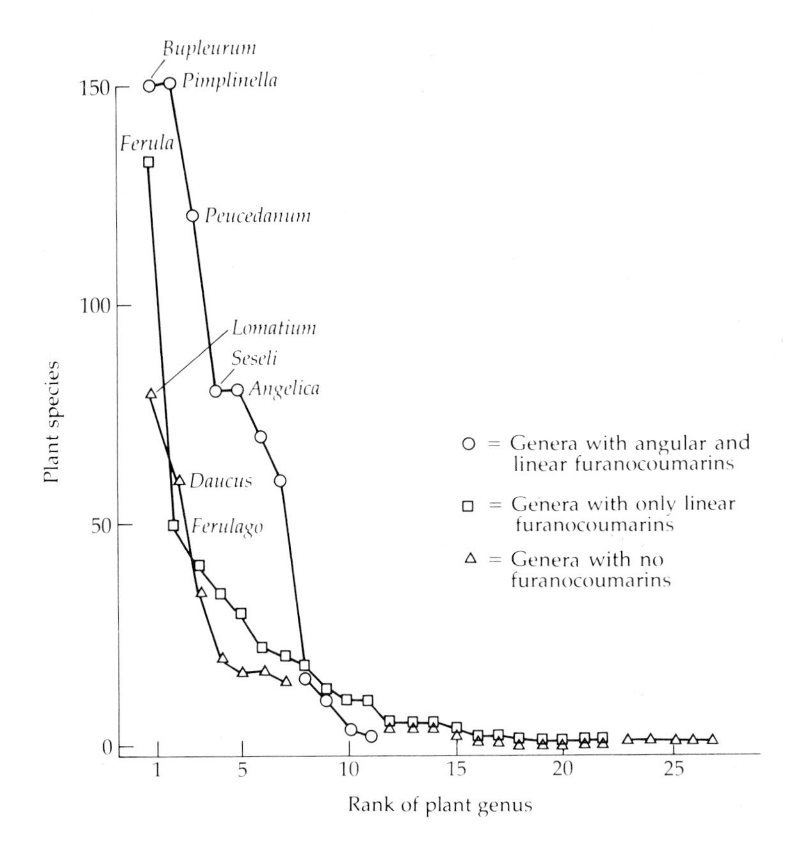

Fig. 7.1. Number of plant species per genus in genera of Umbelliferae with different chemistries. In general, there are more species per genus in those genera whose species contain angular and linear furanocoumarins. (After Berenbaum 1983.)

no coumarins. Courmarin-containing plants are quite consistent with steps 1 and 3 of the Ehrlich and Raven model. Widely occurring precursors gave rise to more specialized molecules, and this process occurred independently in unrelated families. Evolutionary success in terms of number of subtaxa is correlated with the presence of the more derived coumarins, linear and angular. Coumarin-containing plants have certainly radiated in correlation with their novel compounds.

The insects' roles, steps 2, 4 and 6 of the scenario, are corroborated in some aspects by the coumarin story. Plants with only simple coumarins

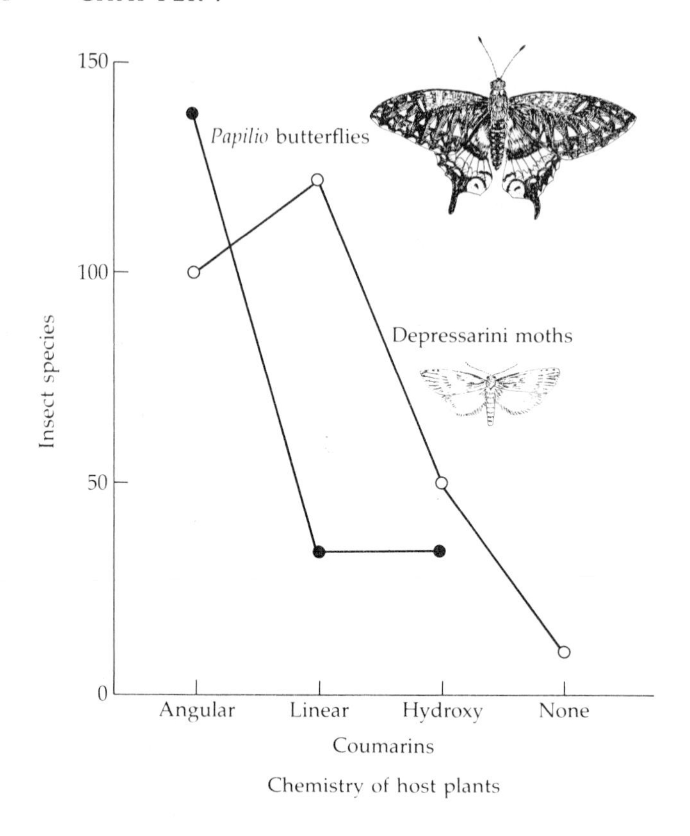

Fig. 7.2. Number of species of Lepidoptera in two groups associated with Umbelliferae differing in their secondary chemistry (after Berenbaum 1983).

and hydroxycoumarins are attacked by a wide variety of polyphagous insects (Berenbaum 1981b). Apparently, these compounds are not particularly toxic to insects, or, at least, many insect groups have species with resistance or tolerance for the compounds. Umbelliferae with linear and angular furanocoumarins have more specialized monophagous or oligophagous faunas (Table 7.2). Caterpillars of some noctuid Lepidoptera, beetles of the genus *Apion*, leaf-mining flies of the genus *Phytomyza*, caterpillars of the moth family Oecophoridae, and caterpillars of butterflies in the genus *Papilio* have all specialized upon these plants. Some adaptations of insects to these chemicals are remarkable. Oecophorid caterpillars protect themselves against the phototoxic properties of linear furanocoumarins by shading themselves

Table 7.2. Degree of host-specialization in insects collected from Umbelliferae differing in chemical composition (from Berenbaum 1981b).

Host plant chemistry	Proportion of		
	Extreme specialists, feeding on only 1–3 genera	Intermediate species, feeding on 4–20 genera	Polyphages, feeding on more than three families
With angular and linear furanocoumarins	0.43	0.285	0.285
With linear furanocoumarins only	0.30	0.30	0.40
Without furanocoumarins	0.00	0.36	0.64

with a rolled-up leaf while feeding. In direct sunlight, linear furanocoumarins cross-link and inactivate DNA, and most plant species containing these compounds grow in direct sunlight rather than shade (Berenbaum 1981b).

In brief, what we know of the insects that feed on plants containing coumarins is consistent with much of Ehrlich and Raven's model, and is difficult to explain without invoking coevolution.

Heliconius Butterflies and Two Types of Vine

In the tropical rain forests and secondary woodlands of Central and South America, some of the most conspicuous insects are the brightly coloured *Heliconius* butterflies, that glide through the air, weaving their way between the lianas and tree trunks (Bates 1862; Gilbert 1971, 1975, 1982, 1983a and b; Mallet & Jackson 1980; Brown 1981). Their larval stages feed on various 'passion-vines' (*Passiflora*): these are scattered throughout the forest and put out new shoots irregularly. The larvae of many species are confined to feeding on the young foliage (though some gregarious species can live on older foliage). In order to lay her full complement of eggs a female butterfly must normally live a long time (several months) and search the forest. Selection for this life-style has led to butterflies with an energy-efficient flight (gliding), an ability almost to hover in front of plants (for 'inspection'), large eyes, and an ability to navigate round a 'trap-line' or regular track in the forest. To survive to do this the butterflies must avoid natural enemies; they are moderately distasteful with warning coloration and are often the major species in a Mullerian mimicry ring (Bates 1862; Moulton 1909; Benson 1972; Gilbert 1983a). Both males and females also require nitrogenous food as adults (Gilbert 1972; Dunlop-Pianka *et al.* 1977; Boggs *et al.* 1981). They obtain this from the third component in this coevolutionary web — curcubitaceous vines, mainly certain members of the genus *Anguria* (= *Psiguria*). Some species of these vines rely on the butterflies for pollination. It is suggested that the regular and excessive supply of male flowers compared with the small number of female flowers is an evolutionary response to the need to ensure regular visits from *Heliconius* as they follow their flight tracks in the forest: the occasional flower here and there might never be discovered. A male inflorescence of these vines may produce around 100 male flowers in succession; as each flower lasts for 1–2 days a particular inflorescence may remain attractive for several months and the butterflies will visit it regularly (Gilbert 1975).

If a female butterfly lays an egg on the young growth of a passion-vine and the resultant larva grows to maturity, it will destroy the shoot.

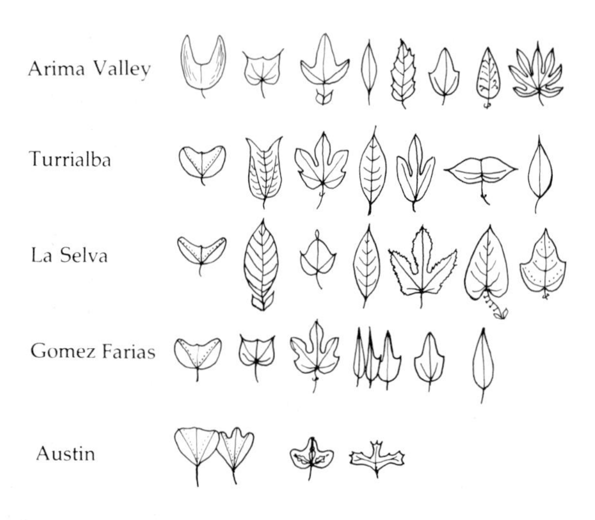

Fig. 7.3. Leaf variation among sympatric species of *Passiflora*. The localities are, from top to bottom: Trinidad, Costa Rica, Costa Rica, Mexico and Texas. (After Gilbert 1975.)

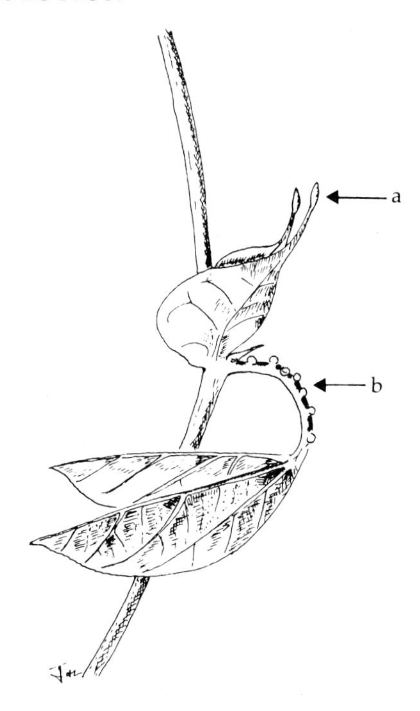

Fig. 7.4. (**a**) Mimetic *Heliconius* eggs on stipule of *Passiflora cyanea*. Several days earlier, the new uncurling growth point was sheltered in the stipule so that the two fake eggs were presented in the area where *Heliconius* usually lays its eggs. Note also the extrafloral nectaries (**b**), the large glands on the petiole. (From a photograph in Gilbert & Raven 1975.)

The patchy and irregular production of young growth in passion-vines may be an evolved response to this herbivory. There are extrafloral nectaries on the shoots; these attract ants (see Chapter 6) and adult egg parasitoids, thereby decreasing the chances of any egg that is laid developing to a full-grown larva. But there are other features of passion-vines that suggest further step-wise coevolved responses.

The leaf form of *Passiflora* species varies greatly (Fig. 7.3) and Gilbert (1975, 1982) has suggested that this prevents females recognizing a range of vines as potential host plants. Support for this hypothesis comes from studies on the swallowtail butterfly *Battus philenor* and its host plants (Rausher 1978; and Rausher, in Gilbert 1982). The most striking coevolved adaptation is the mimic-egg (Fig. 7.4). These structures resemble *Heliconius* eggs, but are part of the vine and have evolved from four different botanical structures (stipules, accessory buds, leaf nectaries and petioles) (Gilbert 1975, 1982); such analogous structures with different morphological origins are evidence of strong selective pressure for their development. *Heliconius* females are reluctant to lay on shoots already with an egg, real or mimetic (Williams & Gilbert 1981). The evolutionary origin of this reluctance is probably associated with the cannabilistic habit of the larvae, which itself may be related to the limited food supply. Shapiro (1981) describes a similar situation in relation to pierid butterflies.

There are other actors in the coevolutionary play (Gilbert 1975, 1977, 1979, 1980), but the taxa discussed above are the major performers. Such reciprocal efforts are the very essence of coevolution.

We must now look for similar evidence in other insect–plant systems.

7.2 Components of Coevolution

7.2.1 *Insect Adaptations to Noxious Phytochemicals and Physical Defences*

Mixed function oxidase (MFO) enzymes provide a major means of insect tolerance to plants containing noxious phytochemicals (step 6 of Table 7.1; Brattsten 1979). All animals have these enzymes and employ them to metabolize foreign, potentially toxic compounds. Vertebrate MFO activity is centred in the liver. Insect MFOs are in cells of the gut, fat bodies and Malphigian tubules. The basic metabolic mechanism of these enzymes is to convert lipophilic foreign chemicals into polar, water-soluble molecules that the animal can dispose of by means of its water-based excretory system. MFOs achieve this by oxidation, reduction, hydrolysis and group transfer and, in some cases, by conjugating the foreign compound with sugars, amino acids, sulphate or other hydrophilic molecules. Although MFOs in insects have been studied as mechanisms of pesticide resistance, phytochemicals occurring naturally in the plant diets of insects are obviously the normal substrates for these enzymes. Current research is turning up a long list of related conjugating enzymes, esterases, reductases, epoxide hydrases, hydrolytic enzymes and group transfer enzymes, which aid and complement the MFOs in the detoxification of noxious phytochemicals by insects.

Table 7.3. Activity of mixed function oxidases in 35 species of caterpillars in relation to feeding habit (from Krieger *et al.* 1971).

Diet specialization (Plant families in diet)		Insect species tested	Aldrin epoxidase activity (pmol min^{-1} mg^{-1} protein)
Monophagous	(1)	8	20.4 ± 9.1
Oligophagous	(2–10)	15	90.0 ± 33.6
Polyphagous	(11 or more)	12	297.4 ± 65.9

MFO activity is inducible within the period that an insect uses a plant containing a noxious substance. For example, MFO levels in the gut of caterpillars of the southern armyworm (*Spodoptera eridania*) depend upon the chemical composition of food plants eaten during the previous 24 hours. The preferred plant for this insect in laboratory tests is lima bean, which does not induce as much MFO activity as plant species that contain phytochemicals less commonly encountered by the insect, such as tomato, basil, coriander, parsley and carrot (Brattsten *et al.* 1977).

Not only do the preferred food plants of polyphagous species induce less MFO activity than less familiar food plants, in general, higher levels of MFO activity may be induced in polyphagous species of phytophagous insects than in oligophagous or monophagous species (Krieger *et al.* 1971; Table 7.2). ('Monophagous', 'oligophagous' and 'polyphagous' here refer to the number of *families* of host plants utilized; cf. Section 6.3.1.) Since synthesis of MFO enzymes is presumably not metabolically cost-free, these data suggest that one ecological advantage of food plant specialization may be lower investment in detoxifying enzymes, leading to greater efficiencies of growth. Paradoxically, data on growth efficiencies show that broadly polyphagous species are not less efficient converters of food into insect biomass than more specialized insect species (Fox & Morrow 1981). These findings throw doubt on the supposed major selective advantage of specialization and should be a stimulus to further work.

MFO activity aside, the behavioural and biochemical defences evolved by phytophages against plant toxins are legion (Rosenthal & Janzen 1979). They include such specialized behaviours as cutting a circular trench in the leaf before feeding, thus preventing the mobilization of feeding deterrents (Carroll & Hoffman 1980), and much more widespread but nonetheless specialized means of exploiting plants such as phloem-feeding and leaf-mining, both methods that may avoid toxins in the palisade cells of the leaf (Feeny 1970).

Insects have also evolved numerous ways of avoiding physical defences in the form of leaf spines, hairs and trichomes (Gilbert 1971; Levin 1973; Johnson 1975; Pillemer & Tingey 1976). An example of this is provided by the caterpillars of an ithomiid butterfly, *Mechanitis isthmia*, which have evolved the novel response to trichomes of spinning a silk scaffolding upon which they can crawl to the spineless edges of leaves, where they feed unhindered (Rathcke & Poole 1974).

Put simply, the evidence for adaptations by insects against plant defences is overwhelming, rich in detail, and there for all to see. However, on their own, such adaptations do not constitute much in the way of evidence for coevolution, although they do clearly illustrate evolution.

7.2.2 *Effects of Phytophages on Plants: Problems for the Coevolution Theory*

Once an insect has adapted to a particular defence mechanism or range of defences, we move to, or return to, stage 2 of Table 7.1. Can we then assume that phytophages generally limit, restrain, depress and reduce the fitness of plants so severely or in such a manner that selection pressure for novel noxious phytochemistry is usually high, and coevolution inevitable? It is this aspect of coevolution that has received least attention and which has least support.

A range of plant genotypes varying in susceptibility to insect attack within one plant species presumably provides the raw material for selection to operate upon, and we have already seen examples of such variability within populations of wild plants in Section 6.3.3. Moreover, there is no doubt that some insect populations have a profound, direct effect upon the abundance of their host plants, revealed particularly clearly in cases of successful biological control of weeds by phytophagous insects (Clausen 1978; De Bach 1964, 1974; Table 7.4). Even marginally more resistant genotypes will be strongly favoured under these circumstances. Indeed it is noteworthy that biological control of plants has been significantly more successful against asexually

Table 7.4. Estimates of the level of depression (q) of plant populations by insect herbivores in biological control programmes or in insect-removal experiments. q is defined as the ratio of the average abundance of the plant in the presence of the herbivore (V^*) to its abundance in the absence of control (K). Most of the estimates are crude, and illustrate only that the effects of a single species of insect on host abundance can sometimes be dramatic. (After Caughley & Lawton 1981.)

Plant species	Main insect herbivore	Method of attack	$\frac{V^*}{K} = q$	Authority
Optunia inermis *O. stricta*	*Cactoblastis cactorum*	Larvae mine the pads	0.002	Caughley & Lawton (1981)
Hypericum perforatum	*Chrysolina quadrigemina* (and also *C. hyperici*)	Larvae attack winter basal growth, with complete loss of flowers	<0.01 0.005	DeBach & Schlinger (1964) Harper (1969)
Senecio jacobaea	*Tyria jacobaeae*	Caterpillars eat leaves	0.003	Harris *et al.* (1978)
Carduus nutans	*Rhinocyllus conicus*	Larvae feed on thistle heads	0.05	Kok & Surles (1975)
Chondrilla juncea	*Cystiphora schmidti* in conjunction with a rust *Puccinia*	Gall midge	0.36 (Wagga) to 0.03 (Tamworth)	Cullen (1978)
Melampyrum lineare	*Atlanticus testaceous*	Eats seedlings, and later foliage of mature plants	0.25	Cantlon (1969)
Machaeranthera canescens	*Hesperotettix viridis*	Eats foliage	0 (excluded completely from desert habitats)	Parker & Root (1981)
Salvinia molesta	*Cyrtobagous singularis*	Eats foliage	$<10^{-5}$	Room *et al.* (1981)

reproducing species than against sexually reproducing, and hence presumably genetically more variable, species (Burdon & Marshall 1981).

A number of other studies reveal major deleterious effects on the components of host plant fitness, for example seed production, by single species of phytophagous insects, without going so far as to demonstrate significant effects on host abundance. Examples, embracing a wide range of insect and plant types, are provided by: Dixon (1971a, b), Thompson (1978), Boscher (1979), Hartnett & Abrahamson (1979), Rausher & Feeny (1980), Newbery (1980), Myers (1981), and Heithaus *et al.* (1982).

Heavy herbivory of this type is exactly what is required to drive coevolution (steps 2 and 3, Table 7.1).

The problems with such examples are twofold. First, how typical of phytophagous insects in general are they? Secondly, how is potential coevolution between a pair of species (a herbivore and its host plant) modified and constrained by whole suites of herbivores acting in concert? We will examine each of these problems in turn.

Interactive and Non-interactive Systems

Caughley and Lawton (1981) following Monro (1967) divided grazing systems into two types, representing the end points of a continuum; they are *non-interactive* systems, where the herbivores have no measurable impact on the performance of their food plants, and *interactive* systems where they do.

Figure 7.5 summarizes population data for the herbivorous insects on two species of plants, namely bracken (*Pteridium aquilinum*) and the grass *Holcus mollis* (Lawton & McNeill 1979). The insect faunas on the two plants share two features typical of many others. First, a small proportion of species may be sufficiently abundant, for part or most of the time, to inflict serious damage upon their hosts. Aphids (*Holcaphis*) and thrips (*Aptinothrips*) on *Holcus*, and broom moths (*Ceramica pisi*) on bracken are good examples. In the main, it is these species that interact strongly with their host plants. Secondly, the majority of species are rare or very rare relative to the abundance of their food plants and remain so over many generations (e.g. Lawton 1982). Low and innocuous rates of naturally occurring herbivory have been noted by

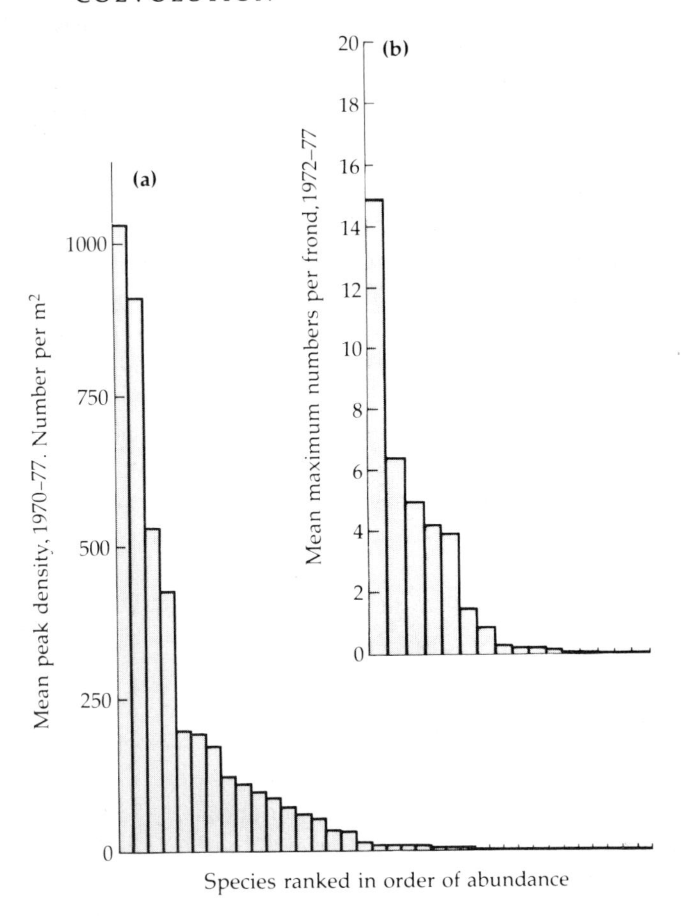

Fig. 7.5. Mean maximum density of phytophagous insects on (**a**) the grass *Holcus mollis* and (**b**) the fern *Pteridium aquilinum* (Lawton & McNeill 1979). Most species are relatively rare most of the time.

other workers for most of the species of herbivores attacking a particular host; examples include Menhinick (1967), Mispagel (1978), Nielsen (1978), Simberloff *et al.* (1978), Tilden (1951) and Zwölfer (1979). We can assume that the majority of these rare species have little or no impact upon the dynamics of their hosts (see Harris 1973) and hence are non-interactive.

The majority of insect species are probably maintained towards the non-interactive end of the spectrum by the vagaries and harshness of the environment, a shortage of shelter sites, or by the impact of natural enemies (Chapter 5; Caughley & Lawton 1981). The importance of natural enemies is revealed particularly clearly by the insects that attack the tropical monocot *Heliconia* and the related, introduced banana (*Musa* spp.) in Central America (Strong 1982b). *Heliconia* and banana have quite similar phytochemistries. Neither plant contains any of the noxious compounds known from other plants. Their main 'secondary' component is wax, which is neither toxic nor repulsive to insects (Gage & Strong 1981).

Herbivory upon *Heliconia* by all species combined averages approximately 1.5 per cent of leaf area in nature. Most of this small amount of damage is caused by the specialised hispine beetles discussed in Section 5.2.2, which are the most abundant insects on *Heliconia*. Under most circumstances, banana suffers little more herbivory than *Heliconia*. Hispine damage is absent, and Lepidoptera are normally so sparse on banana that many leaves mature with no measurable insect damage. The dramatic exception to this picture is in large plantations where bananas are sprayed with insecticides. Wholesale insecticide application to bananas can cause devastating outbreaks of many Lepidoptera species in the plantations, by selective killing of parasitoids. Lepidoptera quickly gain resistance to all insecticides and the only solution to the problem proves to be the cessation of insecticide use. After spraying is stopped, parasitoids return to the plantations, and pest Lepidoptera subside to low population levels. The obvious inference is that wild *Heliconia* is similarly protected by parasitoids from devastating herbivory. The diversity of *Heliconia* and its associated hispines and Lepidoptera is a counter-example to the coevolutionary scenario. In this case, the communities of insects on the plant are profoundly affected by natural enemies and apparently less so by noxious plant chemistry. A similar story of parasitoid protection for other tropical crops is related by Bennett *et al.* (1976).

We do not believe that *Heliconia* and its insects are atypical. Indeed, the data suggest that many, probably the majority, of the species of insects on most host plants are too rare most of the time individually to significantly affect the fitness of their hosts. It is therefore difficult to envisage host plants and most of their herbivores coevolving to any significant extent in the simple manner envisaged by Ehrlich and Raven.

Groups of Phytophages

A second difficulty with Ehrlich and Raven's model centres on the effects of groups of herbivores acting in concert. On some plants, as we have seen, the damage inflicted by individual species of insects is noticeable, sometimes devastating, and is certainly sufficient to select for increasing host resistance. Over and above the effects imposed by single species of herbivores, several recent experiments have shown that if entire suites of herbivores attacking a plant are removed, plant performance markedly improves (e.g. Janzen 1970; Kulman 1971; Gradwell 1974; Morrow & La Marche 1978; Clements *et al.* 1978; Perkins 1978).

The effect of eliminating or markedly reducing herbivores has been well shown by Waloff and Richards' (1977) study on scotch broom (*Sarothamnus scoparius*), a shrub that lives for up to about a decade. Bushes protected by insecticides had higher reproductive rates and, beyond five years of age, survived better than those exposed to the natural level of herbivory (Fig. 7.6). When groups of herbivores act in concert in this way, tight reciprocal coevolution between the host plant and one or two species of phytophagous insects seems particularly unlikely. As soon as plant performance is influenced by several herbivore species, conflicting selection pressures may be generated that restrict or prevent coevolution.

Conflicting selection pressures are clearly revealed by two specialized seed predators of cocklebur (*Xanthium strumarium*). Hare and Futuyma (1978) showed that changes in the structure and chemical composition of the seeds of cocklebur which reduced susceptibility to attack by caterpillars of the tortricid moth *Phaneta imbridana* increased susceptibility to the tephritid fly *Euaresta aequalis*. Although similar examples involving leaf-feeders or stem-miners are lacking, we doubt that Hare and Futuyma's example is unusual or in any sense atypical.

More than One Species of Plant

Finally, the impact of a single species of phytophage may be radically altered by the community in which the plant is growing. Most plants are subject to intense competitive pressure from other plants, both intra- and interspecific, and relatively small amounts of differential defoliation can

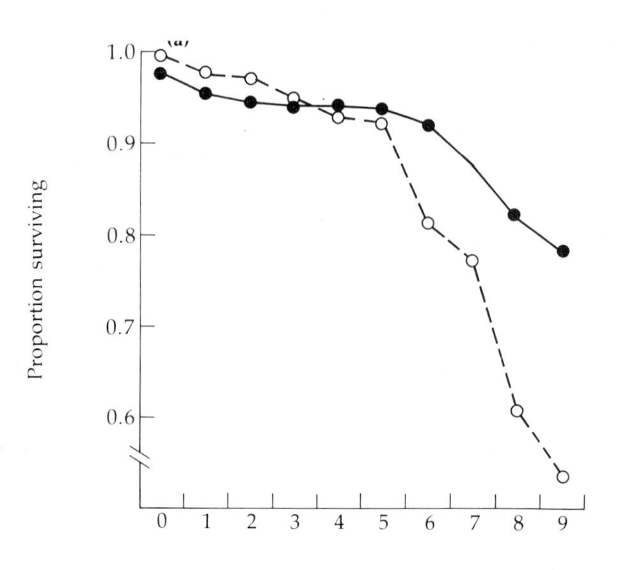

(a)
1.0
0.9
0.8
0.7
0.6
Proportion surviving
0
1
2
3
4
5
6
7
8
9

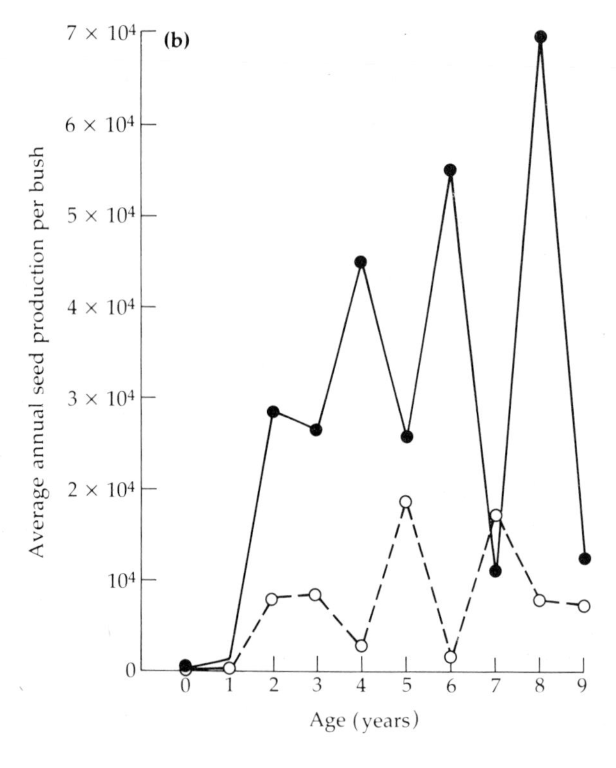

(b)
7 × 10⁴
6 × 10⁴
5 × 10⁴
4 × 10⁴
3 × 10⁴
2 × 10⁴
10⁴
0
Average annual seed production per bush
0
1
2
3
4
5
6
7
8
9
Age (years)

sometimes influence the outcome of this competition (Whittaker 1979). For example, Bentley and Whittaker (1979) suggest that grazing by the chrysomelid beetle *Gastrophysa viridula* limits the distribution of one species of dock *Rumex crispus* particularly when it is growing in the presence of a competitor, *Rumex obtusifolius*. The beetle attacks both dock species, but prefers *R. obtusifolius*. *Rumex crispus* is competitively inferior to *R. obtusifolius*, and grazing by *Gastrophysa* that 'spill over' on to *R. crispus* from nearby *R. obtusifolius* may be sufficient to eliminate *R. crispus* entirely from some habitats. Under these circumstances it is very difficult to know whether grazing by *Gastrophysa* is entirely harmful or partially beneficial to *Rumex obtusifolius*, because it eliminates one of its potential competitors, *R. crispus*.

A Summary of Problems

Drawing these details together, we see that not all insect populations impose significant selection pressures on their hosts; many associations are non-interactive. Some interactive systems involve several species of insects and these may impose conflicting evolutionary demands upon their hosts, whilst the consequences of being grazed may depend upon other plant species growing in the vicinity. All these phenomena pose problems for coevolution.

The notion that coevolution is a widespread and common process in insect–plant interactions is also challenged by a number of other phenomena. First, insects are not the only organisms that attack plants, and it cannot be assumed that plant defences have always evolved to repel insects. Insect resistance to toxic defences evolved by plants for other purposes is an interspecific linkage much looser than that envisaged in the Erlich and Raven model. Secondly, many insects exploit more than one species of host plant (Fig. 6.8). It is unlikely that different

Fig. 7.6. The effect of removing phytophagous insects (by regular spraying with the insecticides dimethoate and malathion) on (**a**) survivorship and (**b**) fecundity of broom bushes *Sarothamnus scoparius* (Waloff & Richards 1977). Sprayed bushes (•——•) apparently survive better and have higher fecundities than unsprayed bushes (○– – –○) (N.B. Only one experimental and one control plot were used in this work; hence differences between treatments may in part reflect differences unrelated to insecticide application. Since the two plots were only 30m apart, such effects are likely to be small.)

species of hosts impose identical selection pressures on oligophagous or polyphagous insects; indeed, the sustained, reciprocal interactions necessary to drive coevolution seem rather unlikely for polyphagous insects. Finally, many plant–insect interactions are variable and unpredictable in time and space. Plants of one species in different parts of a geographic area (Chapter 3) or in different places and habitats (Chapter 6) often support very different insect assemblages, whilst at one place some insects may be common one year and rare or absent another (Chapter 5). It is not at all clear how variation in selection pressures in time and space influence coevolution, but they would seem to make it less likely.

In other words, the known examples of coevolution look increasingly like small vortices in the mainstream of evolution. The sustained, reciprocal and intense interactions that are necessary for coevolution are rare enough that the mainstream is not usually deflected.

7.2.3 Coevolutionary Vortices in an Evolutionary Stream

As this metaphor implies, we believe that although some insect–plant interactions are sucked into intense, escalating whirlpools of reciprocating adaptations, most interactions do not result in special and mutual counteradaptations. Rather, most may be the product of what has been termed 'diffuse coevolution' (Janzen 1980; Fox 1981). Diffuse coevolution implies that plant defences have evolved to cope with attacks from many quarters, not just from one, and that insect adaptations cannot usually be traced to a coevolved arms race with a single species of plant. It may be considered that diffuse coevolution merits a special term only in the historical perspective of enthusiasm for 'focused' coevolution. Diffuse coevolution is evolution in response to a variety of conflicting demands.

In the broadest possible terms, diffuse coevolution may be useful to account for phenomena like the rise of the angiosperms (Section 2.4) and the subsequent diversification of phytophagous insects. Herbivory in general selects for plant defences and among the herbivores are species or groups of species (some of them insects) that significantly depress plant populations, reducing interspecific competition between plants and allowing more plant species to coexist. The result, in the long run, is a rise in insect diversity as insects colonize new species of plants; but many of these insects are unlikely to be closely coevolved with their new hosts (Section 4.1.5).

Coevolution in the form envisaged by Ehrlich and Raven is most likely when relatively few species of host-specific insects exploit short-lived plants scattered in diverse communities. Umbelliferae in 'old fields' are a good example; passion-vines are much longer lived, with much richer faunas, and *a priori* are less obvious candidates for coevolution. Certainly, coevolution as seen by Ehrlich and Raven must be very rare when plants have very long generation times relative to their insect herbivores, and when they sustain very rich and diverse faunas. It is difficult to imagine 'focused' coevolution between oak trees and each of the hundreds of species of phytophagous insects that feed on them (Fox 1981).

Last but not least we must beware of explaining interesting and curious phenomena in communities of phytophagous insects and their hosts as the products of coevolution between contemporary members, without good reason to disregard what has gone on in the past. Much of observed phytochemical diversity, for example, may be 'fossil' adaptation to phytophages long gone, to associations long dissolved. Janzen (1979) terms these adaptations 'anachronisms'. We pick one example from many. Despite considerable effort, it has not proved possible to find any function, metabolic or defensive, for the mimics of insect moulting hormones found in various ferns, including bracken (Kaplanis *et al.* 1967; Lawton 1976; Jones & Firn 1978). Since bracken and other ferns support numerous species of phytophagous insect, it would be tempting to explain the presence of moulting hormone mimics (phytoecdysteroids) as an evolutionary response by ferns to insect herbivores, and to explain the contemporary success of fern-feeding insects as a coevolved response to phytoecdysteroids. Unfortunately, not only are fern-feeding insects totally unaffected by phytoecdysteroids at concentrations normally found in these plants, but so are polyphagous non-fern feeding species, for which coevolution can be entirely ruled out (Jones & Firn 1978). Although it is very difficult to prove, one possible explanation is that phytoecdysteroids are redundant defences; whether they were coevolved, and what organisms they were defences against, is a mystery.

7.3 *Epilogue*

Nothing would have pleased us more than to finish this book with an account of coevolution between insects and their host plants as the major forcc generating patterns in plant and insect diversity, driving the sweep of phytophagous insect evolution (Chapter 2), creating patterns of species richness in contemporary time (Chapter 3), promoting speciation (Chapter 4), and supported by ecological interactions in contemporary time (Chapters 5 and 6). The real ecological world is more complex and less tightly coupled than this. As we noted in Chapter 4, the insect fauna of plants is a *pot pourri* of the coevolved, the pre-adapted and the opportunistic in varied and unpredictable proportions. Coevolution most certainly does not provide a general mechanism to explain the contemporary structure of phytophagous insect communities.

We are encouraged by the patterns that have already emerged from our infant science, and spurred on by considering the processes that now appear to make sense. But we are acutely aware of how little we know, still less understand, about the structure of phytophagous insect communities. Science proceeds by the reciprocal posing and testing of ideas. By summarizing the state of the art, and explaining our ideas, we hope that we will persuade others to take up the study of communities of insects on plants, to identify patterns in their organization, and to test hypotheses about the mechanisms underpinning these patterns.

Appendices

There are two appendices. The first is a thumbnail sketch of the classification and biology of phytophagous insects that are mentioned in our text. The second is a brief consideration of the taxonomic knowledge of some phytophagous insects.

Appendix 1 will be useful to those ecologists who are not entomologists, but it is only a very brief account. Those wishing to know more about the classification and biology of phytophagous insects should consult one of the standard works on entomology; we have followed the classification system used in Imms' classical *General Textbook of Entomology* in its tenth edition, as revised by Richards and Davies (1977).

Appendix 2 deals with the rate of taxonomic discovery of phytophagous insects. Its purpose is to show that these insects are now sufficiently well known taxonomically for a work of this kind not to be seriously biased by taxonomic ignorance.

Appendix 1
A Classification of Phytophagous Insects

Insects are a major class within the Phylum Arthropoda. The 29 living orders of insects are divided into **Apterygota**, comprising four orders of primitive wingless insects, **Exopterygota**, 16 orders of winged insects with incomplete (hemimetabolous) metamorphosis, i.e. lacking a pupal stage, and **Endopterygota** with complete (holometabolous) metamorphosis accompanied by a pupal stage. In general, the juvenile stages of apterygotes and the non-aquatic exopterygotes resemble miniature adults both in form and habits. Immature exopterygotes are often termed 'nymphs', and develop wings continually, through successive instars and moults. Immature endopterygotes are called 'larvae' and are morphologically distinct from adults. Larvae transform to adults and develop wings during a pupal stage.

The nine orders with phytophagous members, listed in Chapter 1, are:

Apterygotes:
1. Collembola: springtails

Exopterygotes:
2. Orthoptera: grasshoppers, locusts
3. Phasmida: stick insects
4. Hemiptera: bugs, cicadas, whiteflies, leaf-hoppers, aphids, etc.
5. Thysanoptera: thrips

Endopterygotes:
6. Coleoptera: beetles
7. Diptera: flies
8. Lepidoptera: butterflies, moths
9. Hymenoptera: sawflies, bees, wasps, ants, etc.

1. ORDER COLLEMBOLA
Springtails

Most Collembola are less than 5 mm long, and feed on small fungi, yeasts and moulds. Those that eat green plants belong to the Family Sminthuridae (Fig. 2.1). *Bourletiella viridescens* (Table 1.1), found on bracken, is a good example. Sminthurids feed externally, chewing the epidermis and surface hairs of plants. They probably also eat microorganisms and detritus on plant surfaces.

2. ORDER ORTHOPTERA

Grasshoppers, locusts, crickets, katydids, etc.

Grasshoppers and locusts are large, active insects, usually with two pairs of wings, though some species have lost the power of flight. All have chewing mouthparts. Not all Orthoptera are phytophagous. Those that are dealt with in this book include representatives of the following taxa.

Superfamily: Tettigonioidea

Family: Tettigoniidae

The tettigonids include bush crickets, long-horned grasshoppers and katydids. Some are entirely or partially predatory, others chew foliage. Typical species are *Orchelimum* (Fig. 1.3) and *Meconema* (Fig. 1.5).

Superfamily: Acridoidea

Family: Acrididae

Acridids are the familiar short-horned grasshoppers and locusts ('short-horned' because they have shorter antennae than tettigonids). With about 9000 species, acridids are the largest group of the Orthoptera. Both nymphs and adults chew plant foliage. A typical grasshopper is illustrated in Fig. 2.1.

Family: Eumastacidae

Eumastids are a small group of relatively primitive, often flightless, Orthoptera. Extensive work has been carried out on speciation in the endemic Australian Subfamily Morabinae (Section 4.2.2).

3. ORDER PHASMIDA

Stick and leaf insects, 'walkingsticks'

Many phasmids are large, and may be up to 20 cm in length. They are tropical, externally chewing folivores, remarkable for their close mimicry of leaves and twigs. A typical stick insect in the genus *Carausius* is shown in Fig. 2.1.

4. ORDER HEMIPTERA

True bugs

In terms of species numbers the Hemiptera are the largest order in the Exopterygota. They are recognized by their highly modified, piercing mouthparts in the form of a needle-like rostrum which is commonly held beneath the head. Most phytophagous bugs exploit their host plants by tapping directly into the vascular system, or by evacuating the contents of cells.

The classification of the Hemiptera is complex, and many families are not phytophagous. Those dealt with in the text can be classified as follows.

4a SUBORDER HOMOPTERA

SERIES: AUCHENORRHYNCHA

All Auchenorrhyncha are phytophagous. The taxon comprises the cicadas and hoppers. Most feed externally on the above-ground parts of their food plants both as nymphs and adults, but some (e.g. cicadas) have subterranean nymphal stages that feed upon roots. They lay their eggs into plant tissues, the females having 'knife-like' ovipositors.

Both nymphal and adult leaf- and plant-hoppers are often capable of rapidly jumping or flying away when disturbed.

English names are confusing, the terms 'leaf-hopper' and 'plant-hopper' being applied rather indiscriminately. Representatives of the following superfamilies, families and subfamilies are mentioned in the text.

Superfamily: Fulgoroidea

Family: Delphacidae

Delphacids are small 'hoppers' found in herbaceous vegetation, particularly on grasses, rushes and related plants. *Prokelisia marginata* (Fig. 1.3) is a typical member. Taxonomically, they are distinguished by a large apical spur on the hind tibia.

Superfamily: Cercopoidea

Family: Cercopidae

Family: Aphrophoridae

Some authorities treat these as two subfamilies of the Cercopidae. The adults are small, jumping Homoptera called frog-hoppers because they are reminiscent of tiny frogs (Fig. 6.7, upper of the pair; the lower is a cicadellid — see below). Frog-hoppers feed externally on their hosts. The nymphs, known as cuckoo-spit insects or spittlebugs, surround themselves with a bubbly froth (Fig. 2.2b). The froth is liquid excreta derived from plant sap, blown into foam by a special valve in the anal cavity, into which a pair of spiracles open. Larval Cercopidae are entirely subterranean, whilst most larval Aphrophoridae live above ground.

Unlike most Homoptera, cercopids and aphrophorids exploit the nutritionally dilute xylem (Fig. 2.6) of their host, rather than the phloem.

Superfamily: Cicadoidea

Family: Cicadidae

Cicadas are large and conspicuous, and the ringing song of the males is a familiar sound on sunny days in many warm parts of the world (e.g. the Mediterranean, Northern India and America). The eggs are laid into the aerial parts of plants, especially twigs; the young nymphs fall to the ground and enter the soil, where they feed from the xylem vessels of small tree roots.

Superfamily: Cicadelloidea

Family: Membracidae

Known as tree-hoppers, ant cows and devil hoppers, these small, jumping Homoptera often have bizarre extensions of the thorax. They live on trees, and are particularly numerous in Central and South America. They are often tended by ants (Section 6.4.1). Members of the *Enchenopa binotota* 'species complex' are the subject of detailed studies on sympatric speciation and host-race formation (Section 4.2.2). A number of species resemble thorns.

Family: Cicadellidae (= Jassidae)

Cicadellid leaf-hoppers or jassids, characterized by spines on their hind tibiae, otherwise resemble small, rather slender frog-hoppers. With the exception of aphids in temperate regions, jassids are probably the most common and abundant of all Homoptera. Most feed on the phloem of their hosts, although members of the important Subfamily Typhlocybinae (Fig. 1.5) evacuate mesophyl cells in leaves, causing small white dots or blotches. *Erythroneura* and *Eupteryx* (Table 5.5b, examples 16 and 17) are typical typhlocybine leaf-hoppers. Figure 5.4 shows a typical grassland, phloem-feeding member of the genus *Elymana*, and Fig. 6.7 (lower of the pair) a member of the North American genus *Comellus*.

SERIES: STERNORRHYNCHA

The sternorrhyncha comprise four distinct superfamilies; all members are phytophagous.

Superfamily: Psylloidea

Family: Psyllidae

Psyllids (jumping plant lice, lerp insects) are small, aphid-sized insects bearing a very superficial resemblance to tiny cicadas. Most species are free-living, but some induce galls on their hosts; nymphs of the many Australian species (e.g. *Cardiaspina* (Fig. 5.6)) are sedentary and secrete a test or 'lerp', often of remarkable complexity, under which they shelter. The family is especially well represented in the Australian region.

Superfamily: Aleyrodoidea

Family: Aleyrodidae

Larvae are oval, flattened, inactive insects that feed externally on their host; often they are blackish with white tufts of wax. Adults are tiny, active winged insects a few millimetres in size, dusted with white powdery wax from which they take their common names of whiteflies or snowflies.

Superfamily: Aphidoidea

Family: Aphididae

Aphids (greenfly, greenbugs) are familiar and important pests, especially in temperate regions. Most feed externally from the phloem of their hosts; this feeding is often referred to as 'sucking', but the fluids usually move through the fine mouthparts (stylet) under positive pressure from inside the phloem. Large volumes of plant vascular fluids can pass through aphids and other Homoptera. The clear fluid discharged from the anus is known as 'honeydew'; it is rich in certain sugars and attracts ants and other insects (Section 6.4.1). Several species are subterranean, living on plant roots, and a few induce galls (e.g. *Pemphigus*, Section 6.3.2). A typical free-living aphid is *Sitobion* (Fig. 5.9a); most aphids may be easily recognized by the pair of small tubes, siphunculae or cornicles, that arise near the end of the abdomen: these secrete a waxy material used in defence against insect predators.

Aphids have fascinating and complex life histories. Many, though by no means all, have alternate summer and winter hosts. The winter host is often a woody bush or tree, the summer host is often a herb, but there are numerous exceptions. This seasonal host alternation* makes aphids very unusual amongst phytophagous insects. Whether or not species

* A few cynipid gall wasps (p. 246) also alternate host plants, but these utilize different species of *Quercus*, not entirely different types of plant as do aphids.

alternate hosts, they usually pass through sexual and asexual reproductive phases at various times of the year, and may have winged (dispersive) phases (as in *Drepanosiphum*, Fig. 2.8c) and unwinged (apterous) phases (as in *Lachnus*, Fig. 1.5). Asexual reproduction allows rapid population increase when host plants are most suitable, for the generations are 'telescoped': embryos form in the mother before her birth, i.e. in the grandmother!

Family: Adelgidae (= Chermidae)

Adelgids are confined to conifers, where they may induce galls. They have complex life cycles and differ from true aphids in lacking cornicles. The related *Phylloscerida*e are serious pests of vines.

Superfamily: Coccoidea

All coccoids (scale insects and mealy bugs) have extremely amorphous apterous females that live more or less immobile on their hosts; they may be scale-like, gall-like or live under a waxy or powdery coat. Male coccids are rarely seen, but as adults resemble small midges with only one pair of wings; immature males resemble females. Like aphids, many coccids secrete honeydew that is attractive to ants.

The following are representative families.

Family: Coccidae

In this group (wax and tortoise scales) the females are flattened, oval and often covered in wax.

Family: Pseudococcidae

The females are covered in a mealy or filamentous waxy secretion from which they derive their common name of mealy bugs.

Family: Diaspididae

Females of these 'armored scales' live under a waxy scale containing the skins of earlier instars. This is the largest family of Coccoidea, and it contains many important pests. The pine-leaf scale *Nuculaspsis* (Section 6.3.3) is an important representative.

4b SUBORDER HETEROPTERA

Technically, the difference between Heteroptera and Homoptera rests on details of wing structure and the position of the rostrum. The forewings of Heteroptera, when present, are clearly divided into two regions (hence their name), a tough, leathery basal area and a membranous apical area. Heteroptera fold their wings flush with the body. In contrast Homoptera have wings of uniform texture that fold to form a 'roof' when the insect is at rest. In Heteroptera, the rostrum arises from the front of the head, in Homoptera it joins the head near the posterior ventral margin.

Unlike Homoptera, which are entirely phytophagous, many Heteroptera are partially or entirely predatory. Those that feed on plants do not 'plug into' the vascular bundles, but after injecting saliva evacuate the contents of several adjacent cells. Many species probably feed on plant tissues that are particularly rich in nitrogen, the growing points of shoots, young flowers or fruits. Thus the feeding of a few Heteroptera may cause far more plant damage than a comparable number of Homoptera. Those that attack flowers, fruits or seeds may be regarded more as plant 'predators' than as plant parasites, like the Homoptera. Some of the more important families of Heteroptera with partially or entirely phytophagous members are as follows. (Other families not mentioned in the text are Tingidae (lace-bugs) and Coreoidea (squash bugs and others.)

Superfamily: Cimicoidea

Family: Miridae (= Capsidae)

The majority of mirids (leaf bugs, plant bugs or capsid bugs) feed on plants, though most also feed on other insects or mites, that is they are partly predatory. A typical mirid is illustrated in Figs 1.5 and 2.1, and the nymph and adults of an entirely grass-feeding species (*Leptoterna*) in Figs. 2.7a and 5.5b. Mirids lay their eggs into plant tissues, with just the egg cap protruding from the plant surface.

Superfamily: Lygaeoidea

Family: Lygaeidae

Lygaeids (seed bugs) are mainly phytophagous, but some are partly predatory. Most of the phytophagous species feed on seeds, either on the plant or after they have been shed. The cotton stainer (*Dysdercus*) belongs to a related family (Pyrrhocoridae) with similar habits.

Superfamily: Pentatomoidea

Family: Acanthosomatidae

Family: Pentatomidae

Acanthosomids are sometimes treated as a subfamily of the pentatomids. Both are rather large, broadly oval or shield-shaped flattened bugs (the Pentatomoidea are commonly called shield bugs), common and widespread on vegetation, especially in the warmer regions of the world. All, except the subfamily Asopinae, are phytophagous and lay their eggs on the surfaces of plants. As their name of stink bugs implies, pentatomids, like most Heteroptera, give off a disagreeable odour when irritated or attacked. *Cyphostethus tristriatus* (Fig. 3.4) feeds on juniper and Cupressus; it is a typical acanthosomid shield bug. Other pentatomoid families are the Cydnidae (burrowing bugs), Scutelleridae (tortoise bugs) and Plataspididae; all are phytophagous.

5. ORDER THYSANOPTERA

Thrips

Thrips are usually small (2–3 mm long) yellow, yellow-brown or black insects; the wings, when present, are very narrow, with greatly reduced venation and long marginal setae (Figs 2.1 and 2.12). Thrips have asymmetrical mouthparts in the form of a conical beak at the base of the head, on the ventral side. Some species are predatory and many fungivorous. The majority of phytophagous species belong to the Family Thripidae and are sap-feeders, scraping and sucking from plant surfaces.

6. ORDER COLEOPTERA

Beetles

Adult beetles are easily recognized by their highly modified hard or leathery forewings (elytra) that meet neatly along the midline of the back. The membranous hindwings, when present, are folded away beneath the elytra until required for flight.

Adult beetles are very different from their larvae, which are often grub-like. Weevil larvae are legless. Adults and larvae may or may not exploit the same kind of food. Many are not phytophagous at any stage of their lives. Phytophages include representatives of the following groups.

6a. SUBORDER ARCHOSTEMATA

Archostemata are primitive, non-phytophagous beetles, included here briefly because they provide clues to the evolutionary origins of the order (Section 2.2.1). Their larvae bore in timber, rotting wood, etc.

6b. SUBORDER POLYPHAGA

This suborder contains the majority of beetle species.

Superfamily: Scarabaeoidea

Family: Scarabaeidae

Most scarab beetles are not phytophagous, but the adults of the Subfamily Melolonthinae, known as June beetles, May beetles and chafers are important defoliators of trees and shrubs. Their subterranean grubs may cause serious damage to plant roots.

Superfamily: Cucujoidea

Family: Languridae

These are elongate, slender beetles (commonly called lizard beetles), with phytophagous larvae that bore in plant stems (e.g. *Languria taedata*, Fig. 1.3)

Family: Coccinellidae

Many species of coccinellids (the familiar ladybirds) have brightly coloured adults, with a distinct oval or nearly spherical shape. The great majority are predatory on aphids or mites, both as adults and as larvae. A few (Subfamily Epilachninae) are phytophagous as adults and larvae; the Mexican bean beetle *Epilachna* (Section 6.4.2) is an example.

Family: Tenebrionidae

Adult and larval tenebrionids (darkling beetles) mainly feed as scavengers on decaying vegetation, fungi, seeds, etc., but a few attack living plants, particularly as adults.

Family: Mordellidae

Mordellids (tumbling flower beetles) are rather small, humpbacked beetles with adults that tumble about in a bizarre fashion when disturbed, making them difficult to pick up. Some species (but not all) have phytophagous, stem-boring larvae (e.g. Fig. 1.3, *Mordellistena splendens*).

Superfamily: Chrysomeloidea

The Chrysomeloidea are a vast assemblage of phytophagous or wood-feeding species. Many of the important phytophagous Coleoptera are in this superfamily.

Family: Cerambicidae

Larval cerambicids mainly bore wood, but a few species are stem- and root-miners. The adults are easily recognized by their long antennae, hence their common name of long-horned beetles.

Family: Chrysomelidae

This family, the leaf beetles, competes very closely with the Curculionidae (below) in the number of described species; over 20 000 are known. All chrysomelid larvae, and many adults, are phytophagous; some larvae feed as external chewers either above or below ground, and others mine leaves. Most adult chrysomelids are fairly small, brightly coloured and often have a metallic lustre. A typical adult chrysomelid (*Chrysolina*) is illustrated in Fig. 2.1, the larva of *Paropsis*, an Australian species, on Fig. 2.7.

The large number of species are organized into several subfamilies. Two specifically dealt with in the text are the Subfamily Halticinae (flea beetles) and the Subfamily Hispinae (hispine beetles). Ecology of flea beetles in the genus *Phyllotreta* is discussed in Sections 5.2.3 and 6.2.4. They are called flea beetles because the adults have swollen hind femora that allow them to jump. Some larval Halticinae are leaf-miners, others are free-living folivores.

The rolled-leaf hispines (Hispinae) are discussed extensively in Sections 5.2.2 and 7.2.2; a typical adult and larva are shown on Fig. 3.13. These hispines are smooth and thus differ from many spiny species. Larvae of rolled-leaf hispines do not mine into plant tissues, as do more characteristic members of this family, but rather 'strip-mine' plant surfaces. Larvae of rolled-leaf hispines are reminiscent of 'waterpennies', the aquatic larvae of Psephenidae (Coleoptera).

Chrysomelinae include the widespread pest *Leptinotarsa declimineata*, the Colorado potato beetle.

Superfamily: Curculionoidea

The Curculionoidea may be the most species-rich group within the Coleoptera, with over 60 000 described species in one family alone, the Curculionidae. The other family we mention, the Apionidae, is small by comparison but still has over 1000 species. Bark-beetles constitute other families within this superfamily. The narrow host plant range of most curculionoids is undoubtedly a major factor in the species richness of the group.

Family: Apionidae

Family: Curculionidae

These two families comprise the weevils. Apionid weevils are sometimes treated as a subfamily of the curculionids. Apionids have mining larvae, inhabiting seeds, stems and plant roots. The adults are free living and differ from Curculionidae in having straight, not elbowed, antennae (Fig. 1.5). They share with Curculionidae the other main hallmarks of weevils, namely clubbed antennae and a head projected into a snout or beak. Snouts of weevils have jaws at the tips.

All curculionid larvae are legless; many are miners of leaves, stems, roots or seeds; others live in shed seeds (including stored grain) and dead plant material. Some species live externally, chewing the surfaces of leaves, sparing only a network of small veins. Many curculionid adults are also phytophagous, feeding externally and generally making small holes in leaves or around their edges. The snout may be used to bore into plants to gain access to softer, more nutritious inner tissues for feeding or to prepare a hole for oviposition. Species whose larvae live in nuts have long fine snouts, sometimes longer than the body.

7. ORDER DIPTERA

Flies

Adult flies have a single pair of membranous wings. The hindwings are modified into small, club-shaped halteres (e.g. Fig. 6.8). Larvae take many forms but are typically grub-like and legless.

Many adult Diptera feed on pollen or nectar, but none feed on plant leaves, roots or stems; a large number of species do not feed at all as adults. Only five families have many species with phytophagous larvae, viz. Cecidomyidae, Anthomyiidae, Tephritidae, Agromyzidae, and Chloropidae. Members of a few other families also occasionally exploit the living tissues of higher plants for food. Detailed studies of phytophagous dipterous larvae (other than gall-formers) often reveal that the ability to feed and survive on plants depends on the presence of microorganisms in the tissues on which they are feeding (e.g. Section 2.1.4).

7a. SUBORDER NEMATOCERA

Family: Cecidomyidae

Adult Cecidomyidae (gall midges) are minute (<3 mm long), delicate flies (Fig. 1.5). Belying their common name, not all cecidomyids live in or induce galls; nor do all have phytophagous larvae. Some species have predatory larvae, some saprophagous. Nonetheless, the majority of species form galls on higher plants in their larval stage. *Dasineura* (Table 1.1) is a typical genus. *Calamomyia alterniflorae* (Fig. 1.3) is a stem-mining cecidomyid that feeds on *Phomopsis* fungi, which it 'cultivates' inside the stems.

7b. SUBORDER BRACHYCERA

Family: Dolichopodidae

Adult dolichopodids (long-legged flies) are predatory, as are many of their larvae, but a small number appear to be phytophagous stem-borers, as *Thrypticus violaceus* (Fig. 1.3).

7c. SUBORDER CYCLORRHAPHA

Superfamily: Drosphiloidea

Family: Drosophilidae

Like *Drosophila melanogaster*, much studied by geneticists, most fruit flies are not phytophagous, but feed as larvae in a variety of places rich in yeasts or other microorganisms, rotting fruit, sap flows, decaying cacti, etc. A number of species, however, appear to be genuine phytophages. Members of the genus *Scaptomyza* for example have leaf-mining larvae.

Family: Ephydridae

Ephydrids (shore flies) are small flies whose larvae feed in a wide variety of ways. Some are phytophagous leaf-miners, e.g. *Hydrellia valida* (Fig. 1.3).

Superfamily: Muscoidea

Family: Anthomyiidae

Most anthomyiids have phytophagous larvae that are leaf- or stem-miners. Important pests are *Delia* (= *Erioischia*) *brassicae*, the cabbage root fly, *Hylemya coarctata*, the wheat bulb fly, and *Delia antiqua*, the onion fly. The genus *Chirosia*, common on bracken (Table 1.1), is also in this group.

Adult anthomyiids resemble small adult house flies, in the closely related family Muscidae.

Superfamily: Agromyzoidea

Family: Agromyzidae

An adult agromyzid is illustrated in Fig. 6.8. Typically they are small flies, inconspicuous and non-phytophagous as adults. Larvae are predominantly leaf-miners (e.g. Fig. 2.1), but some are stem-miners and a few induce galls. Most species are host specific (Fig. 6.8), but some *Liriomyza* species are polyphagous agricultural pests. *Phytomyza* is another genus.

Family: Chloropidae

Small or minute flies with some predatory but many phytophagous larvae. *Oscinella frit*, the frit fly, is an important pest of cereals. *Lipara* induces galls.

Family: Tephritidae (= Trypetidae)

Adult tephritids are small to medium-sized, often quite brightly coloured flies with spotted or banded wings. Their larvae mine into fruits, flower-heads (especially of Compositae), or leaves and plant stems, depending upon the species. Some induce galls, e.g. *Euribia cardui* (Fig. 4.7). The 'med fly' or Mediterranean fruit fly, *Ceratitis capitata*, is a notorious pest that mines fruits of many types. Sympatric speciation (Section 4.2.1) has been studied in *Rhagoletis*.

8. ORDER LEPIDOPTERA

Butterflies and moths

To the layman, butterflies are brightly coloured, fly by day, and close their wings vertically over their backs at rest; moths are dull in colour, fly by night and fold their wings flat across their backs. But taxonomically these distinctions are not strictly valid. Some adult moths are brightly coloured and normally diurnal (e.g. the cinnabar moth (*Tyria jacobaeae*); Arctiidae), and some butterflies are dull brown and can fold their wings (e.g. skippers (Hesperiidae)). In Europe and North America one distinction is that butterflies have clubbed antennae that are never 'hooked' at the tip. The antennae of moths take many forms but are never simply clubbed.

We have followed the common practice of distinguishing between 'macrolepidoptera' and 'microlepidoptera'. This is a distinction of convenience based on size, cutting across modern taxonomic arrangements. Macrolepidoptera include all the butterflies, and the larger moths in the following superfamilies:

'Macro' moths	*Butterflies*
Zygaenoidea	Papilionoidea
Hepialoidea	Hesperioidea
Cossoidea	
Noctuoidea	
Geometroidea	
Sphingoidea	
Bombycoidea	

Adult Lepidoptera either do not feed at all, or more usually are fluid-feeders, exploiting nectar, sap-flows from plant wounds and such apparently unlikely things as urine, sulphur springs and rotting carcasses. A few feed on pollen (Section 7.1.2) or have bizarre blood-sucking habits. Almost all caterpillars are phytophagous, feeding on plants in many ways, especially as external chewing leaf-eaters, leaf-miners and stem-borers.

8a. SUBORDER ZEUGLOPTERA

Family: Micropterigidae

These are primitive Lepidoptera. The adults have functional biting mandibles, not the usual long, flexible, sucking proboscis; they feed on pollen grains. Micropterigid caterpillars appear to eat detritus, mosses and liverworts.

8b. SUBORDER MONOTRYSIA

Superfamily: Hepialoidae

Family: Hepialidae

The hepialids (swift moths) have subterranean, root-feeding larvae.

8c. SUBORDER DITRYSIA

This is large group embracing the great majority of Lepidoptera. The superfamily Tineoidea contains about half the species of microlepidoptera.

Superfamily: Tineoidea

Family: Psychidae

The caterpillars live in cases (inspiring the common name of bag worms) constructed from plant material and silk. The wingless adult female never leaves her larval case.

Family: Gracillariidae

Caterpillars are common leaf-miners, for example the genus *Phyllonorycter* (= *Lithocolletis*) (Fig. 5.9c).

Family: Yponomeutidae

Yponomeuta (Table 4.3) is an example; the larvae feed communally.

Members are known colloquially as ermine moths.

Family: Scythridae

Scythris inspersella is an example (Section 4.1.2).

Family: Oecophoridae

Caterpillars in this and the two previous families frequently live in or under silk webs, or tie leaves with silk. *Drepressaria* (Section 7.1.2) is a member of the Subfamily Depressarini of the Oecophoridae. An adult of these small brown moths is illustrated in Fig. 7.2.

Family: Cosmopterygidae (= Momphidae)

Mostly small moths with pointed wings; larvae of some species mine leaves.

Family: Gelechiidae

Paltodora cytisella (Table 1.1), a bracken stem-galler, is an example of this large family of small moths; the most frequent larval habit is to spin together leaves or seed heads.

Superfamily: Cossoidea

Family: Cossidae

Goat moth caterpillars are generally wood-borers and are often large and smelly; the adults have wing-spans up to 24 cm.

Superfamily: Zygaenoidea

Family: Zygaenidae

These are mostly brightly coloured, day-flying (butterfly-like) moths; they are distasteful to predators and contain compounds sequestered during larval feeding from their host plants. Included in this taxon are burnet moths and foresters.

Superfamily: Tortricoidea

Family: Tortricidae

Familiar tortricids are the codling moth *Laspeyresia* (*Cydia*) *pomonella*, budworms (e.g. *Zeiraphera* and *Choristoneura*) and free-living defoliators like *Tortrix viridana*, common on oak. A leaf rolled up by caterpillars of an *Epinotia* sp. is shown in Fig. 2.2.

Family: Olethreutidae

Olethreutids are closely related to tortricids, and are sometimes grouped with them. The larvae of several species feed on leaves of trees after they have fallen. Others are true phytophages.

Superfamily: Pyraloidea

A large assemblage of small to medium-sized moths, mostly in the one family Pyralidae.

Family: Pyralidae

Many pyralids are not strictly phytophagous, but feed on dried, dead or decaying plant material. The Subfamily Pyraustinae contains the interesting Hawaiian genus *Hedylepta*, discussed in Section 4.2.1. *Chilo* (e.g. Fig. 1.3) is a typical genus of phytophagous pyralids.

Superfamily: Pterophoroidae

Family: Pterophoridae

The wings are divided into lobes, each like a plume or feather (hence their name of plume moths), and are held out when at rest. Caterpillars sometimes mine leaves in the early stages.

Superfamily: Papilionoidea

This and the next superfamily together comprise the butterflies.

Family: Nymphalidae

Nymphalids are the largest group of butterflies, and one of the largest families of Lepidoptera. The forelegs of the adult are reduced and do not function as legs. Subfamilies referred to include the following.

Subfamily Danainae (milkweed butterflies), the group to which the monarch *Danaus plexippus* belongs; the adults are mostly red-brown, white and black marked, and are protected from predators by cardiac-glycosides sequestered from their larval host plants, Asclepiadaceae and Apoeynaceae.

Subfamily Ithomiinae are mostly forest-dwelling South American butterflies, often with transparent patches in their wings.

Subfamily Satyrinae (meadow browns, ringlets, heaths, graylings, marbled whites, etc.) are predominately brownish butterflies with underside patterns that make them very inconspicuous when at rest; the larvae feed on palms, grasses and related plants.

Subfamily Heliconiinae are strikingly handsome South American butterflies; the heliconiids have been the subject of detailed studies of coevolution, discussed in Section 7.1.2.

Subfamily Nymphalinae (fritillaries, tortoiseshells, peacocks, painted ladies, etc.) is a large world-wide group whose larvae mainly chew foliage, often sewing it together with silk. *Ladoga camilla,* a typical nymphalid, is shown in Fig. 3.6.

Family: Riodinidae (= Nemeobiidae)

A largely South American group, some of whose larvae are ant-tended (Section 6.4.1). Some larvae are very hairy (unusual in butterfly caterpillars). Member are known as 'metalmarks'.

Family: Lycaenidae

The major foods of these caterpillars are unusual for Lepidoptera, being more like the foods of mirid bugs: as well as feeding on flowers, fruits and buds, some are partially predatory and take living Homoptera. Caterpillars of many species are tended by ants (Section 6.4.1 and Fig. 6.17), and have a 'honey gland' on the abdomen; and pores on the surface of the body also produce food for ants. Adults often have tails on the hindwings that are used to mimic quivering antennae. Included in this family are the blues, coppers and hairstreaks.

Family: Pieridae

As indicated by their common names of whites and sulphurs, these are mostly white or yellow butterflies, the wing pigments being derived from excretory products. *Pieris* is a familiar genus.

Family: Papilionidae

A world-wide family of about 530 species of mostly large butterflies (swallowtails and birdwings), whose larves feed mainly on Aristolochiaceae, Lauraceae, Rutaceae and Umbelliferae. An adult black swallowtail (*Papilio polyxenes*) is illustrated in Fig. 7.2.

Superfamily: Hesperioidea

Family: Hesperiidae

Fast, darting flyers (hence the common name of 'skippers' for members of the family), these butterflies are mainly small and brown or blackish. The antennal club is less pronounced than in other butterflies and so skippers may be mistaken for moths. The caterpillars mostly feed on grasses.

Superfamily: Geometroidea

Family: Drepanidae

The caterpillar's body tapers posteriorly and lacks the terminal claspers. 'Hooktips' feed on the foliage of trees.

Family: Geometridae

This family contains a very large number of species. The most consistent feature of geometrids is the reduction of the number of pairs of 'claspers' or false legs on the caterpillar's abdomen, so that they move with a characteristic 'looping' gait (Fig. 1.5), giving rise to popular names such as 'inch-worms', 'loopers' and 'earth-measures'. The adults are nocturnal, with a slow fluttering flight. Female adults of some species, such as the winter moth, *Operophtera brumata* (Section 5.3.1), and the fall cankerworm, *Alsophila* (Table 4.3), are flightless. The caterpillars mostly feed externally, and when resting many resemble a twig. A few Hawaiian species have been shown to catch and consume flies that alight on or near these 'twigs':

Superfamily: Bombycoidea

Family: Endromididae

The Kentish glory (*Endromis*) (Fig. 3.5) is probably the sole extant representative of this family. The adult males may fly by day.

Family: Lasiocampidae

The caterpillars are very hairly and some live communally in webs ('tent caterpillars'). They often have a wide range of botanically unrelated food plants. The moths are fairly large and the males may fly by day. Members include the eggar and lappet moths.

Superfamily: Sphingoidea

Family: Sphingidae

The hairless caterpillars of hawk moths have a curved 'horn' at the end of the abdomen, giving them an alarming appearance, but they do not seem to sequester defensive chemicals from their host plants. The adults are often large, streamlined and some can fly as fast as 40 km h^{-1}.

Superfamily: Noctuoidea

Family: Arctiidae

Arctiids (tiger moths, ermines, etc.) are conspicuously marked, often brightly coloured moths. The cinnabar moth *Tyria jacobaeae* is an example (Fig. 6.21). The caterpillars are often hairy, sometimes densely so ('woolly bears'). Many arctiid caterpillars are poisonous, sequestering their host plants' chemicals that are retained into the adult state. The adults of some species, including *Tyria jacobaeae* (unlike most moths), often fly actively during the day.

Family: Noctuidae

There are more species (over 20 000) in the family Noctuidae than in any other group of Lepidoptera. Many of the larvae are important pests, and include cutworms (*Agrotis*), the army worm (*Leucania*) and the southern army worm (*Spodoptera*), all of which are polyphagous and feed externally on foliage and stems. Others are stem-borers. The moths (which include the owlets and underwings) are night-flying, mostly moderate sized and brownish in colour; they can detect the sounds made by the bats which prey on them.

Family: Lymantriidae

Unlike noctuids, the adult tussocks lack a proboscis and do not feed; some females are wingless. The caterpillars often have tufts of hairs and feed mostly on trees. The gipsy moth, *Lymantria* (Fig. 5.7), is an example.

9. ORDER HYMENOPTERA

Bees, ants, wasps, sawflies, gall wasps, etc.

Adult Hymenoptera typically have two pairs of membranous wings with large distinctive 'cells', which are areas enclosed by the wing veins. There are two major suborders: the Symphyta (sawflies), with predominantly phytophagous members, and the Apocrita, of very varied biology and feeding habits.

9a. SUBORDER SYMPHYTA

Sawflies are regarded as the most primitive Hymenoptera. Adults are distinguished from adult Apocrita by the absence of a narrow 'waist' between the abdomen and the thorax. They are called sawflies because the female's ovipositor is adapted for sawing or boring into the substrate, usually plant tissues.

Sawfly caterpillars (Fig. 2.1) are predominantly phytophagous and have independently evolved an appearance remarkably like that of the caterpillars of Lepidoptera (Fig. 2.5a and b). A few are wood-borers (the Siricidae, wood wasps or horn tails).

Superfamily: Xyeloidea

Family: Xyelidae

These are primitive sawflies with larvae that feed on pine flowers or that bore into new pine shoots. An adult of a fossil species is illustrated in Fig. 2.11d.

Superfamily: Cephoidea

Family: Cephidae

Larvae are stem- and shoot-borers, especially in grasses (including cereals). The common name for the group is the stem sawflies.

Superfamily: Tenthredinoidea

Family: Blasticotomidae

The larva of the single European species, *Blasticotoma filiceti* (Section 2.2.2), bores in the stems of ferns.

Family: Diprionidae

As their common name of conifer sawflies implies, the larvae feed on conifers, and are often strongly and conspicuously marked, and gregarious. *Neodiprion* (Table 4.2 and 4.3) is a typical genus. The adults have characteristic feathery antennae (males) or toothed antennae (females).

Family: Tenthredinidae

The great majority of sawflies belong to this family, which includes over 4000 known species, of very variable form and habit, although all the larvae are phytophagous. The majority are external, chewing folivores.

9b. SUBORDER APOCRITA

Included in this suborder are the great majority of the Hymenoptera, all with the abdomen constricted into a 'waist' where it joins the thorax. Bees, wasps, ants and the vast array of parasitic 'wasps' (parasitoids) belong here. For completeness, and because they are such important enemies of phytophagous insects, some of the most important groups of parasitoids are briefly assigned to their taxonomic pigeon-holes in the account that follows.

Superfamily: Ichneumonoidea

Family: Ichneumonidae

Family: Braconidae

Braconid and ichneumonid wasps are parasitoids of phytophagous insects. *Apantales* (Section 6.4.2 and Fig. 6.21) is a braconid; *Cratichneumon* (Section 5.3.1) is an ichneumonid.

Superfamily: Cynipoidea

Family: Cynipidae

The larvae of all cynipids (gall wasps) induce galls on their host plants (e.g. *Diplolepis eglanteriae* (Fig. 5.9b) on *Rosa*). Approximately 85 per cent of known species produce galls on oaks, *Quercus*, and are confined to that genus. A number of species, instead of producing their own gall, take over galls made by other species, and are known as inquilines.

A number of cynipid species have two generations per year, one sexual, the other parthenogenetic. The galls produced by the two generations are usually very different in appearance, and occur on different parts of the host (e.g. catkins and leaves). Some species may even use different species of hosts for each generation.

Superfamily: Chalcidoidea

This superfamily contains a vast number of tiny species. Most are parasitoids or hyperparasitoids, but some are phytophagous. Phytophagous chalcids occur in two families in particular, as follows.

Family: Agaonidae

These species live symbiotically with figs, and the adult fig wasps are essential for pollination of the host; most are specific to a single species of fig. The larvae feed on the ovaries and induce galls.

Family: Eurytomidae

Eurytomids have a diversity of larval feeding habits. Some species from galls in stems or bore into seeds, others are parasitoids of a wide range of insects.

Appendix 2 Taxonomic Knowledge of Phytophagous Insects

The patterns explored in this book are based upon samples of insect–plant diversity, not upon complete censuses. At larger scales the samples will be less complete and gathered by more people than at finer scales. How complete are data at larger scales? How far have taxonomists progressed in documenting the world's phytophagous insects? First, we should recognize that probably not all species of insects will ever be known to science. New species will contiuue to be found and known taxa will continue to be subdivided, grouped and revised. Fortunately for our purposes, exhaustive censuses and complete agreement between taxonomists are not needed to establish the existence of patterns or to test hypotheses to explain such patterns. We require only that representative taxa should be reasonably well known. Here we show why we believe that present taxonomic knowledge is complete enough not to bias our endeavours seriously.

A preliminary picture of global knowledge can be gained from study of historical rates of species discovery. Global histories for representative groups of insects, namely: whiteflies (Aleyrodidae: Hemiptera), weevils (Curculionidae: Coleoptera), phytophagous thrips (Thripidae: Thysanoptera), and papilionoid and danaid butterflies (Papilionoidea and Danainae: Lepidoptera), are shown in Fig. A.1. The rate of description of new species through time is roughly parabolic for most families, commencing with low rates in the first years after publication of Linnaeus' *Systema Naturae*. The two butterfly groups have undergone an extra cycle of discovery. Their initially high discovery rates declined,

Fig. A.1. Histories of species descriptions on a world scale for five groups of phytophagous insects. Data for whiteflies are from Mound and Halsey (1978), weevils from O'Brian and Wimber (1979), thrips from Jacot-Guillarmod (1970–1975), and Papilionoidea and Danainae from the Card Catalogue in the British Museum, Natural History (D.R.S.). Only names with valid status in these references and in the Card Catalogue in May 1979 are included. Open circles are temperate and arctic species; closed circles are tropical species.

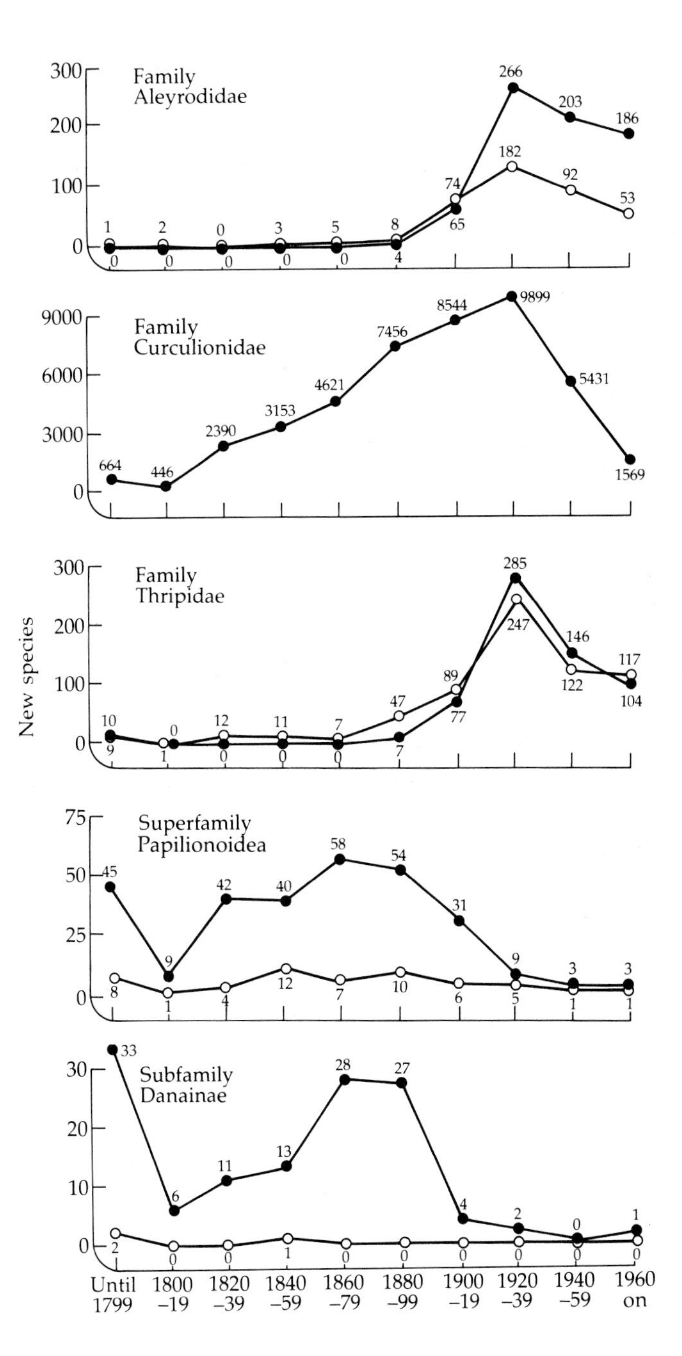
Family
Aleyrodidae
Family
Curculionidae
Family
Thripidae
Superfamily
Papilionoidea
Subfamily
Danainae
New species
Until
1799
1800
–19
1820
–39
1840
–59
1860
–79
1880
–99
1900
–19
1920
–39
1940
–59
1960
on

then climbed again during the latter half of the nineteenth century, and finally declined again recently.

The major lesson of Fig. A.1 is that current rates of species description for these phytophages are relatively low. The largest, most conspicuous insects, such as danaid and papilionid butterflies, now have description rates very close to zero. In fact, danaid butterflies are so completely known globally that any new species is widely discussed by experts. Current description rates for the weevils, which are intermediate in size and conspicuousness, are much higher. At the last count, 1569 new weevil species had been described since 1960; but this is less than four per cent of the total species already known in this group, and present description rates are very low compared to those of the bonanza years in the early twentieth century. Even the very small whiteflies and thrips are not being described at rates as high as those 30 to 50 years ago, though the fraction of total species described recently is much higher for these inconspicuous insects than for weevils or butterflies. Approximately 22 per cent of valid whitefly species and 17 per cent of thrips have been described since 1960.

At finer scales of resolution, we can detect bursts of species descriptions reflecting the labours of a few prolific taxonomists, superimposed upon a gradual decline in the rate of discovery of new species (Fig. A.2). These bursts do not usually occur simultaneously in all regions, so that the overall global curve is smoother than its constituent regional curves. Yet all regional rates of discovery for the weevils have been declining conspicuously for some time, and the total global rate has been declining for at least 60 years.

This fall in the rate of species description is abetted by a change in the working practices of taxonomists as well as a dwindling in the numbers of new species collected. Previously, on spotting a new species in a collection, a taxonomist would describe it; the most frequent modern practice is to set the specimens aside until a complete revision is undertaken of the group in question. This applies in particular to the least known taxa, such as parasitic Hymenoptera, but not to those groups, like butterflies, where the true identity of species already described is well established.

Fortunately for our purposes, we can avoid these problems by focusing upon regional faunas that are extremely well known. Britain has the best known biota of any region (Fig. A.3). Just as occurred on a

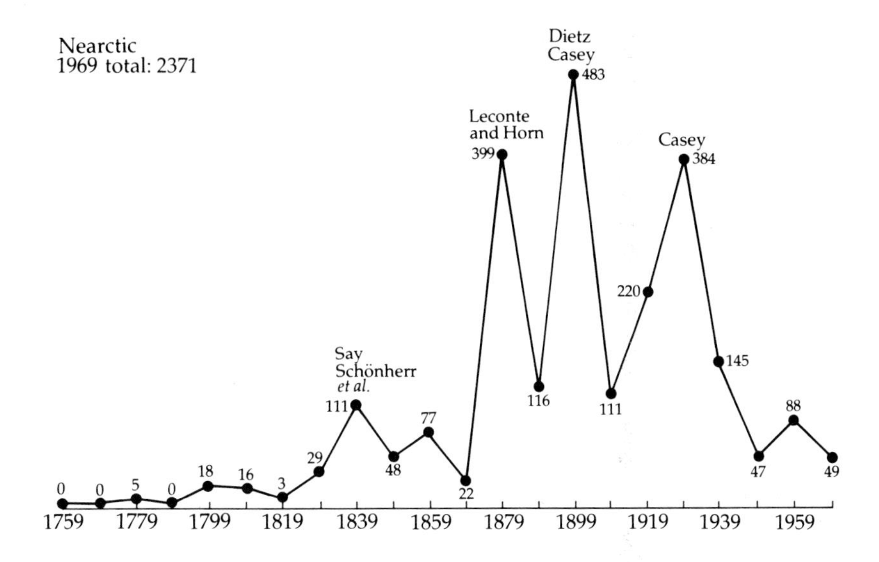

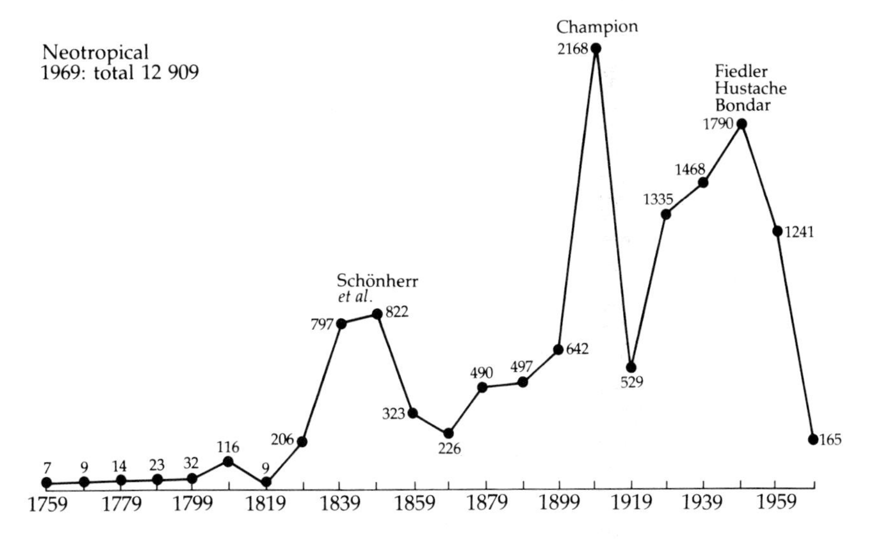

Fig. A.2. Histories of species description for weevils in six regions of the world, from O'Brian and Wimber (1979). Peaks of productivity within regions have usually been the result of the efforts of individuals or small groups of authors, as indicated above each peak. Continued on pages 252–3.

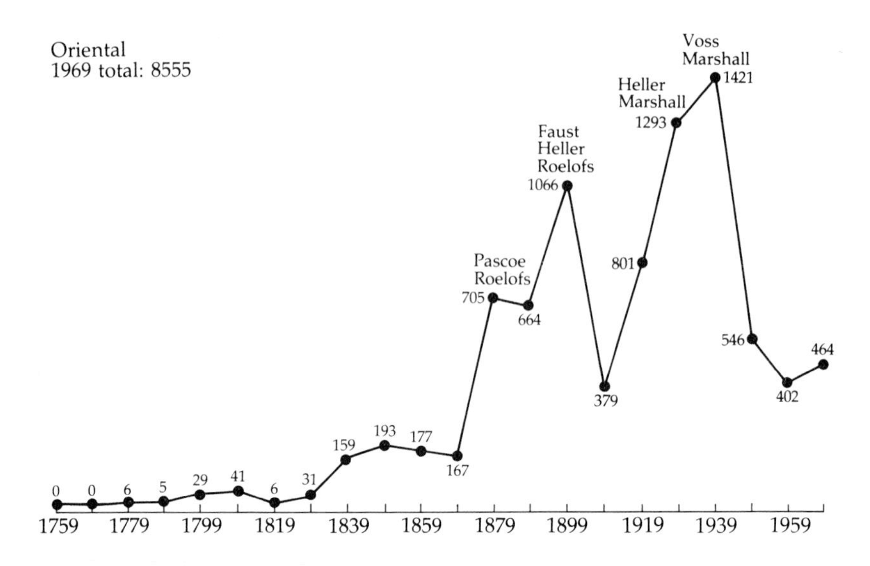
Oriental
1969 total: 8555
Voss
Marshall
1421
Heller
Marshall
1293
Faust
Heller
Roelofs
1066
Pascoe
Roelofs
705
664
801
546
464
402
379
0 0 6 5 29 41 6 31 159 193 177 167
1759 1779 1799 1819 1839 1859 1879 1899 1919 1939 1959

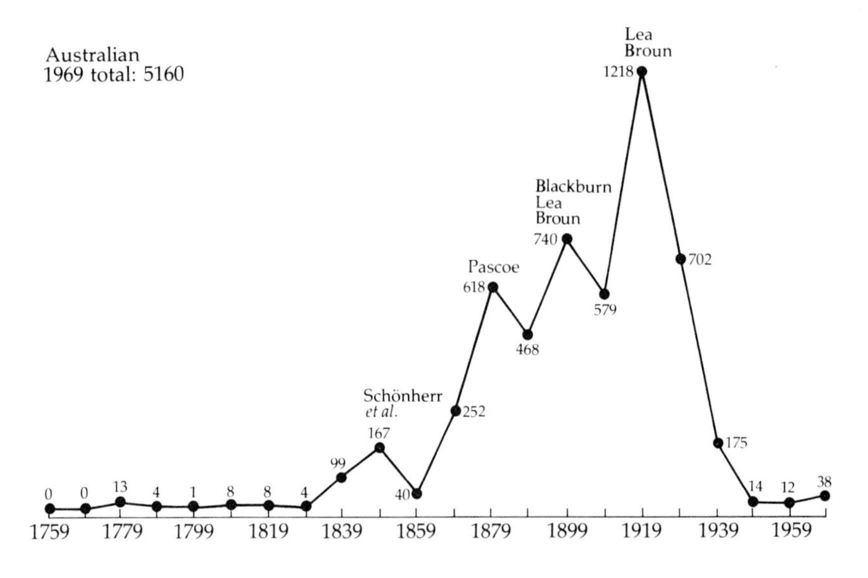
Australian
1969 total: 5160
Lea
Broun
1218
Blackburn
Lea
Broun
740
Pascoe
618
702
579
468
Schönherr
et al.
167
252
175
0 0 13 4 1 8 8 4 99 40 14 12 38
1759 1779 1799 1819 1839 1859 1879 1899 1919 1939 1959

Fig. A.2. *(cont.)*

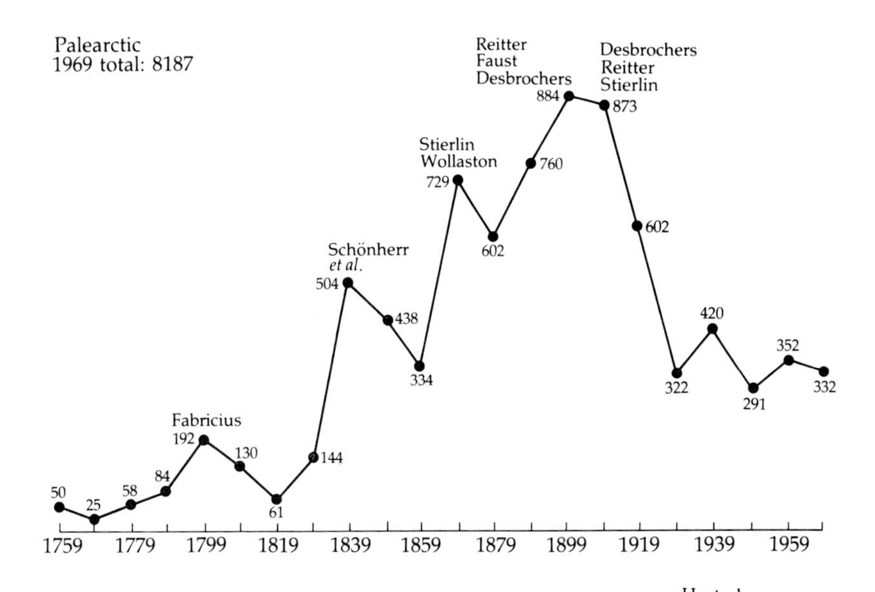
Palearctic
1969 total: 8187
Fabricius
Schönherr
et al.
Stierlin
Wollaston
Reitter
Faust
Desbrochers
Desbrochers
Reitter
Stierlin
50
25
58
84
192
130
61
144
504
438
334
729
602
760
884
873
602
322
420
291
352
332
1759
1779
1799
1819
1839
1859
1879
1899
1919
1939
1959

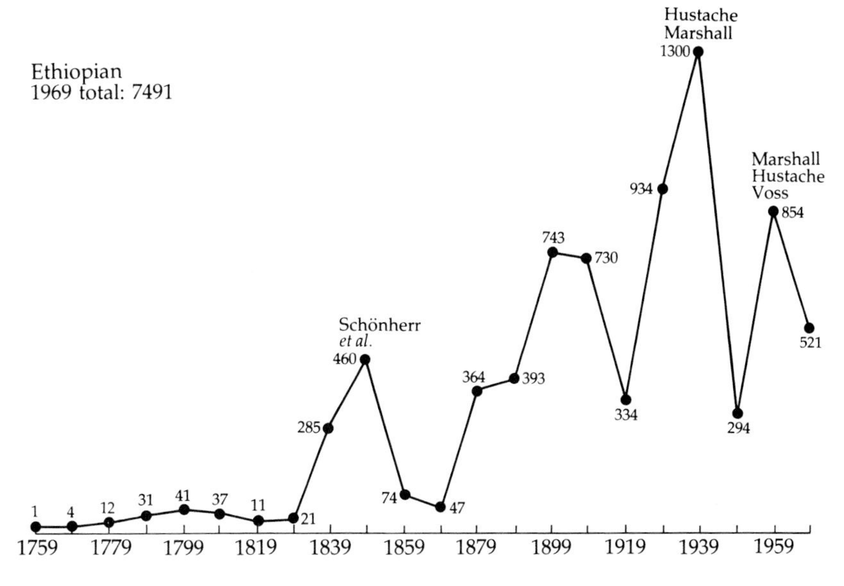
Ethiopian
1969 total: 7491
Schönherr
et al.
Hustache
Marshall
Marshall
Hustache
Voss
1
4
12
31
41
37
11
21
285
460
74
47
364
393
743
730
334
934
1300
294
854
521
1759
1779
1799
1819
1839
1859
1879
1899
1919
1939
1959

Fig. A.2. *(cont.)*

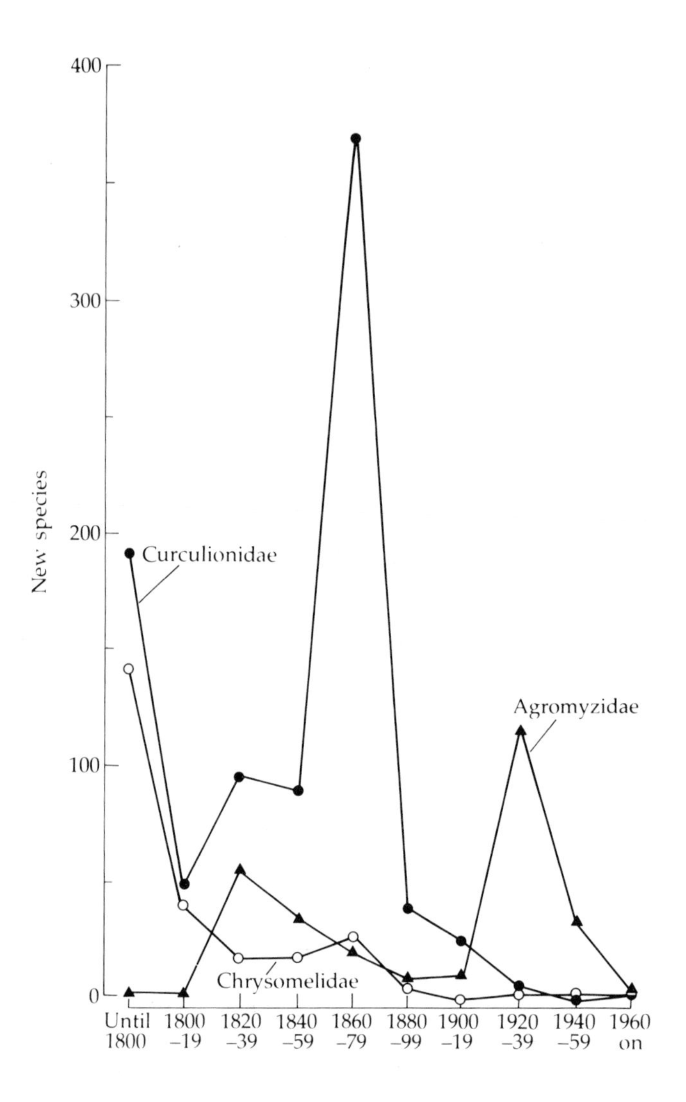

Fig. A.3. Histories of species description for British phytophagous insects, from Kloet and Hinks (1964, 1972, 1976, 1977). Recent rates of description are very close to zero and are a very small proportion of known species for all groups.

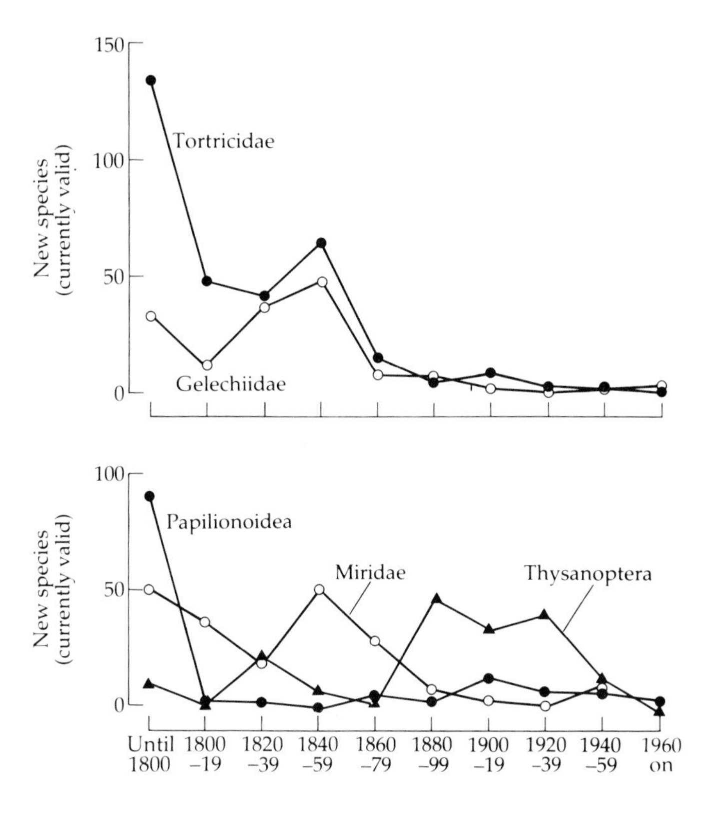

Fig. A.3. (cont.)

global scale, the largest and most conspicuous British species were discovered first. For example, papilionid butterflies, curculionid beetles, and chrysomelid beetles had large fractions of their species described before 1800. The important conclusion is that no group of phytophagous insects in Britain is gaining new species at more than a trickle. Thus it is safe to say that this region's phytophagous entomofauna is close to being completely described at the species level. This is why we have relied repeatedly on British data.

At the other extreme, our ignorance of many groups of insects in tropical forests is vast. Some taxonomists believe there to be millions of undescribed species. We hope that by relying on reasonably well known tropical taxa, like heliconiid butterflies and hispine beetles, we have avoided serious distortions in our tentative search for patterns.

References

Addicott J.F. (1978a) Niche relationships among species of aphids feeding on fireweed. *Can. J. Zool.*, **56**, 1837–41.

Addicott J.F. (1978b) The population dynamics of aphids on fireweed: a comparison of local populations and metapopulations. *Can. J. Zool.*, **56**, 2554–64.

Addicott J.F. (1979) A multispecies aphid–ant association: density dependence and species-specific effects. *Can. J. Zool.*, **57**, 558–69.

Agarwal R.A. & Krishnananda N. (1976) Preference to oviposition and antibiosis mechanism to jassids (*Amrasca devastans* Dist.) in cotton (*Gossypium* sp.). *Symp. Biol. Hung.*, **16**, 13–22.

Ahmad I. & Abror I. (1977) A study of eggs and larvae of two closely related sympatric species of *Oxrhachis* Germar (Homoptera: Membracidae) *Pakistan J. Sci. Ind. Res.*, **19**, 233–8.

Alexander R.D. & Moore T.E. (1962) The evolutionary relationships of 17-year and 13-year cicadas, and three new species (Homoptera, Cicadidae, *Magicicada*). *Misc. Publs. Univ. Mich.*, **121**, 5–57.

Alpin R.T., d'Arcy Ward R. & Rothschild M. (1975) Examination of the large white and small white butterflies (*Pieris* spp.) for the presence of mustard oils and mustard oil glycosides. *J. Entomol.* (A), **50**, 73–8.

Anderson R.M. & May R.M. (1980) Infectious diseases and population cycles of forest insects. *Science*, **210**, 658–61.

Anderson R.M. & May R.M. (1981) The population dynamics of microparasites and their invertebrate hosts. *Phil. Trans. R. Soc.*, **291**, 451–524.

Askew R.R. (1961) On the biology of the inhabitants of oak galls of Cynipidae (Hymenoptera) in Britain. *Trans. Soc. Br. Ent.*, **14**, 237–68.

Askew R.R. (1962) The distribution of galls of *Neuroterus* (Hym: Cynipidae) on oak. *J. Anim. Ecol.*, **31**, 439–55.

Askew R.R. (1980) The diversity of insect communities in leaf-mines and plant galls. *J. Anim. Ecol.*, **49**, 817–29.

Atkinson W.D. & Shorrocks B. (1981) Competition on a divided and ephemeral resource: a simulation model. *J. Anim. Ecol.*, **50**, 461–71.

Atsatt P.R. (1981a) Lycaenid butterflies and ants: selection for enemy-free space. *Am. Nat.*, **118**, 638–54.

Atsatt P.R. (1981b) Ant-dependent food plant selection by the mistletoe butterfly *Ogyris amaryllis* (Lycaenidae). *Oecologia*, **48**, 60–3.

Atsatt P.R. & O'Dowd D.J. (1976) Plant defense guilds. *Science*, **193**, 24–9.

Auerbach M.J. & Hendrix S.D. (1980) Insect–fern interactions: macrolepidopteran utilization and species-area association. *Ecol. Entomol.*, **5**, 99–104.

Auerbach M.J. & Strong D.R. (1981) Nutritional ecology of *Heliconia* herbivores: experiments with plant fertilization and alternative hosts. *Ecol. Monogr.*, **51**, 63–83.

Ausich W.I. & Bottjer D.J. (1982) Tiering in suspension-feeding communities on soft substrata throughout the Phanerozoic. *Science*, **216**, 173–4.

Bach C.E. (1980) Effect of plant density and diversity on the population dynamics of a specialist herbivore, the striped cucumber beetle, *Acalymma vittata* (Fab.). *Ecology*, **61**, 1515–30.

Baltensweiler W. (1971) The relevance of changes in the composition of larch bud moth populations for dynamics of its numbers. In: *Dynamics of Populations* (ed. P.J. der Boer & G.R. Gradwell), pp. 208–19. Centre for Agriculture, Wageningen, Netherlands.

Baltensweiler W., Benz G., Bovey P. & Delucchi V. (1977) Dynamics of larch bud moth populations. *A. Rev. Ent.*, **22**, 79–100.

Banerjee B. (1979) A key-factor analysis of population fluctuations in *Andraca bipunctata* Walker (Lepidoptera: Bombycidae). *Bull. ent. Res.*, **69**, 195–201.

Banerjee B. (1981) An analysis of the effect of latitude, age and area on the number of arthropod pest species of tea. *J. appl. Ecol.*, **18**, 339–42.

Barlow N.D. & Dixon A.F.G. (1980) *Simulation of Lime Aphid Population Dynamics*. PUDOC, Wageningen, Netherlands.

Bartell R.J. (1966) Studies on water loss from eggs of Lepidoptera. Unpublished Ph.D. Thesis, University of London.

Bates H.W. (1862) Contributions to an insect fauna of the Amazon Valley. *Trans. Linn. Soc. Lond.* **23**, 495–566.

Bauchhenb E. & Renner M. (1977) Pulvillus of *Calliphora erythrocephala* Meig. (Diptera: Calliphoridae). *Int. J. Insect Morphol. & Embryol.*, **6**, 225–7.

Beck S.D. & Reese J.C. (1976) Insect–plant interactions: nutrition and metabolism. *Rec. Adv. Phytochem.*, **10**, 41–92.

Beddington J.R., Free C.A. & Lawton J.H. (1978) Characteristics of successful natural enemies in models of biological control of insect pests. *Nature (London)*, **273**, 513–9.

Begon M. & Mortimer M. (1981) *Population Ecology*. Blackwell Scientific Publications, Oxford.

Bennett F.D., Rosen D., Cochereau P. & Wood B.J. (1976) Biological control of pests of tropical fruits and nuts. In *Theory and Practice of Biological Control* (ed. C.B. Huffaker & P.S. Messenger), pp. 359–95. Academic Press, New York.

Benson J.F. (1973) Some problems of testing for density dependence in animal populations. *Oecologia*, **13**, 183–90.

Benson R.B. (1950) An introduction to the natural history of British sawflies (Hymenoptera, Symphyta). *Trans. Soc. Br. Ent.*, **10**, 46–142.

Benson W.W. (1972) Natural selection for Mullerian mimicry in *Heliconius erato* in Costa Rica. *Science*, **176**, 936–9.

Benson W.W. (1978) Resource partitioning in passion vine butterflies. *Evolution*, **32**, 493–518.

Benson W.W., Brown K.S. Jr. & Gilbert L.E. (1976) Coevolution of plants and herbivores: passion flower butterflies. *Evolution*, **29**, 659–80.

Bentley B.L. (1977) Extrafloral nectaries and protection by pugnacious body-guards. *Annu. Rev. Ecol. & Syst.*, **8**, 407–27.

Bentley S. & Whittaker J.B. (1979) Effects of grazing by a chrysomelid beetle, *Gastrophysa viridula*, on competition between *Rumex obtusifolius* and *Rumex crispus*. *J. Ecol.*, **67**, 79–90.

Berenbaum M. (1978) Toxicity of a furanocoumarin to armyworms: a case of biosynthetic escape from insect herbivores. *Science*, **201**, 532–4.

Berenbaum M. (1980) Adaptive significance of midgut pH in larval Lepidoptera. *Am. Nat.*, **115**, 138–46.

Berenbaum M. (1981a) Effects of linear furanocoumarins on an adapted specialist insect (*Papilio polyxenes*). *Ecol. Entomol.*, **6**, 345–51.

Berenbaum M. (1981b) Patterns of furanocoumarin distribution and insect herbivory in the Umbelliferae: plant chemistry and community structure. *Ecology*, **62**, 1254–66.

Berenbaum M.R. (1981c) Patterns of furanocoumarin production and insect herbivory in a population of wild parsnip (*Pastinaca sativa* L.). *Oecologia*, **49**, 236–44.

Berenbaum M. (1983) Coumarins and caterpillars: A case for coevolution. *Evolution*, **37**, 163–179.

Berenbaum M. & Feeny P. (1981) Toxicity of angular furanocoumarins to swallowtails: escalation in the coevolutionary arms race. *Science*, **212**, 927–9.

Bernays E.A. (1981) Plant tannins and insect herbivores: an appraisal: *Ecol. Entomol.*, **6**, 353–60.

Bernays E.A. & Chapman R.F. (1975) The importance of chemical inhibition of feeding in host-plant selection by *Chorthippus parallelus* (Zetterstedt). *Acrida*, **4**, 83–93.

Bigger M. (1976) Oscillations of tropical insect populations. *Nature (London)*, **259**, 207–9.

Billany D.J., Borden J.H. & Brown R.M. (1978) Distribution of *Gilpinia hercyniae* (Hymenoptera, Diprionidae) eggs within sitka spruce trees. *Forestry*, **51**, 67–72.

Birks H.J.B. (1980) British trees and insects: a test of the time hypothesis over the last 13,000 years. *Am. Nat.*, **115**, 600–5.

Blackman R.L. (1967) The effects of different aphid foods on *Adalia bipunctata* L. and *Coccinella 7-punctata* L. *Ann. appl. Biol.*, **59**, 207–19.

Blakley N.R. & Dingle H. (1978) Competition: butterflies eliminate milkweed bugs from a Caribbean island. *Oecologia*, **37**, 133–6.

Boggs C.L., Smiley J.T. & Gilbert L.E. (1981) Patterns of pollen exploitation by *Heliconius* butterflies. *Oecologia*, **48**, 284–9.

Boscher J. (1979) Modified reproductive strategy of leek *Alium porrum* in response to a phytophagous insect, *Acrolepiopsis assectella*. *Oikos*, **33**, 451–6.

Bournier A. (1977) Grape insects. *A. Rev. Ent.*, **22**, 355–76.

Bradley J.D., Tremewan W.G. & Smith A. (1979) *British Tortricoid Moths. Vol. 2. Tortricidae: Olethreutinae*. The Royal Society, London.

Brattsten L.B. (1979) Ecological significance of mixed-function oxydations. *Drug Metabol. Rev.*, **10**, 35–78.

Brattsten L.B., Wilkinson C.F. & Eisner T. (1977) Herbivore–plant interactions: mixed-function oxidases and secondary plant substances. *Science*, **196**, 1349–52.

Brian M.V. (1977) *Ants*. Collins, London.

Brower L.P. & Brower J.V.Z. (1964) Birds, butterflies and plant poisons: a study in ecological chemistry. *Zoologica*, **49**, 137–59.

Brower L.P., Brower J.V.Z. & Corvina J.M. (1967) Plant poisons in a terrestrial food chain. *Proc. natl Acad. Sci., USA*, **57**, 893–8.

Brower L.P. & Glazier S.C. (1975) Localization of heart poisons in the monarch butterfly. *Science*, **188**, 19–25.

Brown E.S. (1959) Immature nut-fall of coconuts in the Solomon Islands. II. *Bull. ent. Res.*, **50**, 523–58.

Brown K.S. (1981) The biology of *Heliconius* and related genera. *A. Rev. Ent.* **26**, 427–456.

Brues C.T. (1946) *Insect Dietary*. Harvard University Press, Cambridge, Mass.

Bruzzese E. (1980) The phytophagous insect fauna of *Rubus* spp. (Rosaceae) in Victoria, a study on the biological control of blackberry (*Rubus fruiticosus* L. Agg.). *J. Aust. ent. Soc.*, **19**, 1–6.

Buckner C.H. & Turnock W.J. (1965) Avian predation on the larch sawfly, *Pristiphora erichsonii* (Htg.) (Hymenoptera: Tenthredinidae). *Ecology*, **46**, 223–36.

Bulmer M.G. (1975) The statistical analysis of density dependence. *Biometrics*, **31**, 901–11.

Burdon J.J. & Marshall D.R. (1981) Biological control and the reproductive mode of weeds. *J. appl. Ecol.*, **18**, 649–58.

Burton J.F. (1975) The effects of recent climatic changes on British insects. *Bird Study*, **22**, 203–4.

Bush G.L. (1975a) Modes of animal speciation. *Annu. Rev. Ecol. & Syst.*, **6**, 339–64.

Bush G.L. (1975b) Sympatric speciation in phytophagous parasitic insects. In: *Evolutionary Strategies of Parasitic Insects and Mites* (ed. P.W. Price), pp. 187–206. Plenum Press, New York.

Cameron E. (1939) The holly leaf-miner (*Phytomyza ilicis*, Curt.) and its parasites. *Bull. ent. Res.*, **30**, 173–208.

Campbell B.C. & Duffy S.S. (1979) Tomatine and parasitic wasps: potential incompatability of plant antibiosis with biological control. *Science*, **205**, 700–2.

Cantlon J.E. (1969) The stability of natural populations and their sensitivity to technology. *Brookhaven Symp. Biol.*, **22**, 197–203.

Carpenter F.M. (1977) Geological history and evolution of the insects. *Proceedings of the XV International Congress on Entomology, Washington D.C.*, pp. 63–70.

Carpenter F.M. & Richardson E.S. Jr (1978) Structure and relationships of the Upper Carboniferous insect, *Prochoroptera calopteryx* (Diaphanopterodea, Prochoropteridae). *Psyche*, **85**, 219–28.

Carroll C.R. & Hoffman C.A. (1980) Chemical feeding deterrent mobilized in response to insect herbivory and counteradaptation by *Epilachna tredecimnotata*. *Science*, **209**, 414–6.

Cassidy M.D. (1978). Development of an induced food plant preference in the Indian stick insect, *Carausius morosus*. *Entomol. Exp. & Appl.*, **24**, 87–93.

Cates R.G. (1980) Feeding patterns of monophagous, oligophagous, and polyphagous insect herbivores: the effect of resource abundance and plant chemistry. *Oecologia*, **46**, 22–31.

Cates R.G. (1981) Host plant predictability and the feeding patterns of monophagous, oligophagous, and polyphagous insect herbivores. *Oecologia*, **48**, 319–26.

Cates R.G. & Rhoades D.F. (1977) Patterns in the production of antiherbivore chemical defenses in plant communities. *Biochem. Syst. Ecol.*, **5**, 185–93.

Caughley G. & Lawton J.H. (1981) Plant–herbivore systems. In: *Theoretical Ecology* (ed. R.M. May), pp. 132–66. Blackwell Scientific Publications, Oxford.

Chapman R.F. & Blaney W.M. (1979) How animals perceive secondary compounds. In: *Herbivores. Their Interaction with Secondary Plant Metabolites* (ed. G.A. Rosenthal & D.H. Janzen), pp. 161–98. Academic Press, New York.

Chew F.S. (1975). Coevolution of pierid butterflies and their cruciferous foodplants. I. The relative quality of available resources. *Oecologia*, **20**, 117–29.

Chew F.S. (1979) Community ecology and *Pieris*–crucifer coevolution. *Jl. N.Y. ent. Soc.*, **87**, 128–34.

Chew F.S. (1981) Coexistence and local extinction in two pierid butterflies. *Am. Nat.*, **118**, 655–72.

Chew F.S. & Rodman J.E. (1979) Plant resources for chemical defense. In: *Herbivores. Their Interaction with Secondary Plant Metabolites* (ed. G.A. Rosenthal & D.H. Janzen), pp. 271–307. Academic Press, New York.

Chiang H.C. (1978) Pest management in corn. *A. Rev. Ent.*, **23**, 101–23.

Chipperfield H.E. (1980) Raspberry as a natural pabulum of the holly blue. *Entomologist's Rec. J. Var.*, **92**, 153.

Churchill G.B., John H.H., Duncan D.P. & Hodson A.C. (1964) Longterm effects of defoliation of aspen by the forest tent caterpillar. *Ecology*, **45**, 630–3.

Claridge M.F. & Reynolds W.J. (1972) Host plant specificity, oviposition behaviour and egg parasitism of some woodland leafhoppers of the genus *Oncopsis* (Hemiptera, Homoptera: Cicadellidae). *Trans. R. ent. Soc. Lon.*, **124**, 149–66.

Claridge M.F. & Wilson M.R. (1976) Diversity and distribution patterns of some mesophyll-feeding leafhoppers of temperate woodland canopy. *Ecol. Entomol.*, **1**, 231–50.

Claridge M.F. & Wilson M.R. (1978) British insects and trees: a study in island biogeography or insect/plant coevolution? *Am. Nat.*, **112**, 451–6.

Claridge M.F. & Wilson M.R. (1981) Host plant associations, diversity and species–area relationships of mesophyll-feeding leafhoppers of trees and shrubs in Britain. *Ecol. Entomol.*, **6**, 217–38.

Claridge M.F. & Wilson M.R. (1982) Insect herbivore guilds and species–area relationships: leafminers on British trees. *Ecol. Entomol.*, **7**, 19–30.

Clark L.R. (1964) The population dynamics of *Cardiaspina albitextura* [Psyllidae]. *Aust. J. Zool.*, **12**, 362–80.

Clark L.R., Geier P.W., Hughes R.D. & Morris R.F. (1967) *The Ecology of Insect Populations in Theory and Practice*. Methuen, London.

Clarke J.F.G. (1971) *The Lepidoptera of Rapa Island*. Smithsonian Contributions to Zoology 56. Smithsonian Institution Press, Washington D.C.

Clausen C.P. (1940) *Entomophagous Insects*. McGraw-Hill, New York.

Clausen C.P. (ed.) (1978) *Introduced Parasites and Predators of Arthropod Pests and Weeds: A World Review*. United States Department of Agriculture. Agricultural Handbook 480. Washington D.C.

Clements R.O., Gibson R.W., Henderson I.F. & Plumb R.T. (1978) Ryegrass: pest and virus problems. *ARC Res. Rev.* 1978. Natnl. Grassland Issue, Whitefriars Press, London. pp. 7–10.

Cody M.L. (1974) *Competition and the Structure of Bird Communities*. Princeton University Press, Princeton, N.J.

Cody M.L. & Diamond J.M. (ed.) (1975) *Ecology and Evolution of Communities*. Belknap Press, Cambridge, Mass.

Cody M.L. & Mooney H.A. (1978) Convergence versus nonconvergence in Mediterranean-climate ecosystems. *Annu. Rev. Ecol. & Syst.*, **9**, 265–321.

Coley R.D. (1980) Effects of leaf age and plant life history patterns on herbivory. *Nature (London)*, **284**, 545–6.

Collins M., Crawley M.J. and McGavin G.C. (1983) Survivorship of the sexual and agamic generations of *Andricus quercuscalicis* on *Quercus cerris* and *Q. robur*. *Ecol. Entomol.*, **8**, 133–8.

Conn E.E. (1979) Cyanide and cyanogenic glycosides. In: *Herbivores. Their Interaction with Secondary Plant Metabolites* (ed. G.A. Rosenthal & D.H. Janzen), pp. 387–412. Academic Press, New York.

Connell J.H. (1975) Some mechanisms producing structure in natural communities: a model and evidence from field experiments. In: *Ecology and Evolution of Communities* (ed. M.L. Cody & J.M. Diamond), pp. 460–90. Belknap Press, Cambridge, Mass.

Connell J.H. (1980) Diversity and the coevolution of competitors, or the ghost of competition past. *Oikos*, **35**, 131–8.

Connor E.F., Faeth S.H., Simberloff D. & Opler P.A. (1980) Taxonomic isolation and the accumulation of herbivorous insects: a comparison of introduced and native trees. *Ecol. Entomol.*, **5**, 205–11.

Connor E.F. & McCoy E.D. (1979) The statistics and biology of the species–area relationship. *Am. Nat.*, **113**, 791–833.

Cornell H.V. & Washburn J.O. (1979) Evolution of the richness–area correlation for cynipid gall wasps on oak trees: a comparison of two geographic areas. *Evolution*, **33**, 257–74.

Courtney S.P. & Courtney S. (1982) The 'edge-effect' in butterfly oviposition: causality in *Anthocharis cardamines* and related species. *Ecol. Entomol.*, **7**, 131–7.

Crane P.R. & Jarzembowski E.A. (1980) Insect leaf mines from the Palaeocene of southern England. *J. Nat. Hist.*, **14**, 629–36.

Cromartie W.J. Jr (1975) The effect of stand size and vegetational background on the colonization of cruciferous plants by herbivorous insects. *J. appl. Ecol.*, **12**, 517–33.

Cronquist A. (1973) Chemical plant taxonomy: a generalist's view of a promising speciality. In: *Chemistry in Botanical Classification* (ed. G. Benz & J. Santesson), pp. 29–39, Academic Press, London.

Crooke M. (1958) Some aspects of forest entomology in Britain. *Proceedings of the X International Congress on Entomology, Montreal 1956*, pp. 233–9.

Cross J.R. (1975) Biological flora of the British Isles. *Rhododendron ponticum* L. *J. Ecol.*, **63**, 345–64.

Crowson R.A. (1981) *The Biology of the Coleoptera*. Academic Press, New York.

Cullen J.M. (1978) Evaluating the success of the programme for the biological control of *Chondrilla juncea* L. *Proceedings of the IV International Symposium on Biological Control of Weeds, Gainesville, Florida*, pp. 117–21.

Cummins K.W. & Wuycheck J.C. (1971) Caloric equivalents for investigations in ecological energetics. *Int. Ver. theor. ang. Limnol.*, **18**, 1–158.

Darlington A. (1974) The galls on oak. In: *The British Oak. Its History and Natural History* (ed. M.G. Morris & F.H. Perring), pp. 298–311. E.W. Classey, Faringdon, Oxon.

Davis B.N.K. (1973) The Hemiptera and Coleoptera of stinging nettle (*Urtica dioica* L.) in East Anglia. *J. appl. Ecol.*, **10**, 213–37.

Davis B.N.K. (1975) The colonization of isolated patches of nettles (*Urtica dioica* L.) by insects. *J. appl. Ecol.*, **12**, 1–14.

Day P.R. (1972) Crop resistance to pests and pathogens. In: *Pest Control; Strategies for the Future*, pp. 257–71. National Academy of Sciences, Washington DC.

DeBach P. (1964) *Biological Control of Insect Pests and Weeds*. Chapman & Hall, London.

DeBach P. (1971) The use of imported natural enemies in insect pest management ecology. *Proceedings of the Tall Timbers Conference on Ecology of Animal Control by Habitat Management.*, *3*, pp. 211–33. Tallahassee, Florida.

DeBach P. (1974) *Biological Control by Natural Enemies*. Cambridge University Press, Cambridge.

DeBach P. & Schlinger E.I. (ed.) (1964) *Biological Control of Insect Pests and Weeds*. Chapman & Hall, London.

Dempster J.P. (1975) *Animal Population Ecology*. Academic Press, London.
Dempster J.P. & Hall M.L. (1980) An attempt at re-establishing the swallowtail butterfly at Wicken Fen. *Ecol. Entomol.*, **5**, 327–34.
Dempster J.P. & Lakhani K.H. (1979) A population model for cinnabar moth and its food plant, ragwort. *J. Anim. Ecol.*, **48**, 143–63.
Dennis R.L.H. (1977) *The British Butterflies. Their Origin and Establishment*. E.W. Classey, Faringdon, Oxon.
Denno R.F. (1980) Ecotope differentiation in a guild of sap-feeding insects on the salt marsh grass, *Spartina patens*. *Ecology*, **61**, 702–14.
Denno R.F., Raup M.J. & Tallamy D.W. (1981) Organization of a guild of sap-feeding insects: equilibrium vs nonequilibrium coexistence. In: *Life History Patterns: Habitat and Geographic Variation* (ed. R.F. Denno & H. Dingle), pp. 151–81. Springer-Verlag, New York.
Dickson J.G., Conner R.N., Fleet R.R., Kroll J.C. & Jackson J.A. (ed.) (1979) *The Role of Insectivorous Birds in Forest Ecosystems*. Academic Press, New York.
Dixon A.F.G. (1970) Quality and availability of food for a sycamore aphid population. *Symp. Brit. Ecol. Soc.*, **10**, 271–87.
Dixon A.F.G. (1971a) The role of aphids in wood formation. I. The effect of the sycamore aphid, *Drepanosiphum platanoides* (Schr.) (Aphididae), on the growth of sycamore, *Acer pseudoplatanus* (L.). *J. appl. Ecol.*, **8**, 165–79.
Dixon A.F.G. (1971b) The role of aphids in wood formation II. The effect of the lime aphid, *Eucallipterus tiliae* L. (Aphididae), on the growth of lime, *Tilia* x *vulgaris* Hayne. *J. appl. Ecol.*, **8**, 393–9.
Dixon A.F.G. (1979) Sycamore aphid numbers: the role of weather, host and aphid. *Symp. Brit. Ecol. Soc.*, **20**, 105–21.
Dolinger P.M., Ehrlich P.R., Fitch W.L. & Breedlove D.E. (1973) Alkaloid and predation patterns in Colorado lupine populations. *Oecologia*, **13**, 191–204.
Doutt R.L. (1959) The biology of parasitic Hymenoptera. *A. Rev. Ent.*, **4**, 161–82.
Downey J.C. & Fuller W.C. (1961) Variation in *Plebejus icariodes* (Lycaenidae). I. Foodplant specificity. *J. Lepid. Soc.*, **15**, 34–42.
Doyle J.A. (1978) Origin of angiosperms. *Annu. Rev. Ecol. & Syst.*, **9**, 365–92.
Dugdale J.S. (1975) The insects in relation to plants. In: *Biogeography and Ecology in New Zealand* (ed. G. Kuschel), pp. 561–89. *Monographiae Biologicae* 27. Junk, The Hague.
Dunlap-Pianka H., Boggs C.L. & Gilbert L.E. (1977) Ovarian dynamics in heliconiine butterflies: programmed senescence versus eternal youth. *Science* **197**, 487–90.
Eastop V.F. (1973) Deductions from the present day host plants of aphids and related species. *Symp. R. ent. Soc. Lond.*, **6**, 157–77.
Eastop V. (1979) Sternorrhyncha as angiosperm taxonomists. *Symp. Bot. Upsal.*, **22**, 120–34.
Edmunds G.F. Jr & Alstad D.N. (1978) Coevolution in insect herbivores and conifers. *Science*, **199**, 941–5.
Edwards J.S. & Tarkanian M. (1970) The adhesive pads of Heteroptera: a re-examination. *Proc. R. ent. Soc. Lond. (A)*, **45**, 1–5.

Ehrlich P.R. and Birch L.C. (1967) The 'balance of nature' and 'population control'. *Am. Nat.*, **101**, 97 107.

Ehrlich P.R. & Raven P.H. (1964) Butterflies and plants: a study in coevolution. *Evolution*, **18**, 586 608.

Eisner T., Johnessee J.S., Carrel J., Hendry L.B. & Meinwald J. (1974) Defensive use by an insect of a plant resin. *Science*, **184**, 996 9.

Elseth G.D. & Baumgardner K.D. (1981) *Population Biology*. Van Nostrand, New York.

Elton C.S. (1958) *The Ecology of Invasions by Animals and Plants*. Chapman & Hall, London.

Embree D.G. (1971) The biological control of the winter moth in eastern Canada by introduced parasites. In: *Biological Control* (ed. C.B. Huffaker), pp. 217–26. Plenum Press, New York.

Emden H.F. van (1978) Insects and secondary plant substances – an alternative viewpoint with special reference to aphids. In: *Biochemical Aspects of Plant and Animal Coevolution* (ed. J.B. Harborne), pp. 309–23. Academic Press, London.

Emden H.F. van & Way M.J. (1973) Host plants in the population dynamics of insects. *Symp. R. Entomol. Soc. Lond.*, **6**, 181 99.

Endler J.A. (1977) *Geographic Variation, Speciation and Clines*. Princeton University Press, Princeton.

Entwistle P.F. (1972) *Pests of Cocoa*. Longmans, London.

Entwistle P.F. (1978) Twenty years of *Eriozona syrphoides* (Fallén) (Dipt., Syrphidae) in Britain. *Entomologist's mon. Mag.*, **113**, 146.

Entwistle P.F., Johnson C.G. & Dunn E. (1959) New pests of cocoa (*Theobroma cacao* L.) in Ghana following applications of insecticides. *Nature (London)*, **184**, 2040.

Etten C.H. van & Tookey H.L. (1979) Chemistry and biological effects of glucosinolates. In: *Herbivores. Their Interaction with Secondary Plant Metabolites*. (ed. G.A. Rosenthal & D.H. Janzen), pp. 471 500. Academic Press, New York.

Faeth S.H. (1980) Invertebrate predation of leaf-miners at low densities. *Ecol. Entomol.*, **5**, 111 114.

Faeth S.H. & Simberloff D. (1981a) Population regulation of a leaf-mining insect, *Cameraria* sp. nov., at increased field densities. *Ecology*, **62**, 620 24.

Faeth S.H. & Simberloff D. (1981b) Experimental isolation of oak host plants: effects on mortality, survivorship, and abundances of leaf-mining insects. *Ecology*, **62**, 625 35.

Feeny P. (1970) Seasonal changes in oak leaf tannins and nutrients as a cause of spring feeding by winter moth caterpillars. *Ecology*, **51**, 565 81.

Feeny P. (1975) Biochemical coevolution between plants and their insect herbivores. In: *Coevolution of Animals and Plants* (ed. L.E. Gilbert & P.H. Raven), pp. 3–19. University of Texas Press, Austin and London.

Feeny P. (1976) Plant apparency and chemical defense. *Rec. Adv. Phytochem.*, **10**, 1 40. *Coevolution* (ed. J.B. Harborne), pp. 163 206. Academic Press, London.

Fellows D.P. & Heed W.B. (1972) Factors affecting host plant selection in desert-adapted cactiphilic *Drosophila*. *Ecology*, **53**, 850–8.

Fennah R.G. (1963) Nutritional factors associated with seasonal population increase of cacao thrips *Selenothrips rubrocinctus* (Giard) (Thysanoptera), on cashew, *Anacardium occidentale*. *Bull. ent. Res.*, **53**, 681–713.

Fischlin A. & Baltensweiler W. (1979) Systems analysis of the larch budmoth system, part I. *Bull. Soc. Ent. Suisse*, **52**, 273–89.

Flanders S.E. (1942) Abortive development in parasitic Hymenoptera, induced by the food-plant of the insect host. *J. Econ. Entomol.*, **35**, 834–5.

Force D.C. (1974) Ecology of insect host–parasitoid communities. *Science*, **184**, 624–32.

Ford E.B. (1945) *Butterflies*. Collins, London.

Fowden L. & Lea P.J. (1979) Mechanisms of plant avoidance of autotoxicity by secondary metabolites, especially by nonprotein amino acids. In: *Herbivores. Their Interaction with Secondary Plant Metabolites* (ed. G.A. Rosenthal & D.H. Janzen), pp. 135–60. Academic Press, New York.

Fowler S.V. & Lawton J.H. (1982) The effects of host-plant distribution and local abundance on the species richness of agromyzid flies attacking British umbellifers. *Ecol. Entomol.*, **7**, 257–65.

Fox L.R. (1975) Cannibalism in natural populations. *Annu. Rev. Ecol. & Syst.*, **6**, 87–106.

Fox L.R. (1981) Defense and dynamics in plant–herbivore systems. *Amer. Zool.*, **21**, 853–64.

Fox L.R. & Macauley B.J. (1977) Insect grazing on *Eucalyptus* in response to variation in leaf tannins and nitrogen. *Oecologia*, **29**, 145–62.

Fox L.R. & Morrow P.A. (1981) Specialization: species property or local phenomenon? *Science*, **211**, 887–93.

Fritz R.S. (1982) An ant–treehopper mutualism: effects of *Formica subsericea* on the survival of *Vanduzea arquata*. *Ecol. Entomol.*, **7**, 267–76.

Futuyma D.J. (1976) Food plant specialization and environmental predictability in Lepidoptera. *Am. Nat.*, **110**, 285–92.

Futuyma D.J. (1983) Evolutionary interactions among herbivorous insects and plants. In: *Coevolution* (ed. D.J. Futuyma & M. Slatkin), pp. 207–31. Sinauer, New York.

Futuyma D.J. & Gould F. (1979) Associations of plants and insects in a deciduous forest. *Ecol. Monogr.*, **49**, 33–50.

Futuyma D.J. & Mayer G.C. (1980) Non-allopatric speciation in animals. *Syst. Zool.*, **29**, 254–71.

Gage D. & Strong D.R. Jr (1981) The chemistry of *Heliconia imbricata* and *H. latispatha* and the slow growth of a hispine beetle herbivore. *Biochem. Syst. Ecol.*, **9**, 79–81.

Gallun R.L. & Khush G.S. (1980) Genetic factors affecting expression and stability of resistance. In: *Breeding Plants Resistant to Insects* (ed. F.G. Maxwell & P.R. Jennings), pp. 64–85. John Wiley & Sons, New York.

Gatz A.J. Jr (1980) Phenetic packing and community structure: a methodological comment. *Am. Nat.*, **116**, 147–9.

Geier P.W., Clark L.R. & Briese D.T. (1983) Principles for the control of arthropod pests 1. Elements and functions involved in pest control. *Protection Ecol.*, **5**, 1–96.

Gibson C.W.D. (1980) Niche use patterns among some Stenodemini (Heteroptera: Miridae) of limestone grassland, and an investigation of the possibility of interspecific competition between *Notostira elongata* Geoffroy and *Megaloceraea recticornis* Geoffroy. *Oecologia*, **47**, 352–64.

Gibson C. & Visser M. (1982) Interspecific competition between two field populations of grass-feeding bugs. *Ecol. Entomol.*, **7**, 61–7.

Gilbert L.E. (1971) Butterfly–plant coevolution: Has *Passiflora adenopoda* won the selectional race with heliconiine butterflies? *Science*, **172**, 585–6.

Gilbert L.E. (1972) Pollen feeding and the reproductive biology of *Heliconius* butterflies. *Proc. natl Acad. Sci., USA*, **69**, 1403–7.

Gilbert L.E. (1975) Ecological consequences of a coevolved mutualism between butterflies and plants. In: *Coevolution of Animals and Plants* (ed. L.E. Gilbert & P.R. Raven), pp. 210–240. University of Texas Press, Austin and London.

Gilbert L.E. (1977) The role of insect–plant coevolution in the organization of ecosystems. *Coll. Int. CNRS*, **265**, 399–413.

Gilbert L.E. (1979) Development of theory in the analysis of insect–plant interactions. In: *Analysis of Ecological Systems* (ed. D.J. Horn & R.D. Mitchell), pp. 117–54. Ohio State University Press, Columbus, Ohio.

Gilbert L.E. (1980) Food web organization and the conservation of neotropical diversity. In: *Conservation Biology. An Evolutionary–Ecological Perspective*. (ed. M.E. Soulé & B.A. Wilcox), pp. 11–33. Sinauer, Sunderland, Mass.

Gilbert L.E. (1982) The coevolution of a butterfly and a vine. *Scient. Am.* **247**, August 1982, 102–7.

Gilbert L.E. (1983a) Coevolution and mimicry. In: *Coevolution* (ed. D.J. Futuyma & M. Slatkin), pp. 263–81. Sinauer, New York.

Gilbert L.E. (1983b) The biology of butterfly communities. In: *Biology of Butterflies, Symp. R. Entomol. Soc. Lond.* (in press).

Gilbert L.E. & Raven P.H. (ed.) (1975) *Coevolution of Animals and Plants*. University of Texas Press, Austin.

Gilbert L.E. & Singer M.C. (1975) Butterfly ecology. *Annu. Rev. Ecol. & Syst.*, **6**, 365–97.

Gilbert L.E. & Smiley J.T. (1978) Determinants of local diversity in phytophagous insects: host specialists in tropical environments. *Symp. R. Entomol. Soc. Lond.*, **9**, 89–104.

Gilbert N. (1980) Comparative dynamics of a single-host aphid. I. The evidence. *J. Anim. Ecol.*, **49**, 351–69.

Gillett J.D. & Wigglesworth V.B. (1932) The climbing organ of an insect, *Rhodnius prolixus* (Hemiptera, Reduviidae). *Proc. R. Soc. B.*, **3**, 364–76.

Goeden R.D. (1971) The phytophagous insect fauna of milk thistle in Southern California. *J. Econ. Entomol.*, **64**, 1101–4.

Goeden R.D. (1974) Comparative survey of the phytophagous insect faunas of

Italian thistle, *Carduus pycnocephalus*, in southern California and southern Europe relative to biological weed control. *Environ. Entomol.*, **3**, 464–74.

Goeden R.D. & Louda S.M. (1976) Biotic interference with insects imported for weed control. *A. Rev. Ent.*, **21**, 325–42.

Goeden R.D. & Ricker D.W. (1968) The phytophagous insect fauna of Russian thistle (*Salsola kali* var. *tenuifolia*) in southern California. *Ann. ent. Soc. Am.*, **61**, 67–72.

Golterman H.L. (ed.) (1969) *Methods for Chemical Analysis of Fresh Waters*. Blackwell Scientific Publications, Oxford.

Gould F. (1979) Rapid host range evolution in a population of the phytophagous mite *Tetranychus urticae* Koch. *Evolution*, **33**, 791–802.

Gould S.J. (1981) Palaeontology plus ecology as palaeobiology. In: *Theoretical Ecology* (ed. R.M. May), pp. 295–317. Blackwell Scientific Publications, Oxford.

Gradwell G.R. (1974) The effect of defoliators on tree growth. In: *The British Oak. Its History and Natural History* (ed. M.G. Morris & F.H. Perring), pp. 182–93. E.W. Classey, Faringdon, Oxon.

Greenblatt J.A. & Barbosa P. (1981) Effects of host diet on two pupal parasitoids of the gypsy moth: *Brachymeria intermedia* (Nees.) and *Coccygomimus turionellae* (L.). *J. appl. Entomol.*, **18**, 1–10.

Greenslade P.J.M. (1971) Interspecific competition and frequency changes amongst ants in Solomon Islands cocoa plantations. *J. appl. Ecol.*, **8**, 323–52.

Gross S.W. & Fritz R.S. (1982) Differential stratification, movement and parasitism of sexes of the bagworm, *Thyridopteryx ephemeraeformis* on red cedar. *Ecol. Entomol.*, **7**, 149–54.

Guttman S.I., Wood T.K. & Karlin A.A. (1981) Genetic differentiation along host plant lines in the sympatric *Enchenopa binotata* Say complex (Homoptera: Membracidae). *Evolution*, **35**, 205–17.

Hairston N.G., Smith F.E. & Slobodkin L.B. (1960) Community structure, population control, and competition. *Am. Nat.*, **44**, 421–5.

Hammond P.M. (1974) Changes in the British coleopterous fauna. In: *The Changing Flora and Fauna of Britain* (ed. D.L. Hawksworth), pp. 323–69. Academic Press, London.

Hammond P.M. (1980) Speciation in the face of gene flow — sympatric-parapatric speciation. In: *The Evolving Biosphere* (ed. P.L. Forey), pp. 37–48. British Museum (Natural History), Cambridge University Press.

Harborne J.B. (1982) *Introduction to Ecological Biochemistry*. Academic Press, London.

Hare J.D. & Futuyma D.J. (1978) Different effects of variation in *Xanthium strumarium* L. (Compositae) in two insect seed predators. *Oecologia*, **37**, 109–20.

Harland W.B. *et al.* (ed.) (1967) *The Fossil Record*. Geological Society of London, London.

Harper J.L. (1969) The role of predation in vegetational diversity. *Brookhaven Symp. Biol.* **22**, 48–61.

Harper J.L. (1977) *Population Biology of Plants*. Academic Press, London.

Harris P. (1973) Insects in the population dynamics of plants. *Symp. R. Entomol. Soc. Lond.*, **6**, 201–9.

Harris P., Wilkinson A.T.S., Thompson L.S. & Neary M. (1978) Interaction between the cinnabar moth, *Tyria jacobaeae* L. (Lep.: Arctiidae) and ragwort, *Senecio jacobaea* L. (Compositae) in Canada. *Proceedings of the IV International Symposium on Biological Control of Weeds, Gainesville, Florida*, pp. 174–80.

Harrison J.O. (1964) Factors affecting the abundance of Lepidoptera in banana plantations. *Ecology*, **45**, 508–19.

Harrison J.W. (1927) Experiments on the egg-laying instincts of the saw-fly *Pontania salicis* Christ, and their bearing on the inheritance of acquired characters; with some remarks on a new principle in evolution. *Proc. R. Soc. B.*, **101**, 115–26.

Hartnett D.C. & Abrahamson W.G. (1979) The effects of stem gall insects on life history patterns in *Solidago canadensis*. *Ecology*, **60**, 910–17.

Hassell M.P. (1976) *The Dynamics of Competition and Predation*. Edward Arnold, London.

Hassell M.P. (1978) *The Dynamics of Arthropod Predator–Prey Systems*. Princeton University Press, Princeton, N.J.

Hassell M.P. (1980) Foraging strategies, population models and biological control: a case study. *J. Anim. Ecol.*, **49**, 603–28.

Hassell M.P. (1981) Arthropod predator–prey systems. In: *Theoretical Ecology* (ed. R.M. May), pp. 105–31. Blackwell Scientific Publications, Oxford.

Hassell M.P. (1982) Patterns of parasitism by insect parasitoids in patchy environments. *Ecol. Entomol.*, **7**, 365–77.

Hassell M.P., Lawton J.H. & May R.M. (1976) Patterns of dynamical behaviour in single-species populations. *J. Anim. Ecol.*, **45**, 471–86.

Haukioja E. & Hakala T. (1975) Herbivore cycles and periodic outbreaks. Formulation of a general hypothesis. *Rep. Kevo Subarctic Res. Stat.*, **12**, 1–9.

Haukioja E. & Niemela P. (1979) Birch leaves as a resource for herbivores: seasonal occurrence of increased resistance in foliage after mechanical damage of adjacent leaves. *Oecologia*, **39**, 151–9.

Haukioja E., Niemela P., Iso-livari L., Ojala H. & Aro E-M. (1978) Birch leaves as a resource for herbivores. I. Variation in the suitability of leaves. *Rep. Kevo Subarctic Res. Stat.*, **14**, 5–12.

Hayes J.L. (1981) The population ecology of a natural population of the pierid butterfly *Colias alexandra*. *Oecologia*, **49**, 188–200.

Heads P.A. & Lawton J.H. (1983) Studies on the natural enemy complex of the holly leaf-miner: the effects of scale on the detection of aggregative responses and the implications for biological control. *Oikos*, **40**, 267–76.

Heath J. (ed.) (1973) *Provisional Atlas of the Insects of the British Isles. Part 2. Lepidoptera (Moths–part one)*. Biological Records Centre, NERC.

Heath J. (1974) A century of change in the Lepidoptera. In: *The Changing Flora and Fauna of Britain* (ed. D.L. Hawksworth), pp. 275–92. Academic Press, London.

Heath J. (ed.) (1983). *Atlas of the Butterflies of the British Isles*. Biological Records Centre, ITE (in press).

Hedin P.A., Thompson A.C. & Gueldner R.C. (1976) Cotton plant and insect constituents that control boll weevil behavior and development. *Rec. Adv. Phytochem.*, **10**, 271–350.

Heithaus E.R., Stashko E. & Anderson P.K. (1982) Cumulative effects of plant–animal interactions on seed production by *Bauhinia ungulata*, a neotropical legume. *Ecology*, **63**, 1294–1302.

Hennig W. (1981) *Insect Phylogeny*. John Wiley & Sons, Chichester.

Hickey L.J. & Hodges R.W. (1975) Lepidopteran leaf mine from the early Eocene Wind River Formation of northwestern Wyoming. *Science*, **189**, 718–20.

Higgins L.G. & Riley N.D. (1970) *A Field Guide to the Butterflies of Britain and Europe*. Collins, London.

Hille Ris Lambers D. (1979) Aphids as botanists? *Symb. Bot. Upsal.*, **22**, 114–9.

Hirose Y., Suzuki Y., Takagi M., Hiehata K., Yamasaki M., Kimo H., Yamanaka M., Iga M. & Yamaguchi K. (1980) Population dynamics of the citrus swallowtail, *Papilio xuthus* Linné (Lepidoptera: Papilionidae): mechanisms stabilizing its numbers. *Res. Popul. Ecol.*, **21**, 260–85.

Hocking B. (1970) Insect associations with swollen thorn acacias. *Trans. R. ent. Soc. Lond.*, **122**, 211–55.

Hodek I. (ed.) (1966) *Ecology of Aphidophagous Insects*. Academia, Prague.

Holling C.S., Jones D.D. & Clark W.C. (1976) Ecological policy design: a case study of forest and pest management. *International Institute of Applied Systems Analysis Conference*, **1**, 139–58.

Holloway J.D. & Hebert P.D.N. (1979) Ecological and taxonomic trends in macrolepidopteran host plant selection. *Biol. J. Linn. Soc.*, **11**, 229–51.

Holmes R.T., Scultz J. C. & Nothnagle P. (1979) Bird predation on forest insects: an exclosure experiment. *Science*, **206**, 462–3.

Holt R.D. (1977) Predation, apparent competition, and the structure of prey communities. *Theoret. Pop. Biol.*, **12**, 197–229.

Holway R.T. (1935) Preliminary note on the structure of the pretarsus and its possible phyolgenetic significance. *Psyche*, **42**, 1–24.

Hopkins M.J.G. & Whittaker J.B. (1980) Interactions between *Apion* species (Coleoptera: Curculionidae) and Polygonaceae. II. *Apion violaceum* Kirby and *Rumex obtusifolius L. Ecol. Entomol.*, **5**, 241–7.

Horwood A.R. (1919) *British Wild Flowers in their Natural Haunts*. 6 volumes. Gresham, London.

Hsiao T.H. (1978) Host plant adaptations among geographical populations of the Colorado potato beetle. *Entomol. Exp. & Appl.*, **24**, 437–47.

Huffaker C.B. (ed.) (1971) *Biological Control*. Plenum Press, New York.

Huffaker C.B. & Kennett C.E. (1969) Some aspects of assessing efficiency of natural enemies. *Can. Ent.*, **101**, 425–47.

Hughes N.F. & Smart J. (1967) Plant–insect relationships in Palaeozoic and later time. In: *The Fossil Record.* (ed. W.B. Harland *et al.*), pp. 107–117. Geological Society of London, London.

Hutchinson G.E. (1959) Homage to Santa Rosalia or why are there so many kinds of animals? *Am. Nat.*, **93**, 145–59.

Hutchinson G.E. (1965) *The Ecological Theater and the Evolutionary Play.* Yale University Press, New Haven.

Hutchinson G.E. (1978) *An Introduction to Population Ecology.* Yale University Press, New Haven.

Huxley C. (1980) Symbiosis between ants and epiphytes. *Biol. Rev.*, **55**, 321–40.

Ikeda T., Matsumara F. & Benjamin D.M. (1977) Chemical basis for feeding adaptation of pine sawflies *Neodiprion rugifrons* and *Neodiprion swainei. Science*, **197**, 497–9.

Inouye D.W. & Talor O.R.Jr (1979) A temperate region plant–ant–seed predator system: consequences of extra floral nectar secretion by *Helianthella quinqenervis. Ecology*, **60**, 1–7.

Ito Y. (1961) Factors that affect the fluctuations of animal numbers, with special reference to insect outbreaks. *Bull. natn. Inst. agric. Sci., Tokyo*, **13**, 57–89.

Jacot-Guillarmod C.F. (1970–1975) Catalogue of the Thysanoptera of the World. *Ann. Cape prov. Mus.* **7** (1), 1–216; (2), 217–515; (3), 517–976; (4), 977–1255.

Jaenike J. (1978a) Resource predictability and niche breadth in the *Drosophila quinaria* species group. *Evolution*, **32**, 676–8.

Jaenike J. (1978b) On optimal oviposition behavior in phytophagous insects. *Theoret. Pop. Biol.*, **14**, 350–6.

Jaenike J. (1981) Criteria for ascertaining the existence of host races. *Am. Nat.*, **117**, 830–4.

Janzen D.H. (1966) Coevolution of mutualism between ants and acacias in Central America. *Evolution*, **20**, 249–75.

Janzen D.H. (1967) Interaction of the bull's horn acacia (*Acacia cornigera* L.) with an ant inhabitant (*Pseudomyrmex ferruginea* F. Smith) in eastern Mexico. *Kansas Univ. Sci. Bull.*, **47**, 315–558.

Janzen D.H. (1968a) Host plants as islands in evolutionary and contemporary time. *Am. Nat.*, **102**, 592–5.

Janzen D.H. (1968b) Allelopathy by myrmecophytes: The ant *Azteca* as an allelopathic agent of *Cecropia. Ecology*, **50**, 147–53.

Janzen D.H. (1970) Herbivores and the number of tree species in tropical forests. *Am. Nat.*, **104**, 501–28.

Janzen D.H. (1972) Protection of *Barteria* (Passifloraceae) by *Pachysima* ants (Pseudomyrmecinae) in a Nigerian rain forest. *Ecology*, **53**, 885–92.

Janzen D.H. (1973) Host plants as islands. II. Competition in evolutionary and contemporary time. *Am. Nat.*, **107**, 786–90.

Janzen D.H. (1974) Epiphytic myrmecophytes in Sarawak: Mutualism through the feeding of plants by ants. *Biotropica*, **6**, 237–59.

Janzen D.H. (1978) The ecology and evolutionary biology of seed chemistry as relates to seed predation. In: *Biochemical Aspects of Plant and Animal*

Janzen D.H. (1979) New horizons in the biology of plant defenses. In: *Herbivores. Their Interaction with Secondary Plant Metabolites* (ed. G.A. Rosenthal & D.H. Janzen), pp. 331–50. Academic Press, New York.

Janzen D.H. (1980) When is it coevolution? *Evolution*, **34**, 611–2.

Jeanne R.L. (1979) A latitudinal gradient in rates of ant predation. *Ecology*, **60**, 1211–24.

Jermy T. (1976) Insect–host plant relationship — co-evolution or sequential evolution? *Symp. Biol. Hung.*, **16**, 109–113.

Joern A. (1979a) Resource utilization and community structure in assemblages of arid grassland grasshoppers (Orthoptera: Acrididae). *Trans. Am. ent. Soc.*, **105**, 253–300.

Joern A. (1979b) Feeding patterns in grasshoppers (Orthoptera: Acrididae): factors influencing diet specialization. *Oecologia*, **38**, 325–47.

Johnson H.B. (1975) Plant pubescence: an ecological perspective. *Bot. Rev.*, **41**, 233–58.

Jones C.G. (1983) Microorganisms as mediators of resource exploitation. In: *A New Ecology: Novel Approaches to Interactive Systems* (ed. P.W. Price, C.N. Slobodchikoff & W.S. Gaud), pp. 53–99. John Wiley & Sons, New York.

Jones C.G., Aldrich J.R. & Blum M.S. (1981) Baldcypress allelochemics and the inhibition of silkworm enteric microorganisms. Some ecological considerations. *J. Chem. Ecol.*, **7**, 103–14.

Jones C.G. & Firn R.D. (1978) The role of phytoecdysteroids in bracken fern, *Pteridium aquilinum* (L.) Kuhn as a defense against phytophagous insect attack. *J. Chem. Ecol.*, **4**, 117–38.

Jones D.A. (1979) Chemical defense: primary or secondary function? *Am. Nat.*, **113**, 445–51.

Jones F.G.W. (1977) Pests, resistance and fertilizers. *12th Colloquium of the International Potash Institute*, 111–35.

Jones J.S. (1981) Models of speciation — the evidence from *Drosophila*. *Nature (London)*, **289**, 743–4.

Jones R.E. (1977) Movement patterns and egg distribution in cabbage butterflies. *J. Anim. Ecol.*, **46**, 195–212.

Jones T. (1954) The external morphology of *Chirothrips hamatus* (Trybom) (Thysanoptera). *Trans. R. ent. Soc. Lond.*, **105**, 163–187.

Journet A.R.P. (1980) Intraspecific variation in food plant favourability to phytophagous insects: psyllids on *Eucalyptus blakelyi* M. *Ecol. Entomol.*, **5**, 249–61.

Kaplanis J.N., Thompson M.J., Robbins W.E. & Bryce B.M. (1967) Insect hormones: alpha-ecdysone and 20-hydroxyecdysone in bracken fern. *Science*, **157**, 1436–8.

Karban R. & Ricklefs R.E. (1983) Host characteristics, sampling intensity, and species richness of Lepidoptera larvae on broad-leaved trees in southern Ontario. *Ecology*, **64**, 636–41.

Kareiva P. (1982a) Exclusion experiments and the competitive release of insects feeding on collards. *Ecology*, **63**, 696–704.

Kareiva P. (1982b) Influence of vegetation texture on herbivore populations: resource concentration and herbivore movement. In: *Variable Plants and Herbivores in Natural and Managed Ecosystems* (ed. R.F. Denno & M.S. McClure), pp. 259–89. Academic Press, New York.

Kayumbo H. (1963) The water relations of eggs of some Hemiptera-Heteroptera. Unpublished Ph. D. Thesis, University of London.

Keeler K.H. (1979) Distribution of plants with extrafloral nectaries and ants at two elevations in Jamaica. *Biotropica*, **11**, 152–4.

Keeler K.H. (1981) A model of selection for facultative nonsymbiotic mutualism. *Am. Nat.*, **118**, 488–98.

Kendall P. (1981) *Bromius obscurus* (L.) in Britain (Col., Chrysomelidae). *Entomol. Mon. Mag.*, **117**, 233–4.

Kevan P.G., Chaloner W.G. & Savile D.B.O. (1975) Interrelationships of early terrestrial arthropods and plants. *Palaeontology*, **18**, 391–417.

Key K.H.L. (1974) Speciation in the Australian morabine grasshoppers. Taxonomy and ecology. In: *Genetic Mechanisms of Speciation of Insects* (ed. M.J.D. White), pp.43–56. Australia and New Zealand Book Co., Sydney.

Key K.H.L. (1981) Species, parapatry, and the morabine grasshoppers. *System. Zool.*, **30**, 425–58.

Klausner E., Miller E.R. & Dingle H. (1980) *Nerium oleander* as an alternative host plant for south Florida milkweed bugs, *Oncopeltus fasciatus*. *Ecol. Entomol.*, **5**, 137–42.

Kloet G.S. & Hincks W.D. (1964) *A Check List of British Insects. Part 1: Small Orders and Hemiptera*. Royal Entomological Society of London.

Kloet G.S. & Hincks W.D. (1972) *A Check List of British Insects. Part 2: Lepidoptera*. Royal Entomological Society of London.

Kloet G.S. & Hincks W.D. (1976) *A Check List of British Insects. Part 5: Diptera and Siphonaptera*. Royal Entomological Society of London.

Kloet G.S. & Hincks W.D. (1977) *A Check List of British Insects. Part 3: Coleoptera and Strepsiptera*. Royal Entomological Society of London.

Knerer G. & Atwood C.E. (1973) Diprionid sawflies: polymorphism and speciation. *Science*, **179**, 1090–99.

Knoll A.H., Niklas K.J. & Tiffney B.H. (1979) Phanerozoic land-plant diversity in North America. *Science*, **206**, 1400–2.

Knoll A.H. & Rothwell G.W. (1981) Paleobotany: perspectives in 1980. *Paleobiology*, **7**, 7–35.

Kogan M. (1981) Dynamics of insect adaptations to soybean: impact of integrated pest management. *Environ. Entomol.*, **10**, 363–71.

Kok L.T. & Surles W.W. (1975) Successful biocontrol of musk thistle by an introduced weevil, *Rhinocyllus conicus*. *Environ. Entomol.*, **4**, 1025–7.

Koptur S. (1979) Facultative mutualism between weedy vetches bearing extrafloral nectaries and weedy ants in California. *Am. J. Bot.*, **66**, 1016–20.

Koptur S., Smith A.R. & Baker I. (1982) Nectaries in some neotropical species of

Polypodium (Polypodiaceae): preliminary observations and analysis. *Biotropica*, **14**, 108–113.

Kowalski R. (1977) Further elaboration of the winter moth population models. *J. Anim. Ecol.*, **46**, 471–82.

Kozhanchikov I.V. (1950) The conditions under which the cabbage moth (*Barathra brassicae* L.) feeds on new plants. (In Russian.) *Dokl. Akad. Nauk. SSR. (NS)*, **73**, 385–7.

Krieger R.I., Feeny P.P. & Wilkinson C.F. (1971) Detoxification enzymes in the guts of caterpillars: an evolutionary answer to plant defenses? *Science*, **172**, 579–81.

Kulman H.M. (1971) Effects of insect defoliation on growth and mortality of trees. *A. Rev. Ent.*, **16**, 289–324.

Kuris A.M., Blaustein A.R. & Alio J.J. (1980) Hosts as islands. *Am. Nat.*, **116**, 570–86.

Lack D. (1971) *Ecological Isolation in Birds*. Blackwell Scientific Publications, Oxford.

Laine K.J. & Niemela P. (1980) The influence of ants on the survival of mountain birches during an *Oporinia autumnata* (Lep., Geometridae) outbreak. *Oecologia*, **47**, 39–42.

Lamb R.J. (1980) The rise and decline of a local population of the aphid *Aphis barberae* (Homoptera: Aphididae). *Can. Ent.*, 112, 1285–9.

Latimer W. (1981) Acoustic competition in bush crickets. *Ecol. Entomol.*, **6**, 35–45.

Lawton J.H. (1976) The structure of the arthropod community on bracken. *Bot. J. Linn. Soc.*, **73**, 187–216.

Lawton J.H. (1978) Host-plant influences on insect diversity: the effects of space and time. *Symp. R. Entomol. Soc. London*, **9**, 105–25.

Lawton J.H. (1982) Vacant niches and unsaturated communities: a comparison of bracken herbivores at sites on two continents. *J. Anim. Ecol.*, **51**, 573–95

Lawton J.H. (1983a) Plant architecture and the diversity of phytophagous insects. *A. Rev. Ent.*, **28**, 23–29.

Lawton J.H. (1983b) Non-competitive populations, non-convergent communities, and vacant niches: the herbivores on bracken. In: *Ecological Communities: Conceptual Issues and the Evidence* (ed. D.R. Strong Jr, D. Simberloff & L.G. Abele), in press. Princeton University Press, Princeton, N.J.

Lawton J.H. (1983c) Herbivore community organisation: general models and specific tests with phytophagous insects. In: *A New Ecology: Novel Approaches to Interactive Systems* (ed. P.W. Price, C.N. Slobodchikoff & W.S. Gaud), in press. John Wiley & Sons, New York.

Lawton J.H., Cornell H., Dritschilo W. & Hendrix S.D. (1981) Species as islands: comments on a paper by Kuris *et al. Am. Nat.*, **117**, 623–7.

Lawton J.H. & Hassell M.P. (1981) Asymmetrical competition in insects. *Nature (London)*, **289**, 793–5.

Lawton J.H. & Hassell M.P. (1983) Interspecific competition in insects. In:

Ecological Entomology (ed. C.B. Huffaker & R.L. Rabb), in press. John Wiley & Sons, New York.

Lawton J.H. & McNeill S. (1979) Between the devil and the deep blue sea: on the problem of being a herbivore. *Symp. Brit. Ecol. Soc.*, **20**, 223–44.

Lawton J.H. & Price P.W. (1979) Species richness of parasites on hosts: agromyzid flies on the British Umbelliferae. *J. Anim. Ecol.*, **48**, 619–37.

Lawton J.H. & Schröder D. (1977) Effects of plant type, size of geographical range and taxonomic isolation on number of insect species associated with British plants. *Nature (London)*, **265**, 137–40.

Lawton J.H. & Schröder D. (1978) Some observations on the structure of phytophagous insect communities: the implications for biological control. *Proceedings of the IVth International Symposium on Biological Control of Weeds, Gainesville, Florida*, pp. 57–73. University of Florida.

Lawton J.H. & Strong D.R.Jr (1981) Community patterns and competition in folivorous insects. *Am. Nat.*, **118**, 317–38.

Lemen C. (1981) Elm trees and elm leaf beetles: patterns of herbivory. *Oikos*, **36**, 65–7.

Le Quesne W.J. (1965) The establishment of the relative status of sympatric forms, with special reference to cases among Hemiptera. *Zool. Beitr. (N.F.)*, **11**, 117–28.

Le Quesne W.J. (1972) Studies on the coexistence of three species of *Eupteryx* (Hemiptera: Cicadellidae) on nettle. *J. Entomol. (A)*, **47**, 37–44.

Leston D. (1971) Ants, capsids and swollen shoot in Ghana: interactions and the implications for pest control. *Proceedings of the 3rd International Cocoa Research Conference, Accra 1969*, 205–21.

Leston D. (1973) The ant mosaic — tropical tree crops and the limiting of pests and diseases. *Proc. natl Acad. Sci., USA*, **19**, 311–41.

Levin D.A. (1973) The role of trichomes in plant defense. *Q. Rev. Biol.*, **48**, 3–15.

Levin D.A. (1976) The chemical defenses of plants to pathogens and herbivores. *Annu. Rev. Ecol. & Syst.*, **7**, 121–59.

Levins R. & MacArthur R.H. (1969) An hypothesis to explain the incidence of monophagy. *Ecology*, **50**, 910–1.

Lewis T. (1969) Factors affecting primary patterns of infestation. *Ann. appl. Biol.*, **63**, 315–44.

Lindroth C.H. (1974) *Carabidae. Handbooks for the Identification of British Insects. 4(2)*. Royal Entomological Society, London.

Luck R.F. (1971) An appraisal of two methods of analyzing insect life tables. *Can. Ent.*, **103**, 1261–71.

MacArthur R.H. (1972) *Geographical Ecology. Patterns in the Distribution of Species*. Harper & Row, New York.

MacArthur R.H. & Wilson E.O. (1967) *The Theory of Island Biogeography*. Princeton University Press, Princeton, N.J.

MacGarvin M. (1982) Species–area relationships of insects on host plants: herbivores on rosebay willowherb. *J. Anim. Ecol.*, **51**, 207–23.

MacGarvin M. (1983) Species–area relationships of insects on rosebay willowherb. Unpublished D. Phil. Thesis, University of York.

MacKay D.A. & Singer M.C. (1982) The basis of an apparent preference for isolated host plants by ovipositing *Euptychia libye* butterflies. *Ecol. Entomol.*, **7**, 299–303.

McClure M.S. (1974) Biology of *Erythroneura lawsonii* (Homoptera: Cicadellidae) and coexistence in the sycamore leaf-feeding guild. *Environ. Entomol.*, **3**, 59–68.

McClure M.S. (1980) Competition between exotic species: scale insects on hemlock. *Ecology*, **61**, 1391–1401.

McClure M.S. (1981) Effects of voltinism, interspecific competition and parasitism on the population dynamics of the hemlock scales, *Fiorinia externa* and *Tsugaspidiotus tsugae* (Homoptera: Diaspididae). *Ecol. Entomol.*, **6**, 47–54.

McClure M.S. & Price P.W. (1975) Competition and coexistence among sympatric *Erythroneura* leafhoppers (Homoptera: Cicadellidae) on American sycamore. *Ecology*, **56**, 1388–97.

McClure M.S. & Price P.W. (1976) Ecotope characteristics of coexisting *Erythroneura* leafhoppers (Homoptera: Cicadellidae) on sycamore. *Ecology*, **57**, 928–40.

McKey D. (1974) Ant-plants: selective eating of an unoccupied *Barteria* by a colobus monkey. *Biotropica*, **6**, 269–70.

McKey D. (1979) The distribution of secondary compounds within plants. In: *Herbivores. Their Interaction with Secondary Plant Metabolites* (ed. G.A. Rosenthal & D.H. Janzen), pp. 55–133. Academic Press, New York.

McLain D.K. (1981) Resource partitioning by three species of hemipteran herbivores on the basis of host plant density. *Oecologia*, **48**, 414–7.

McNeill S. (1973) The dynamics of a population of *Leptoterna dolabrata* (Heteroptera: Miridae) in relation to its food resources. *J. Anim. Ecol.*, **42**, 495–507.

McNeill S. & Southwood T.R.E. (1978) The role of nitrogen in the development of insect/plant relationships. In: *Biochemical Aspects of Plant and Animal Coevolution* (ed. J.B. Harborne), pp. 77–98. Academic Press, London.

Majer J.D. (1976a) The ant mosaic in Ghana cocoa farms: further structural considerations. *J. appl. Ecol.*, **13**, 145–55.

Majer J.D. (1976b) The influence of ants and ant manipulation on the cocoa farm fauna. *J. appl. Ecol.*, **13**, 157–75.

Malicky H., Sobhian R. & Zwölfer H. (1970) Investigations on the possibilities of a biological control of *Rhamnus cathartica* L. in Canada: host ranges, feeding sites, and phenology of insects associated with European Rhamnaceae. *Zeit. angew. Ent.*, **65**, 77–97.

Mallet J.L.B. & Jackson D.A. (1980) The ecology and social behaviour of the Neotropical butterfly *Heliconius xanthocles* Bates in Colombia. *Zool. J. Linn. Soc.*, **70**, 1–13.

Malyshev S.I. (1968) *Genesis of the Hymenoptera and the Phases of their Evolution*. Methuen, London.

Mamaev B.M. (1968) *Evolution of Gall Forming Insects — Gall Midges*. British Library Lending Division, Boston Spa. Translated by A. Crozy (1975).

Marten J.L. (1966) The insect ecology of red pine plantations in central Ontario. IV. The crown fauna. *Can. Ent.*, **98**, 10–27.

Martin M.H. (1982) Notes on the biology of *Andricus quercuscalicis* (Burgsdorff) (Hymenoptera: Cynipidae), the inducer of knopper galls on the acorns of *Quercus robur*. *Entomologist's. mon. Mag.*, **118**, 121–3.

Mattson W.J. Jr (1980) Herbivory in relation to plant nitrogen cotent. *Annu. Rev. Ecol. & Syst.*, **11**, 119–61.

May R.M. (1972) Limit cycles in predator–prey communities. *Science*, **177**, 900–2.

May R.M. (1973) Time-delay versus stability in population models with two and three trophic levels. *Ecology*, **54**, 315–25.

May R.M. (1974a) *Stability and Complexity in Model Ecosystems*. Princeton University Press, Princeton. N.J.

May R.M. (1974b) Biological populations with nonoverlapping generations: stable points, stable cycles and chaos. *Science*, **186**, 645–7.

May R.M. (1975) Deterministic models with chaotic dynamics. *Nature (London)*, **256**, 165–6.

May R.M. (ed.) (1981) *Theoretical Ecology*. Blackwell Scientific Publications, Oxford.

May R.M. & MacArthur R.H. (1981) Niche overlap as a function of environmental variability. *Proc. natl Acad. Sci., USA*, **69**, 1109–13.

May R.M., Conway G.R., Hassell M.P. & Southwood T.R.E. (1974) Time delays, density dependence and single-species oscillations. *J. Anim. Ecol.*, **43**, 747–70.

Maynard Smith J. (1966) Sympatric speciation. *Am. Nat.*, **100**, 637–50.

Mayr E. (1963) *Animal Species and Evolution*. Harvard University Press, Cambridge, Mass.

Meijden E. van der (1979) Herbivore exploitation of a fugitive plant species: local survival and extinction of the cinnabar moth and ragwort in a heterogenous environment. *Oecologia*, **42**, 307–23.

Meijden E. van der (1980) Can hosts escape from their parasitoids? The effects of food shortage on the braconid parasitoid *Apanteles popularis* and its host *Tyria jacobaeae*. *Neth. J. Zool.*, **30**, 382–92.

Mellanby K. & French R.A. (1958) The importance of drinking water to larval insects. *Entomol. Exp. & Appl.*, **1**, 116–24.

Menhinick E.F. (1967) Structure, stability and energy flow in plants and arthropods in a *Sericea lespedeza* stand. *Ecol. Monogr.*, **37**, 255–72.

Menken S.B.J. (1981) Host races and sympatric speciation in small ermine moths, Yponomeutidae. *Entomol. Exp. & Appl.*, **30**, 280–92.

Messenger P.S. (1975) Parasites, predators, and population dynamics. In: *Insects, Science and Society* (ed. D. Pimentel), pp. 201–223. Academic Press, New York.

Messina F.J. (1981) Plant protection as a consequence of an ant–membracrid mutualism: interactions on goldenrod (*Solidago* sp.). *Ecology*, **62**, 1433–40.

Miller C.A. & Renault T.R. (1976) Incidence of parasitoids attacking endemic spruce budworm (Lepidoptera: Tortricidae) populations in New Brunswick. *Can. Ent.*, **108**, 1045–52.

Milne A. (1957) The natural control of insect populations. *Can. Ent.*, **89**, 193–213.

Mispagel M.E. (1978) The ecology and bioenergetics of the acridid grasshopper, *Bootettix punctatus*, on creosotebush, *Larrea tridentata*, in the northern Mojave desert. *Ecology*, **59**, 779–88.

Mitter C., Futuyma D.J., Schneider J.C. & Hare J.D. (1979) Genetic variation and host plant relations in a parthenogenetic moth. *Evolution*, **33**, 777–90.

Monro J. (1967) The exploitation and conservation of resources by populations of insects. *J. Anim. Ecol.*, **36**, 531–47.

Mook L.J. (1963) Birds and the spruce budworm. *Mem. ent. Soc. Can.*, **31**, 268–71.

Moran N. (1981) Intraspecific variability in herbivore performance and host quality: a field study of *Uroleucon caligatum* (Homoptera: Aphididae) and its *Solidago* hosts (Asteraceae). *Ecol. Entomol.*, **6**, 301–6.

Moran V.C. (1980) Interactions between phytophagous insects and their *Opuntia* hosts. *Ecol. Entomol.*, **5**, 153–64.

Moran V.C. & Southwood T.R.E. (1982) The guild composition of arthropod communities in trees. *J. Anim. Ecol.*, **51**, 289–306.

Morris M.G. (1974) Oak as a habitat for insect life. In: *The British Oak. Its History and Natural History* (ed. M.G. Morris & F.H. Perring), pp. 274–97. E.W. Classey, Faringdon, Oxon.

Morris R.F., Chesire W.F., Miller C.A. & Mott D.G. (1958) The numerical response of avian and mammalian predators during a gradation of the spruce budworm. *Ecology*, **39**, 487–94.

Morrison G. & Strong D.R.Jr (1980) Spatial variations in host density and the intensity of parasitism: some empirical examples. *Environ. Entomol.*, **9**, 149–52.

Morrison G. & Strong D.R.Jr (1981) Spatial variations in egg density and the intensity of parasitism in a neotropical chrysomelid (*Cephaloleia consanguinea*). *Ecol. Entomol.*, **6**, 55–61.

Morrow P.A. (1977) The significance of phytophagous insects in the *Eucalyptus* forests of Australia. In: *The Role of Arthropods in Forest Ecosystems* (ed. W.J. Mattson), pp. 19–29. Springer-Verlag, New York.

Morrow P.A., Bellas T.E. & Eisner T. (1976) *Eucalyptus* oils in defensive regurgitate of sawfly larvae (Hymenoptera: Pergidae). *Oecologia*, **19**, 293–302.

Morrow P.A. & Fox L.R. (1980) Effects of variation in *Eucalyptus* essential oil yield on insect growth and grazing damage. *Oecologia*, **45**, 209–19.

Morrow P.A. & LaMarche V.C.Jr (1978) Tree ring evidence for chronic insect suppression of productivity in subalpine *Eucalyptus*. *Science*, **201**, 1244–6.

Moulton J.C. (1909) On some of the principal mimetic (Mullerian) combinations of tropical butterflies. *Trans. ent. Soc. Lond.*, 1909, 585–606.

Mound L.A. & Halsey S.H. (1978) *Whitefly of the World. A Systematic*

Catalogue of the Aleyrodidae (Homoptera) with Host Plant and Natural Enemy Data. British Museum (Natural History) and John Wiley & Sons, Chichester.

Murdoch W.W. (1966) 'Community structure, population control, and competition' — a critique. *Am. Nat.*, **100**, 219–26.

Murdoch W.W., Evans F.C. & Peterson C.H. (1972) Diversity and pattern in plants and insects. *Ecology*, **53**, 819–29.

Myers J.H. (1981) Interactions between western tent caterpillars and wild rose: a test of some general plant herbivore hypotheses. *J. Anim. Ecol.*, **50**, 11–25.

Myers J.H. & Post B.J. (1981) Plant nitrogen and fluctuations of insect populations: A test with the cinnabar moth — tansy ragwort system. *Oecologia*, **48**, 151–6.

Myerscough P.J. (1980) Biological flora of the British Isles. *Epilobium angustifolium L. (Chamaenerion angustifolium* (L.) Scop). *J. Ecol.*, **68**, 1047–74.

Nakamura K. & Ohgushi T. (1981) Studies on the population dynamics of a thistle-feeding lady beetle, *Henosephilachna pustulosa* (Kôno) in a cool temperate climax forest II. Life tables, key-factor analysis, and detection of regulatory mechanisms. *Res. Popul. Ecol.*, **23**, 210–31.

Neuvonen S. & Niemela P. (1981) Species richness of macrolepidoptera on Finnish deciduous trees and shrubs. *Oecologia*, **51**, 364–70.

Newbery D.McC. (1980) Interactions between the coccid, *Icerya seychellarum* (Westw.), and its host tree species on Aldabra Atoll. I. *Euphorbia pyrifolia*. *Oecologia*, **46**, 171–9.

Nielsen B.O. (1978) Above ground food resources and herbivory in a beech forest ecosystem. *Oikos*, **31**, 273–9.

Nielsen B.O., & Ejlersen A. (1977) The distribution pattern of herbivory in a beech canopy. *Ecol. Entomol.*, **2**, 293–9.

Niemela P. & Haukioja E. (1982) Seasonal patterns in species richness of herbivores: macrolepidopteran larvae on Finnish deciduous trees. *Ecol. Entomol.*, **7**, 169–75.

Niemela P., Tahvanainen J., Sorjonen J., Hokkanen T. & Neuvonen S. (1982) The influence of host plant growth form and phenology on the life strategies of Finnish macrolepidopterous larvae. *Oikos*, **39**, 164–70.

Niemela P., Tuomi J. & Haukioja E. (1980) Age-specific resistance in trees: defoliation of tamaracks (*Larix laricina*) by larch bud moth (*Zeiraphera improbana*). *Rep. Kevo Subarctic Res. Stat.*, **16**, 49–57.

Nisbet R.M. & Gurney W.S.C. (1982) *Modelling Fluctuating Populations*. John Wiley & Sons, Chichester.

O'Brian C.W. & Wimber G.J. (1979) The use of trend curves of rates of species descriptions: Examples from the Curculionidae (Coleoptera). *Coleopts. Bull.*, **33**, 151–66.

O'Connor B.A. (1950) Premature nutfall of coconuts in the British Solomon Islands Protectorate. *Fiji Agric. J.*, **21**, 1–22.

Onuf C.P. (1978) Nutritive value as a factor in plant–insect interactions with an emphasis on field studies. In: *The Ecology of Arboreal Folivores* (ed. G.G.

Montgomery), pp. 85–96. Smithsonian Institution Press, Washington D.C.
Onuf C.P., Teal J.M. & Valiela I. (1977) Interactions of nutrients, plant growth and herbivory in a mangrove ecosystem. *Ecology*, **58**, 514–26.
Opler P.A., (1973) Fossil lepidopterous leaf mines demonstrate the age of some insect–plant relationships. *Science*, **179**, 1321–23.
Opler P.A. (1974) Oaks as evolutionary islands for leaf-mining insects. *Am. Scient.*, **62**, 67–73.
Orians G.H. & Solbrig O.T. (ed.) (1977) *Convergent Evolution in Warm Deserts*. Academic Press, London.
Otte D. & Joern A. (1977) On feeding patterns in desert grasshoppers and the evolution of specialized diets. *Proc. Acad. nat. Sci. Philad.*, **128**, 89–126.
Owen D.F. & Whiteway W.R. (1980) *Buddleia davidii* in Britain: history and development of an associated fauna. *Biol. Conserv.*, **17**, 149–55.
Paine R.T. (1980) Food webs: linkage, interaction strength and community infrastructure. *J. Anim. Ecol.*, **49**, 667–85.
Painter R.H. (1951) *Insect Resistance in Crop Plants*. Macmillan, New York.
Papavizas G.C. (ed.) (1981) *Biological Control in Crop Production*. Allanheld and Osmun, New York.
Parker M.A. & Root R.B. (1981) Insect herbivores limit habitat distribution of a native composite *Machaeranthera canescens*. *Ecology*, **62**, 1390–2.
Perkins D.B. (1978) Enhancement of effect of *Neochetina eichhorniae* for biological control of waterhyacinth. *Proceedings of the IVth International Symposium on the Biological Control of Weeds, Gainesville, Florida*, pp. 87–97. University of Florida.
Perrin R.M. (1980) The role of environmental diversity in crop protection. *Protection Ecol.*, **2**, 77–114.
Perring F.H. & Walters S.M. (ed.) (1962) *Atlas of the British Flora*. B.S.B.I., Nelson, London.
Peterman R.M., Clark W.C. & Holling C.S. (1979) The dynamics of resilience. Shifting stability domains in fish and insect systems. *Symp. Brit. Ecol. Soc.*, **20**, 321–41.
Phillips P.A. & Barnes M.M. (1975) Host race formation among sympatric apple, walnut and plum populations of the codling moth, *Laspeyresia pomonella*. *Ann. ent. Soc. Am.*, **68**, 1053–60.
Pianka E.R. (1972) r- and K- selection or b- and d-selection? *Am. Nat.*, **106**, 581–8.
Pickett C.H. & Clark W.D. (1979) The function of extrafloral nectaries in *Opuntia acanthocarpa* (Cactaceae). *Am. J. Bot.*, **66**, 618–25.
Pictet A. (1911) Un nouvel exemple de l'heredite des caracteres acquis. *Arch. Soc. Phyc. Nat. Geneve*, **31**, 561–3.
Pielou E.C. (1974) Biogeographic range comparisons and evidence of geographic variation in host–parasite relations. *Ecology*, **55**, 1359–67.
Pielou E.C. (1979) *Biogeography*. Wiley-Interscience, New York.
Pierce N.E. & Mead P.S. (1981) Parasitoids as selective agents in the symbiosis

between lycaenid butterfly larvae and ants. *Science*, **211**, 1185–7.

Pillemer E.A. & Tingey W.M. (1976) Hooked trichomes: a physical plant barrier to a major agricultural pest. *Science*, **193**, 482–4.

Pimentel D. (1961) The influence of plant spatial patterns on insect populations. *Ann. ent. Soc. Am.*, **54**, 61–9.

Pimm S.L. & Lawton J.H. (1980) Are food webs divided into compartments? *J. Anim. Ecol.*, **49**, 879–98.

Pipkin S.B., Rodriquez R.L. & Leon J. (1966) Plant host specificity among flower feeding neotropical *Drosophila*. *Am. Nat.*, **100**, 135–55.

Podoler H. & Rogers D. (1975) A new method for the identification of key factors from life-table data. *J. Anim. Ecol.*, **44**, 85–114.

Polis G.A. (1981) The evolution and dynamics of intraspecific predation. *Annu. Rev. Ecol. & Syst.*, **12**, 225–51.

Pollard E. (1979) Population ecology and change in range of the white admiral butterfly *Ladoga camilla* L. in England. *Ecol. Entomol.*, **4**, 61–74.

Popov Y.A. & Wootton R.J. (1977) The Upper Liassic Heteroptera of Mecklengburg and Saxony. *Syst. Entomol.*, **2**, 333–51.

Powell J.A. (1980) Evolution of larval food preferences in microlepidoptera. *A. Rev. Ent.*, **25**, 133–59.

Prestidge R.A. (1982a) Instar duration, adult consumption, oviposition and nitrogen utilization efficiencies of leafhoppers feeding on different quality food (Auchenorrhyncha: Homoptera). *Ecol. Entomol.*, **7**, 91–101.

Prestidge R.A. (1982b) The influence of nitrogenous fertilizer on the grassland Auchenorrhyncha (Homoptera). *J. appl. Ecol.*, **19**, 735–49.

Prestidge R.A. & McNeill S. (1983) The importance of nitrogen in the ecology of grassland Auchenorryncha. *Symp. Brit. Ecol. Soc.*, **22**, in press.

Preston F.W. (1960) Time and space and the variation of species. *Ecology*, **41**, 611–27.

Preston F.W. (1962) The canonical distribution of commonness and rarity. *Ecology*, **43**, 185–215, 410–32.

Price P.W. (1971) Niche breadth and dominance of parasitic insects sharing the same host species. *Ecology*, **52**, 587–96.

Price P.W. (1976) Colonization of crops by arthropods: non-equilibrium communities in soybean fields. *Environ. Entomol.*, **5**, 605–11.

Price P.W. (1977) General concepts on the evolutionary biology of parasites. *Evolution*, **31**, 405–20.

Price P.W. (1980) *Evolutionary Biology of Parasites*. Princeton University Press, Princeton, N.J.

Price P.W. (1983) Alternative paradigms in community ecology. In: *A New Ecology: Novel Approaches to Interactive Systems*. (ed. P.W. Price, C.N. Slobodchikoff & W.S. Gaud), in press. John Wiley & Sons, New York.

Price P.W., Bouton C.E., Gross P., McPheron B.A., Thompson J.N. & Weis A.E. (1980) Interactions among three trophic levels: influence of plants on interactions between insect herbivores and natural enemies. *Annu. Rev.*

Ecol. & Syst., **11**, 41–65.

Price P.W., Slobodchikoff C.N. & Gaud W.S. (ed.) (1983) *A New Ecology: Novel Approaches to Interactive Systems*. John Wiley & Sons, New York.

Price P.W. & Willson M.F. (1976) Some consequences for a parasitic herbivore, the milkweed longhorn beetle, *Tetraopes tetrophthalmus*, of a host-plant shift from *Asclepias syriaca* to *A. verticillata*. *Oecologia*, **25**, 331–40.

Pschorn-Walcher H. (1977) Biological control of forest insects. *A. Rev. Ent.*, **22**, 1–22.

Pschorn-Walcher H. & Zwölfer H. (1956) The predator complex of the white-fir woolly aphids (Genus *Dryfusia*. Adelgidae). *Z. agnew. Ent.*, **39**, 63–75.

Pyle R., Bentzien J. & Opler P. (1981) Insect conservation. *A. Rev. Ent.*, **26**, 233–58.

Rainey R.C. (ed.) (1976) *Insect Flight. Symp. R. Entomol. Soc. Lond.*, **7**.

Ralph C.P. (1977) Effect of host plant density on populations of a specialized seed sucking bug, *Oncopeltus fasciatus*. *Ecology*, **58**, 799–809.

Ramsay J.A., Butler C.G. & Sang J.H. (1938) The humidity gradient at the surface of a transpiring leaf. *J. exp. Biol.*, **15**, 255–65.

Rathcke B.J. (1976) Competition and coexistence within a guild of herbivorous insects. *Ecology*, **57**, 76–87.

Rathcke B.J. & Poole R.W. (1974) Coevolutionary race continues: butterfly larval adaptation to plant trichomes. *Science*, **187**, 175–6.

Raup D.M. (1972) Taxonomic diversity during the Phanerozoic. *Science*, **177**, 1065–71.

Rausher M.D. (1978) Search image for leaf shape in a butterfly. *Science*, **200**, 1071–3.

Rausher M.D. (1981) The effect of native vegetation on the susceptibility of *Aristolochia reticulata* (Aristolochiaceae) to herbivore attack. *Ecology*, **62**, 1187–95.

Rausher M.D. (1982) Population differentiation in *Euphydryas editha* butterflies: larval adaptation to different hosts. *Evolution*, **36**, 581–90.

Rausher M.D. & Feeny P. (1980) Herbivory, plant density and plant reproductive success: the effect of *Battus philenor* on *Aristolochis reticulata*. *Ecology*, **61**, 905–17.

Rausher M.D., MacKay D.A. & Singer M.C. (1981) Pre- and post-alighting host discrimination by *Euphydryas editha* butterflies: the behavioural mechanisms causing clumped distributions of egg clusters. *Anim. Behav.*, **29**, 1220–8.

Reader P.M. & Southwood T.R.E. (1981) The relationship between palatability to invertebrates and the successional status of a plant. *Oecologia*, **51**, 271–5.

Readshaw J.L. (1964) A theory of phasmatid outbreak release. *Aust. J. Zool.*, **13**, 475–90.

Redfern M. & Cameron R.A.D. (1978) Population dynamics of the yew gall midge *Taxomyia taxi* (Inchbald) (Diptera: Cecidomyiidae). *Ecol. Entomol.*, **3**, 251–63.

Rey, J.R. (1981) Ecological biogeography of arthropods on *Spartina* islands in northwest Florida. *Ecol. Monogr.*, **51**, 237–65.

Rey J.R., McCoy E.D. & Strong D.R. Jr, (1981) Herbivore pests, habitat islands, and the species–area relationship. *Am. Nat.*, **117**, 611–22.

Rhoades D.F. (1979) Evolution of plant chemical defense against herbivores. In: *Herbivores. Their Interaction with Secondary Plant Metabolites* (ed. G.A. Rosenthal & D.H. Janzen), pp. 3–54. Academic Press, New York.

Rhoades D.F. & Cates R.G. (1976) Toward a general theory of plant antiherbivore chemistry. *Rec. Adv. Phytochem.*, **10**, 168–213.

Richards O.W. (1940) The biology of the small white butterfly (*Pieris rapae*), with special reference to the factors controlling its abundance. *J. Anim. Ecol.*, **9**, 243–88.

Richards O.W. & Davies R.G. (1977) *Imm's General Textbook of Entomology*. Vols I and II. Chapman & Hall, London.

Rigby C. & Lawton J.H. (1981) Species–area relationships of arthropods on host plants: herbivores on bracken. *J. Biogeog.*, **8**, 125–33.

Risch S.J. (1981) Insect herbivore abundance in tropical monocultures and polycultures: an experimental test of two hypotheses. *Ecology*, **62**, 1325–40.

Risch S.J. & Carroll C.R. (1982) Effect of a keystone predaceous ant, *Solenopsis geminata*, on arthropods in a tropical agroecosystem. *Ecology*, **63**, 1979–83.

Risch S.J. & Rickson F.R. (1981) Mutualism in which ants must be present before plants produce food bodies. *Nature (London)*, **291**, 149–50.

Robertson T.S. (1981) The decline of *Carterocephalus palaemon* (Pallas) and *Maculinea arion* (L.) in Great Britain. *Entomologist's. Gaz.*, **32**, 5–12.

Robinson A.G. (1980) A new species of *Aphis* L. (Homoptera: Aphididae) from nasturtium. *Can. Ent.*, **112**, 123–5.

Robinson T., (1974) Metabolism and function of alkaloids in plants. *Science*, **184**, 430–5.

Rockwood L. (1974) Seasonal changes in the susceptibility of *Crescentia alata* leaves to the flea beetle, *Oedionychus* sp. *Ecology*, **55**, 142–8.

Rohdendorf B.B. & Raznitsin A.P. (ed.) (1980) The historical development of the class Insecta. (In Russian.) *Trudy Paleont. Inst. (Moscow)*, **175**, 1–268.

Room P.M. (1971) The relative distribution of ant species in Ghana's cocoa farms. *J. Anim. Ecol.*, **40**, 735–51.

Room P.M., Harley K.L.S., Forno I.W. & Sands D.P.A. (1981) Successful biological control of the floating weed *Salvinia*. *Nature (London)*, **294**, 78–80.

Root R.B. (1973) Organization of a plant–arthropod association in simple and diverse habitats: the fauna of collards (*Brassica oleracea*). *Ecol. Monogr.*, **43**, 95–124.

Root R.B. & Tahvanainen J.O. (1969) Role of winter cress, *Barbarea vulgaris*, as a temporal host in the seasonal development of the crucifer fauna. *Ann. Ent. Soc. Am.*, **62**, 852–5.

Rosenthal G.A. & Bell E.A. (1979) Naturally occurring, toxic nonprotein amino acids. In: *Herbivores. Their Interaction with Secondary Plant Metabolites* (ed. G.A. Rosenthal & D.H. Janzen), pp. 353–85. Academic Press, New York.

Rosenthal G.A. & Janzen D.H. (ed.) (1979) *Herbivores. Their Interaction with*

Secondary Plant Metabolites. Academic Press, New York.

Ross G.N. (1966) Life-history studies on Mexican butterflies. IV. The ecology and ethology of *Anatole rossi*, a myrmecophilous metalmark (Lepidoptera: Riodinidae). *Ann. ent. Soc. Am.*, **59**, 985–1004.

Rothschild G.H.L. (1971) The biology and ecology of rice-stem borers in Sarawak (Malaysian Borneo). *J. appl. Ecol.*, **8**, 287–322.

Rothschild M. (1973) Secondary plant substances and warning colouration in insects. *Symp. R. Entomol. Soc. Lond.*, **6**, 59–79.

Royama T. (1981a) Fundamental concepts and methodology for the analysis of animal population dynamics, with particular reference to univoltine species. *Ecol. Monogr.*, **51**, 473–93.

Royama T. (1981b) Evaluation of mortality factors in insect life table analysis. *Ecol. Monogr.*, **51**, 495–505.

Ryan C.A. & Green T.R. (1974) Proteinase inhibitors in natural plant protection. *Rec. Adv. Phytochem.*, **8**, 123–40.

Sale P.F. (1977) Maintenance of high diversity in coral reef fish communities. *Am. Nat.*, **111**, 337–59.

Sanders C.J. & Knight F.B. (1968) Natural regulation of the aphid *Pterocomma populifoliae* on bigtooth aspen in northern lower Michigan. *Ecology*, **49**, 234–44.

Sattler K. (1981) *Scythris inspersella* (Huebner, (1817)) new to the British fauna (Lepidoptera: Scythrididae). *Entomologist's Gaz.*, **32**, 13–17.

Schemske D.W. (1980) The evolutionary significance of extrafloral nectar production by *Costus woodsonii* (Zingiberaceae): an experimental analysis of ant protection. *J. Ecol.*, **68**, 959–67.

Schemske D.W. (1982) Ecological correlates of a neotropical mutualism: ant assemblages at *Costus* extrafloral nectaries. *Ecology*, **63**, 932–41.

Schoener T.W. (1974) Resource partitioning in ecological communities. *Science*, **185**, 27–39.

Schoonhoven L.M. (1973) Plant recognition by lepidopterous larvae. *Symp. R. Entomol. Soc. London*, **6**, 87–99.

Schroder C. (1903) Uber experimentell erzielte instinktuariationen. *Verh. deutsch. Zool. Ges.*, p. 158.

Schultz J.C. & Baldwin I.T. (1982) Oak leaf quality declines in response to defoliation by gypsy moth larvae. *Science*, **217**, 149–51.

Schultz J.C., Otte D. & Enders F. (1977) *Larrea* as a habitat component for desert arthropods. In: *Creosote Bush: Biology and Chemistry of* Larrea *in New World Deserts* (ed. T.J. Mabry, J.H. Hunzicker & D.R. Difeo), pp. 176–208. Dowden, Huchinson & Ross, Stroudsburg, Penn.

Schulz CA. & Meijer J. (1978) Migration of leafhoppers (Homoptera: Auchenorrhyncha) into a new polder. *Holarctic Ecol.*, **1**, 73–8.

Scorer A.G. (1913) *The Entomologist's Log-book*. Routledge, London.

Scriber J.M. (1979) Effects of leaf-water supplementation upon post-ingestive nutritional indices of forb-, shrub-, vine-, and tree-feeding Lepidoptera. *Entomol. exp. Appl.*, **25**, 240–52.

Scriber J.M. & Feeny P. (1979) Growth of herbivorous caterpillars in relation to feeding specialization and to the growth form of their food plants. *Ecology*, **60**, 829–50.

Scriber J.M. & Slansky F. Jr (1981) The nutritional ecology of immature insects. *A. Rev. Ent.*, **26**, 183–211.

Seifert R.P. (1982) Neotropical *Heliconia* insect communities. *Q. Rev. Biol.*, **57**, 1–28.

Seifert R.P. (1983) Does competition structure communities? Field studies on neotropical *Heliconia* insect communities. In: *Ecological Communities: Conceptual Issues and the Evidence* (ed. D.R. Strong, D. Simberloff & L.G. Abele), in press. Princeton University Press, Princeton, N.J.

Seifert R.P. & Seifert F.H. (1976) A community matrix analysis of *Heliconia* insect communities. *Am. Nat.*, **110**, 461–83.

Seifert R.P. & Seifert F.H. (1979a) A *Heliconia* insect community in a Venezuelan cloud forest. *Ecology*, **60**, 462–7.

Seifert R.P. & Seifert F.H. (1979b) Utilization of *Heliconia* (Musaceae) by the beetle *Xenarescus monocerus* (Oliver) (Chrysomelidae: Hispinae) in a Venezuelan forest. *Biotropica*, **11**, 51–9.

Seigler D. & Price P.W. (1976) Secondary compounds in plants: primary functions. *Am. Nat.*, **110**, 101–5.

Shapiro A.M. (1974) Partitioning of resources among lupine-feeding Lepidoptera. *Am. Mid., Nat.*, **91**, 243–8.

Shapiro A.M. (1981) The pierid red-egg syndrome. *Am. Nat.*, **117**, 276–94.

Sharov A.G. (1972) On the phylogenetic relations of the order of thripses (Thysanoptera). (In Russian.) *Ent. Obozr.*, **51**, 854–8.

Shields O. (1976). Fossil butterflies and the evolution of Lepidoptera. *J. Res. Lep.*, **15**, 132–43.

Shorrocks B., Atkinson W. & Charlesworth P. (1979) Competition on a divided and ephemeral resource. *J. Anim. Ecol.*, **48**, 899–908.

Shotton F.W. (ed.) (1977) *British Quaternary Studies. Recent Advances*. Oxford University Press. Oxford.

Side K.C. (1955) A study of the insects living on the wayfaring tree. *Bull. amat. Ent. Soc.*, **14**, 3–5, 11–14, 19–22, 28–31, 42–3, 47–50.

Siegel S. (1956) *Nonparametric Statistics for the Behavioral Sciences*. McGraw Hill, New York.

Simberloff D.S. (1976) Experimental zoogeography of islands: Effects of island size. *Ecology*, **57**, 629–48.

Simberloff D.S. (1978) Colonisation of islands by insects: immigration, extinction and diversity. *Symp. R. Entomol. Soc. Lond.*, **9**, 139–53.

Simberloff D.S., Brown B.J. & Lowrie S. (1978) Isopod and insect root borers may benefit Florida mangroves. *Science*, **210**, 630–2.

Singer M.C. (1971) Evolution of food-plant preference in the butterfly *Euphydryas editha*. *Evolution*, **25**, 383–9.

Skalski A.W. (1979) A new member of the family Micropterigidae (Lepidoptera) from the Lower Cretaceous of Transbaikalia. *Paleontol. Zh.*, 1979, 90–7.

Skinner G.J. (1980) The feeding habits of the wood-ant, *Formica rufa* (Hymenoptera: Formicidae) in limestone woodland in North-west England. *J. Anim. Ecol.*, **49**, 417–33.

Skinner G.J. & Whittaker J.B. (1981) An experimental investigation of inter-relationships between the wood-ant (*Formica rufa*) and some tree-canopy herbivores. *J. Anim. Ecol.*, **50**, 313–26.

Slade N.A. (1977) Statistical detection of density dependence from a series of sequential censuses. *Ecology*, **58**, 1094–1102.

Slansky F. Jr (1974) Relationship of larval food-plants and voltinism patterns in temperate butterflies. *Psyche*, **81**, 243–53.

Slansky F. Jr (1976) Phagism relationships among butterflies. *Jl. N. Y. Ent. Soc.* **84**, 91–105.

Slobodkin L.B., Smith F.E. & Hairston N.G. (1967) Regulation in terrestrial ecosystems, and the implied balance of nature. *Am. Nat.*, **101**, 109–24.

Smart J. & Hughes N.F. (1973) The insect and the plant: progressive palaeoecological integration. *Symp. R. Entomol. Soc. Lond.*, **6**, 143–55.

Smiley J. (1978) Plant chemistry and the evolution of host specificity: new evidence from *Heliconius* and *Passiflora*. *Science*, **201**, 745–7.

Smith D.A.S. (1978) Cardiac glycosides in *Danaus chrysippus* (L.) provide some protection against an insect parasitoid. *Experientia*, **34**, 844–5.

Sogawa K. (1982) The rice brown planthopper: feeding physiology and host plant interactions. *A. Rev. Ent.*, **27**, 49–74.

Solomon B.P. (1981) Response of a host-specific herbivore to resource density, relative abundance, and phenology. *Ecology*, **62**, 1205–14.

Solomon M.E. & Glen D.M. (1979) Prey density and rate of predation by tits (*Parus* spp.) on larvae of codling moth (*Cydia pomonella*) under bark. *J. appl. Ecol.*, **16**, 49–59.

Southwood T.R.E. (1953) The morphology and taxonomy of the genus *Orthotylus* Fieber (Hem., Miridae), with special reference to the British species. *Trans. R. ent. Soc. Lond.*, **104**, 415–449.

Southwood T.R.E. (1955) Some studies on the systematics and ecology of Heteroptera. Unpublished Ph.D. Thesis, University of London.

Southwood T.R.E. (1957) The zoogeography of the British Hemiptera Heteroptera. *Proc. S. Lond. ent. Nat. Hist. Soc.*, 1956, 111–36.

Southwood T.R.E. (1960a) The abundance of the Hawaiian trees and the number of their associated insect species. *Proc. Hawaiian ent. Soc.*, **17**, 299–303.

Southwood T.R.E. (1960b) The flight activity of Heteroptera. *Trans. R. Entomol. Soc. Lond.*, **112**, 173–220.

Southwood T.R.E. (1961a) The number of species of insect associated with various trees. *J. Anim. Ecol.*, **30**, 1–8.

Southwood T.R.E. (1961b) The evolution of the insect–host tree relationship — a new approach *Proceedings of the XIth International Congress on Entomology., Vienna 1960*, pp. 651–4.

Southwood T.R.E. (1973) The insect/plant relationship — an evolutionary perspective. *Symp. R. Entomol. Soc. Lond.*, **6**, 3–30.

Southwood T.R.E. (1975) The dynamics of insect populations. In: *Insects, Science and Society* (ed. D. Pimentel), pp. 151–99. Academic Press, New York.

Southwood T.R.E. (1977a) The stability of the trophic milieu, its influence on the evolution of behaviour and of responsiveness to trophic signals. *Coll. Int. CNRS*, **265**, 471–93.

Southwood T.R.E. (1977b) Habitat, the templet for ecological strategies. *J. Anim. Ecol.*, **46**, 337–65.

Southwood T.R.E. (1978a) The components of diversity. *Symp. R. Entomol. Soc. Lond.*, **9**, 19–40.

Southwood T.R.E. (1978b) *Ecological Methods*. Chapman & Hall, London.

Southwood T.R.E., Brown V.K. & Reader P.M. (1979) The relationships of plant and insect diversities in succession. *Biol. J. Linn. Soc.*, **12**, 327–48.

Southwood T.R.E. & Comins H.N. (1976) A synoptic population model. *J. Anim. Ecol.*, **45**, 949–65.

Southwood T.R.E., May R.M., Hassell M.P. & Conway G.R. (1974) Ecological strategies and population parameters. *Am. Nat.*, **108**, 791–804.

Southwood T.R.E., Moran V.C. & Kennedy C.E.J. (1982a) The assessment of arboreal insect fauna: comparisons of knockdown sampling and faunal lists. *Ecol. Entomol.*, **7**, 331–40.

Southwood T.R.E., Moran V.C. & Kennedy C.E.J. (1982b) The richness, abundance and biomass of the arthropod communities on trees. *J. Anim. Ecol.*, **51**, 635–49.

Southwood T.R.E. & Reader P.M. (1976) Population census data and key factor analysis for the viburnum whitefly, *Aleurotrachelus jelinekii* (Frauenf.), on three bushes. *J. Anim. Ecol.*, **45**, 313–25.

Spencer K.A. (1972) *Handbooks for the Identification of British Insects. Diptera, Agromyzidae. Vol. X, Part 5(g)*. Royal Entomological Society, London.

Sporne K.R. (1975) *The Morphology of Pteridophytes*. Hutchinson, London.

Stark R.W. (1965) Recent trends in forest entomology. *A. Rev. Ent.*, **10**, 303–24.

Starks K.J., Muniappan R. & Eikenbary R.D. (1972) Interaction between plant resistance and parasitism against the greenbug on barley and sorghum. *Ann. ent. Soc. Am.*, **65**, 650–5.

Starmer W.T., Heed W.B., Miranda M., Miller M.W. & Phaff H.J. (1976) The ecology of yeast flora associated with cactiphilic Drosophila and their host plants in the Sonoran Desert. *Microb. Ecol.*, **3**, 11–30.

Starý P. & Rejmánek M. (1981) Number of parasitoids per host in different systematic groups of aphids: the implications for introduction strategy in biological control (Homoptera: Aphidoidea; Hymenoptera: Aphididae). *Entomol. Scandinav. (Suppl.)*, **15**, 341–51.

Stiling P.D. (1980) Competition and coexistence among *Eupteryx* leafhoppers (Hemiptera: Cicadellide) occurring on stinging nettles (*Urtica dioica*). *J. Anim. Ecol.*, **49**, 793–805.

Stiling P.D., Broadbeck B.V. & Strong D.R. (1982) Foliar nitrogen and larval parasitism as determinants of leafminer distribution patterns on *Spartina alterniflora. Ecol. Entomol.*, **7**, 447–52.

Stiling P.D. & Strong D.R. (1982) Egg density and the intensity of parasitism in *Prokelesia marginata* (Homoptera: Delphacidae). *Ecology*, **63**, 1630–5.

Stiling P.D. & Strong D.R. (1983) Weak competition among *Spartina* stem borers, by means of murder. *Ecology*, **64**, 770–8.

Stork N.E. (1980) A scanning electron microscope study of tarsal adhesive setae in the Coleoptera. *Zool. J. Linn. Soc.*, **68**, 173–306.

Straatman R. (1962) Notes on certain Lepidoptera ovipositing on plants which are toxic to their larvae. *J. Lepid. Soc.*, **16**, 99–103.

Strong D.R. Jr (1974a) Rapid asymptotic species accumulation in phytophagous insect communities: the pests of Cacao. *Science*, **185**, 1064–6.

Strong D.R. Jr (1974b) Nonasymptotic species richness models and the insects of British trees. *Proc. natl Acad. Sci., USA*, **71**, 2766–9.

Strong D.R. Jr (1977) Insect species richness: hispine beetles on *Heliconia latispatha. Ecology*, **58**, 573–82.

Strong D.R. Jr (1979) Biogeographic dynamics of insect–host plant communities. *A. Rev. Ent.*, **24**, 89–119.

Strong D.R. Jr (1982a) Potential interspecific competition and host specificity: hispine beetles on *Heliconia. Ecol. Entomol.*, **7**, 217–20.

Strong D.R. Jr (1982b) Harmonious coexistence of hispine beetles on *Heliconia* in experimental and natural communities. *Ecology*, **63**, 1039–49.

Strong D.R. Jr (1983a) Density-vague ecology and liberal population regulation in insects. In: *A New Ecology: Novel Approaches to Interactive Systems* (ed. P.W. Price, C.N. Slobodchikoff & W.S. Gaud), in press. John Wiley & Sons, New York.

Strong D.R. Jr (1983b) Exorcising the ghost of competition past from insect communities. In: *Ecological Communities: Conceptual Issues and the Evidence* (ed. D.R. Strong, D. Simberloff & L.G. Abele), in press. Princeton University Press, Princeton, N.J.

Strong D.R. Jr & Levin D.A. (1975) Species richness of the parasitic fungi of British trees. *Proc. natl Acad. Sci., USA*, **72**, 2116–9.

Strong D.R. Jr & Levin D.A. (1979) Species richness of plant parasites and growth form of their hosts. *Am. Nat.*, **114**, 1–22.

Strong D.R. Jr, McCoy E.D. & Rey J.R. (1977) Time and the number of herbivore species: the pests of sugarcane. *Ecology*, **58**, 167–75.

Strong D.R. Jr, Simberloff D. & Abele L.G. (ed.) (1983) *Ecological Communities: Conceptual Issues and the Evidence*. Princeton University Press, Princeton, N.J. In press.

Stubbs M. (1977) Density dependence in the life-cycles of animals and its important in K- and r-strategies. *J. Anim. Ecol.*, **46**, 677–88.

Sunderland K.D. & Vickerman G.P. (1980) Aphid feeding by some polyphagous predators in relation to aphid density in cereal fields. *J. appl. Ecol.*, **17**, 389–96.

Swain T. (1978) Plant–animal coevolution: a synoptic view of the Paleozoic and

Mesozoic. In: *Biochemical Aspects of Plant and Animal Coevolution* (ed. J.B. Harborne), pp. 3–19. Academic Press, London.

Tahvanainen J.O. & Root R.B. (1972) The influence of vegetation diversity on the population ecology of a specialized herbivore, *Phyllotreta cruciferae* (Coleoptera: Chrysomelidae). *Oecologia*, **10**, 321–46.

Taksdal G. (1965) Hemiptera (Heteroptera) collected on ornamental trees and shrubs at the agricultural college of Norway. *Ås. Norsk ent. Tidsskrift*, **13**, 5–10.

Tanton M.T. (1962) The effect of leaf 'toughness' on the feeding of larvae of the mustard beetle. *Ent. Exp. & Appl.*, **5**, 74–78.

Tauber C.A. & Tauber M.J. (1981) Insect seasonal cycles: genetics and evolution. *Annu. Rev. Ecol. & Syst.*, **12**, 281–308.

Tavormina S.J. (1982) Sympatric genetic divergence in the leaf-mining insect *Liriomyza brassicae* (Diptera: Agromyzidae): *Evolution*, **36**, 523–34.

Taylor L.R. (1959) Abortive feeding behaviour in a black aphid of the *Aphis fabae* group. *Entomol. Exp. & Appl.*, **2**, 143–153.

Taylor L.R., Woiwood I.P. & Perry J.N. (1980) Variance and the large scale spatial stability of aphids, moths and birds. *J. Anim. Ecol.*, **49**, 831–54.

Taylor R.A.J. & Taylor L.R. (1979) A behavioural model for the evolution of spatial dynamics. *Symp. Brit. Ecol. Soc.*, **20**, 1–27.

Tempel A.S. (1981) Field studies on the relationship between herbivore damage and tannin concentration in bracken (*Pteridium aquilinum* Kuhn). *Oecologia*, **51**, 97–106.

Templeton A.R. (1979) Genetics of colonization and establishment of exotic species. In: *Genetics in Relation to Insect Management* (ed. M.A. Hoy & J.J. McKelvey Jr), pp. 41–9. Rockefeller Foundation, New York.

Tepedino V.J. & Stanton N.L. (1976) Cushion plants as islands. *Oecologia*, **25**, 243–56.

Thielges B.A. (1968) Altered polyphenol metabolism in the foliage of *Pinus silvestris* associated with European pine sawfly attack. *Can. J. Bot.*, **46**, 724–5.

Thompson J.N. (1978) Within-patch structure and dynamics in *Pastinaca sativa* and resource availability to a specialized herbivore. *Ecology*, **59**, 443–8.

Thompson J.N. (1981) Reversed animal–plant interactions: the evolution of insectivorous and ant-fed plants. *Biol. J. Linn. Soc.*, **16**, 147–55.

Thompson J.N. (1982) *Interaction and Coevolution*. John Wiley & Sons, New York.

Thompson J.N. & Price P.W. (1977) Plant plasticity, phenology and herbivore dispersion: wild parsnip and the parsnip webworm. *Ecology*, **58**, 1112–9.

Thompson W.R. (continued by F.J. Simmonds & B. Herting) (1943–1971) *A Catalogue of the Parasites and Predators of Insect Pests* (Section 1, parts 1–11). Commonwealth Agricultural Bureau, Belleville, Ontario.

Tilden J.W. (1951) The insect associates of *Baccharis pilularis* De Candolle. *Microentomology*, **16**, 149–85.

Tilman D. (1978) Cherries, ants and tent caterpillars: timing of nectar production in relation to susceptibility of caterpillars to ant predation. *Ecology*, **59**, 686–92.

Tinbergen N. (1974) *Curious Naturalists*. Penguin, London.

Turnipseed S.G. & Kogan M. (1976) Soybean entomology. *A. Rev. Ent.*, **21**, 247–82.

Varley G.C. (1949) Population changes in German forest pests. *J. Anim. Ecol.*, **18**, 117–22.

Varley G.C. & Gradwell G.R. (1960) Key factors in population studies. *J. Anim. Ecol.*, **26**, 251–61.

Varley G.C. & Gradwell G.R. (1970) Recent advances in insect population dynamics. *A. Rev. Ent.* **15**, 1–24.

Varley G.C., Gradwell G.R. & Hassell M.P. (1973) *Insect Population Ecology. An Analytical Approach*. Blackwell Scientific Publications, Oxford.

Vince S.W., Valiela I. & Teal J.M. (1981) An experimental study of the structure of herbivorous insect communities in a salt marsh. *Ecology*, **62**, 1662–78.

Voûte A.D. (1946) Regulation of the density of insect-populations in virgin forests and cultivated woods. *Arch. néerl. Zool.*, **7**, 435–70.

Wallner W.E. & Walton G.S. (1979) Host defoliation: a possible determinant of gypsy moth population quality. *Ann. ent. Soc. Am.*, **72**, 62–7.

Waloff N. (1966) Scotch broom (*Sarothamnus scoparius* (L.) Wimmer) and its fauna introduced into the Pacific Northwest of America. *J. appl. Ecol.*, **3**, 293–311.

Waloff N. (1968a) Studies on the insect fauna on Scotch broom *Sarothamnus scoparius* (L.) Wimmer. *Adv. Ecol. Res.*, **5**, 87–208.

Waloff N. (1968b) A comparison of factors affecting different insect species on the same host plant. *Symp. R. Entomol. Soc. Lond.*, **4**, 76–87.

Waloff N. & Richards O.W. (1977) The effect of insect fauna on growth, mortality and natality of broom, *Sarothamnus scoparius*. *J. appl. Ecol.*, **14**, 787–98.

Waloff N. & Thompson P. (1980) Census data of populations of some leafhoppers (Auchenorrhyncha, Homoptera) of acid grassland. *J. Anim. Ecol.*, **49**, 395–416.

Ward L.K. (1977) The conservation of juniper: the associated fauna with special reference to southern England. *J. appl. Ecol.*, **14**, 81–120.

Ward L.K. & Lakhani K.H. (1977) The conservation of juniper: the fauna of food-plant island sites in southern England. *J. appl. Ecol.*, **14**, 121–35.

Washburn J.O. & Cornell H.V. (1981) Parasitoids, patches, and phenology: their possible role in the local extinction of a cynipid gall wasp population. *Ecology*, **62**, 1597–1607.

Watanabe M. (1981) Population dynamics of the swallowtail butterfly, *Papilio xuthus* L., in a deforested area. *Res. Popul. Ecol.*, **23**, 74–93.

Watanabe M. & Omata K (1978) On the mortalilty factors of the Lycaenid butterfly, *Artopoetes pryeri* M. (Lepidoptera, Lycaenidae). *Jap. J. Ecol.*, **28**, 367–70.

Waterhouse D.F. (ed). (1970) *The Insects of Australia*. Melbourne University Press.

Way M.J. (1953) The relationship between certain ant species with particular reference to biological control of the coreid, *Theraptus* sp. *Bull. ent. Res.*, **44**, 669–91.

Way M.J. (1963) Mutualism between ants and honeydew-producing Homoptera. *A. Rev. Ent.*, **8**, 307–44.

Way M.J. & Murdie G. (1965) An example of varietal variations in resistance of Brussels sprouts. *Ann. appl. Biol.*, **56**, 326–8.

Whalley P. (1977) Lower Cretaceous Lepidoptera. *Nature (London)*, **266**, 526.

Whalley P. & Jarzembowski E.A. (1981) A new assessment of *Rhyniella*, the earliest known insect, from the Devonian of Rhynie, Scotland. *Nature (London)*, **291**, 317.

Wheeler A.G. Jr. (1974) Phytophagous arthropod fauna of crownvetch in Pennsylvania. *Can. Ent.*, **106**, 897–908.

White M.G. (1966) Ecological and chemical control of damage by the *Phytolyma* gall bug (Hemiptera: Psyllidae) on the iroko tree (*Chlorophora* species) in Africa. *Proc. R. Entomol. Soc. (C)*, **31**, 43–4.

White M.J.D. (1974) Speciation in the Australian morabine grasshoppers: The cytogenetic evidence. In: *Genetic Mechanisms of Speciation in Insects* (ed. M.J.D. White), pp. 57–68. Australia and New Zealand Book Co., Sydney.

White M.J.D. (1978) *Modes of Speciation*. W.H. Freeman, San Francisco.

Whitham T.G. (1978) Habitat selection by *Pemphigus* aphids in response to resource limitation and competition. *Ecology*, **59**, 1164–76.

Whitham T.G. (1981) Individual trees as heterogeneous environments: adaptation to herbivory or epigenetic noise? In: *Insect Life History Patterns: Habitat and Geographic Variation* (ed. R.F. Denno & H. Dingle), pp. 9–27. Springer-Verlag, New York.

Whitham T.G. & Slobodchikoff C.N. (1981) Evolution by individuals, plant–herbivore interactions, and mosaics of genetic variability: the adaptive significance of somatic mutations. *Oecologia*, **49**, 287–92.

Whittaker J.B. (1979) Invertebrate grazing, competition and plant dynamics. *Symp. Brit. Ecol. Soc.*, **20**, 207–222.

Whittaker R.H. (1969) Evolution of diversity in plant communities. *Brookhaven Symp. Biol.*, **22**, 178–98.

Whittaker R.H. (1977) Evolution of species diversity in land communities. *Evol. Biol.*, **10**, 1–67.

Whittaker R.H. & Feeny P.P. (1971) Allelochemicals: chemical interactions between species. *Science*, **171**, 757–70.

Whittaker R.H. & Woodwell G.M. (1972) Evolution of natural communities. In: *Ecosystem Structure and Function* (ed. J.A. Wiens), pp. 137–56. Oregon State University Press, Corvallis.

Wigglesworth V.B. (1972) *Principles of Insect Physiology*. Chapman & Hall, London.

Wiklund C. (1973) Host plant suitability and the mechanism of host selection in larvae of *Papilio machaon*. *Entomol. Exp. & Appl.*, **16**, 232–42.

Wiklund C. (1974) The concept of oligophagy and the natural habits and host plants of *Papilio machaon* L. in Fennoscandia. *Entomol. Scand.*, **5**, 151–60.

Wiklund C. (1977) Oviposition, feeding and spatial separation of breeding and foraging habitats in a population of *Leptidea sinapis* (Lepidoptera), *Oikos*, **28**, 56–68.

Williams C.B. (1964) *Patterns in the Balance of Nature*. Academic Press, New York.

Williams K.S. & Gilbert L.E. (1981) Insects as selective agents on plant vegetative morphology: egg mimicry reduces egg laying by butterflies. *Science*, **212**, 467–9.

Williamson M.H. (1972) *The Analysis of Biological Populations*. Edward Arnold, London.

Williamson M.H. (1981) *Island Populations*. Oxford University Press, Oxford.

Williamson M.H. (1957) An elementary theory of interspecific competition. *Nature (London)*, **180**, 422–5.

Willmer P.G. (1980) The effects of a fluctuating environment on the water relations of larval Lepidoptera. *Ecol. Entomol.*, **5**, 271–92.

Wilson M.V.H. (1978) Evolutionary significance of North American Paleocene insect faunas. *Quaest. Entomol.*, **14**, 35–2.

Winter T.G. (1974) New host plant records of Lepidoptera associated with conifer afforestation in Britain. *Entomologists' Gaz.*, **25**, 247–58.

Wise D.H. (1981) A removal experiment with darkling beetles: lack of evidence for interspecific competition. *Ecology*, **62**, 727–38.

Wolda H. (1978a) Fluctuations in abundance of tropical insects. *Am. Nat.*, **112**, 1017–45.

Wolda H. (1978b) Seasonal fluctuations in rainfall, food and abundance of tropical insects. *J. Anim. Ecol.*, **47**, 369–81.

Wood T.K. (1980) Divergence in the *Enchenopa binotata* Say complex (Homoptera: Membracidae) effected by host plant adaptation. *Evolution*, **34**, 147–60.

Wood T.K. & Guttman S.I. (1981) The role of host plants in the speciation of treehoppers: an example from the *Enchenopa binotata* complex. In: *Insect Life History Patterns: Habitat and Geographic Variation* (ed. R.F. Denno & H. Dingle), pp. 39–54. Springer-Verlag, New York.

Wood T.K. & Guttman S.I. (1982) Ecological and behavioral basis for reproductive isolation in the sympatric *Enchenopa binotata* complex (Homoptera: Membracidae). *Evolution*, **36**, 233–42.

Woodhead S. (1981) Environmental and biotic factors affecting the phenolic content of different cultivars of *Sorghum bicolor*. *J. Chem. Ecol.*, **7**, 1035–47.

Wootton R.J. (1981) Palaeozoic insects. *A. Rev. Ent.*, **26**, 319–44.

Wratten S.D., Goddard P. & Edwards P.J. (1981) British trees and insects: the role of palatability. *Am. Nat.*, **118**, 916–9.

Yasumatsu K. (1976) Rice stem-borers. In: *Studies in Biological Control* (ed. V.L. Delucchi), pp. 121–37. Cambridge University Press, Cambridge.

Zaret T.M. (1980) *Predation and Freshwater Communities*. Yale University Press, New Haven.

Zelazny B. & Pacumbaba E. (1982) Phytophagous insects associated with cadang-cadang infected and healthy coconut palms in South-eastern Luzon, Phillipines. *Ecol. Entomol.*, **7**, 113–20.

Zimmerman E.C. (1958) *Insects of Hawaii. Vol. 8. Lepidoptera: Pyraloidea.* University of Hawaii Press, Honolulu.

Zimmerman E.C. (1960) Possible evidence of rapid evolution in Hawaiian moths. *Evolution*, **14**, 137–8.

Zwölfer H. (1965) Preliminary list of phytophagous insects attacking wild Cynareae (Compositae) in Europe. *Comm. Inst. Biol. Contr. Tech. Bull.*, **6**, 81–153.

Zwölfer H. (1979) Strategies and counterstrategies in insect population systems competing for space and food in flower heads and plant galls. *Fortschr. Zool.*, **25**, 331–53.

Author Index

Organism Index

Subject Index

KING'S
COLL. LIBR.
CAMB.